High-Temperature Oxidation and Sulphidation Processes

Titles of Related Interest—

Ashby ENGINEERING MATERIALS 1
Ashby ENGINEERING MATERIALS 2
Ashby PERSPECTIVES IN HYDROGEN IN METALS
Brook IMPACT OF NON-DESTRUCTIVE TESTING
Koppel AUTOMATION IN MINING, MINERAL AND METAL PROCESSING 1989
Ruhle METAL-CERAMIC INTERFACES
Taya METAL MATRIX COMPOSITES

Other CIM Proceedings Published by Pergamon

Bergman FERROUS AND NON-FERROUS ALLOY PROCESSES
Bickert REDUCTION AND CASTING OF ALUMINUM
Bouchard PRODUCTION, REFINING, FABRICATION AND RECYCLING OF LIGHT METALS
Chalkley TAILING AND EFFLUENT MANAGEMENT
Closset PRODUCTION AND ELECTROLYSIS OF LIGHT METALS
Dobby PROCESSING OF COMPLEX ORES
Jaeck PRIMARY AND SECONDARY LEAD PROCESSING
Jonas DIRECT ROLLING AND HOT CHARGING OF STRAND CAST BILLETS
Kachaniwsky IMPACT OF OXYGEN ON THE PRODUCTIVITY OF NON-FERROUS METALLURGICAL PROCESSES
Lait F. WEINBERG INTERNATIONAL SYMPOSIUM ON SOLIDIFICATION PROCESSING
Macmillan QUALITY AND PROCESS CONTROL IN REDUCTION AND CASTING OF ALUMINUM AND OTHER LIGHT METALS
Mostaghaci PROCESSING OF CERAMIC AND METAL MATRIX COMPOSITES
Plumpton PRODUCTION AND PROCESSING OF FINE PARTICLES
Purdy FUNDAMENTALS AND APPLICATIONS OF TERNARY DIFFUSION
Rigaud ADVANCES IN REFRACTORIES FOR THE METALLURGICAL INDUSTRIES
Ruddle ACCELERATED COOLING OF ROLLED STEEL
Salter GOLD METALLURGY
Thompson COMPUTER SOFTWARE IN CHEMICAL AND EXTRACTIVE METALLURGY
Twigge-Molecey MATERIALS HANDLING IN PYROMETALLURGY
Twigge-Molecey PROCESS GAS HANDLING AND CLEANING
Tyson FRACTURE MECHANICS
Wilkinson ADVANCED STRUCTURAL MATERIALS

Related Journals

(Free sample copies available upon request)

ACTA METALLURGICA
CANADIAN METALLURGICAL QUARTERLY
MATERIALS RESEARCH BULLETIN
MINERALS ENGINEERING
SCRIPTA METALLURGICA

PROCEEDINGS OF THE INTERNATIONAL SYMPOSIUM ON HIGH-TEMPERATURE OXIDATION AND SULPHIDATION PROCESSES, HAMILTON, ONTARIO, CANADA, AUGUST 26-30, 1990

High-Temperature Oxidation and Sulphidation Processes

Editor
J.D. Embury
Department of Materials Science and Engineering
McMaster University
Hamilton, Ontario, Canada

Symposium organized by The Metallurgical Society of CIM

29th ANNUAL CONFERENCE OF METALLURGISTS OF CIM
29e CONFÉRENCE ANNUELLE DES MÉTALLURGISTES DE L'ICM

Pergamon Press
Member of Maxwell Macmillan Pergamon Publishing Corporation
New York Oxford Beijing Frankfurt São Paulo Sydney Tokyo Toronto

Pergamon Press Offices:

U.S.A.	Pergamon Press, Inc., Maxwell House, Fairview Park, Elmsford, New York 10523, U.S.A.
U.K.	Pergamon Press plc, Headington Hill Hall, Oxford OX3 0BW, England
PEOPLE'S REPUBLIC OF CHINA	Pergamon Press, 0909 China World Tower, No. 1 Jian Guo Men Wai Avenue, Beijing 1000004, People's Republic of China
FEDERAL REPUBLIC OF GERMANY	Pergamon Press GmbH, Hammerweg 6, D-6242 Kronberg, Federal Republic of Germany
BRAZIL	Pergamon Editora Ltda, Rua Eça de Queiros, 346 CEP 04011, Paraiso, São Paulo, Brazil
AUSTRALIA	Pergamon Press Australia Pty Ltd., P.O. Box 544, Potts Point, NSW 2011, Australia
JAPAN	Pergamon Press, 8th Floor, Matsuoka Central Building, 1-7-1 Nishishinjuku, Shinjuku-ku, Tokyo 160, Japan
CANADA	Pergamon Press Canada Ltd., Suite 271, 253 College Street, Toronto, Ontario M5T 1R5 Canada

Library of Congress Cataloging in Publication Data

ISBN 0-08-040415-4

Printing: 1 2 3 4 5 6 7 8 9 Year: 0 1 2 3 4 5 6 7 8 9

Printed in the United States of America

The paper used in this publication meets the minimum requirements of American National Standard for Information Sciences-Permanence of Paper for Printed Library Materials, ANSI Z 39.48-1984

Foreword

The present volume has been produced to mark the retirement of Professor W.W. Smeltzer. It is appropriate that the volume deals with the oxidation and sulphidation resistance of a wide range of metals, alloys, and intermetallic compounds not only because it is in this area of the science of materials that Professor Smeltzer has made his outstanding contributions, but because in the development of materials for high-temperature applications the environmental stability is of paramount importance.

The papers contained in this volume give a valuable overview of a field which is critical in the development of new materials. They present analytical models based on the applications of thermodynamic and kinetic descriptions of the processes controlling high-temperature stability together with detailed experimental studies which utilize a wide variety of analytical techniques. In addition, the volume contains papers which deal with the understanding of oxidation, and sulphidation in complex engineering materials. Thus, the volume encompasses the essential relationship between the understanding of the stability of materials, and their selection for high-temperature application, and the design of structures based on environmental stability. This relationship is one which is crucial to the development of materials for high-temperature application including intermetallic compounds and ceramics.

It is fitting that the book is dedicated to Professor Walter Smeltzer by his colleagues at McMaster, his students, and his colleagues in the broad fields of high-temperature chemistry, and materials science. In the past four decades Professor Smeltzer has provided much intellectual leadership to the understanding and analysis of oxidation and sulphidation. This volume serves as a token of the respect, and esteem of Professor Smeltzer's many colleagues.

J.D. Embury
Hamilton, Ontario

August 1990

Table of Contents

Interactive mechanisms in the high-temperature oxidation of metals

G.H. Meier, F.S. Pettit and N. Birks
Materials Science and Engineering Department, University of Pittsburgh, Pittsburgh, Pennsylvania, U.S.A.

INTRODUCTION

The subject of the high temperature oxidation of metals was given a firm scientific foundation in 1933 when Carl Wagner(1) published his theory of metal tarnishing or scaling reactions. This brought together the defect theory of ionic solids,the electrochemical nature of most metal oxidation reactions and the theory of transport through ionic solids, to produce a model for high temperature oxidation of metals that not only explained the phenomenon but allowed behavior to be predicted in certain limiting cases. In the ensueing years, this model has been applied widely and has been found to have general validity over a very wide range of metal systems. The theory of course is based on several limiting assumptions that restrict the breadth of its strict application to some specific cases. Nevertheless, the basic principles on which the theory is founded can be applied in every case, even if the final solutions are somewhat less complete than that provided by Wagner.

The purpose of this paper is to examine several situations in which the interpretation of the behavior of systems under high temperature oxidation requires the consideration of factors extra to those included in the original, general, treatment provided by Wagner.

In general , such situations will arise when several processes are acting simultaneously to affect the progress of the oxidation reaction. These processes can be either physical or chemical in nature and may proceed together but independently or may interact with and modify each other.

1. SPECIMEN SHAPE

The formation of porosity

Wagner's original model(1),shown in Figure 1, assumed that the growing scale was perfectly compact and adherent, which is easy to achieve when a simple flat surface is considered. In this case the scale that forms from the reaction

$$2M + O2 = 2MO \qquad (1)$$

can exist on the metal surface and grow as the reaction proceeds without the introduction of growth stresses(2-4). Such stresses are introduced by the difficulty of accomodating a scale of volume Vmo that forms from an equivalent volume of metal Vm. In general these are different , but for a planar scale there is no difficulty in accomodating the thickness ,which is taken up by the scale having a thickness different to that of the metal from which it was formed.

In the situation of a real component or specimen however , the size is finite and there are , consequently, edges and corners to define the geometrical limits of the piece. In the oxidation of a flat surface, the scale metal interface is able to relax towards the center of the specimen without interference . In the case of a finite specimen,as the scale grows it is restricted in the extent to which it can relax, particularly in the region of an edge and ,especially, in the region of a corner. In these areas ,the geometry makes it extremely difficult for the scale to relax in both directions at once. Consequently, the scale can only maintain contact with the metal on all fronts in these areas if it can deform plastically by creep under forces that are defined by and limited by the magnitude of the scale-metal adhesion.

It is in such areas that the scale separates from the metal and the formation of a porous zone begins. This situation can spread from the edges and corners, eventually leading to the development of a scale that shows porosity at the metal scale interface throughout.

The development of this structure depends on the relative values of factors such as the growth rate of the scale, the force of the metal-scale adhesion and the creep properties of the scale.

The Reaction Temperature

Most metal oxidation reactions are exothermic. There is therefore a tendency for the specimen to superheat during the reaction as the enthalpy of the reaction is absorbed or dissipated. This situation is most acute at the start of a reaction when the reaction proceeds most rapidly.

The rate of heat evolution is given by the product of the reaction rate and the reaction enthalpy

$$dh/dt = dm/dt.\Delta H \qquad (2)$$

In this equation the rate of reaction, dm/dt, is expressed in terms of a parabolic scaling constant that is itself a function of temperature

$$Kp = Ko\exp(-Q/RT) \qquad (3)$$

Thus, the rate of heat liberation will fall as the reaction proceeds. The factor that is most important in the design of an experiment is the temperature of the specimen and the rate of increase of the specimen temperature, due to absorbtion of the heat of reaction, which can be approximated by

$$dT/dt = (dh/dt - dhl/dt)/MsCp \qquad (4)$$

where dhl/dt represents the rate at which heat is lost from the specimen,Ms and Cp are the specimen original weight and its specific heat, respectively. The only terms over which the experimentalist has control are the specimen weight and thickness. Clearly ,for a given weight , a specimen of thin section will increase the term dT/dt and lead to high specimen over temperatures . If a large, thick, specimen is chosen to minimise this effect, then the experiment runs into the problem that it takes an increasingly longer time for the specimen to reach the reaction temperature. Thus a compromise must be found between uncertainty in the specimen temperature and uncertainty in the time at which the specimen reached the set temperature in the start of a reaction. These considerations become more critical with systems that show high reaction rates and which have high enthalpies of reaction. The implied uncertainty in the starting conditions of a reaction become less important as the duration of the reaction is increased.

There are practical cases of metal overheating by oxidation, as in the self sustained burning of magnesium ribbon,the thermit processes for the reduction of metal oxides by more reactive metals,titanium fires and the autoignition of cargos and piles of direct reduced iron pellets.

Additional effects of specimen size occur when transport through an oxide is not rate controlling, e.g. when volatile products are produced by the oxidation reaction. These will be discussed subsequently.

2.THERMAL CYCLING

If ,instead of being maintained constant, the temperature is allowed to vary during high temperature oxidation, then a system of stresses will be generated within the scale and at the scale metal interface due to the different coefficients of thermal expansion of the reacting metal and the growing scale(5).

More severe thermal stresses are imposed on the system as the high and low temperatures are more widely separated.

The stresses that are induced by thermal cycling may cause mechanical changes to arise in the scales which range from brittle fracture of the scale , loss of adhesion at the scale-metal interface, introduction of physical defects into the scale and spallation of pieces of scale from the specimen. It is also possible that a scale that is sufficiently ductile may absorb the stresses by plastic deformation, although this is less likely as the temperature is reduced in its cycle, and the scale becomes less ductile.

The damage to a scale following thermal cycling,indicated in Figure 2, almost always results in a scale-metal situation in which the scale is rendered less protective towards subsequent oxidation .

In many commercial processes where metals are exposed to thermal cycling and are in addition designed to support structural stresses, the resistance of the system to thermal stresses is an important consideration in design and in the selection of materials.

The main criteria in designing systems to resist thermal cycling are to reduce the rate of scale growth and to increase the adhesion of the scale to the metal(6-8) . The first is accomplished by selecting, from the range of alloys with suitable mechanical properties, those which develop a scale that is adherent and grows very slowly. Techniques to achieve superior scale metal adhesion currently incude the addition of small amounts of active elements such as yttrium and to minimise the concentration of acutely suface active elements such as sulfur that are suspected of segregating to the scale metal interface and compromising the adhesion between the two.

Apart from the more obvious damage to the scale that is caused , thermal cycling that leads to frequent spallation ecourages the reformation of the protective layer which in the case of an alloy causes the denudation of the protective alloying element and eventually the mechanism of oxidation must change to include the formation of oxides of the other alloy components, which usually leads to more rapid deterioration and signals the beginning of the end of the useful life of the component.

3.ALLOY OXIDATION

The oxidation of alloys follows the same principles as the oxidation of pure metals except that several additional factors introduce complications that are not relevent to pure metals.

Such factors include the possibility of forming several different oxides, the formation of mixed oxides and the possibility of diffusion within the metal phase.

These factors lead to the occurence of phenomena such as internal oxidation, where the more stable oxide forms as particles within the metal phase below a scale composed of a less stable oxide.

Many complications arise from the exposure of alloys to oxidation(9). Basically the aim in producing a heat resistant alloy is to cause the formation of a passivating layer of a very stable ,highly stoichiometric oxide on the alloy surface. If this could be achieved , the alloy would be expected to behave reasonably in accordance with Wagner's mechanisms. Unfortunately,the duration of such behavior, which just about defines the life time of the alloy,is limited and sooner or later the denudation of the passivating element in the alloy at the metal surface,removal of the passivating scale by spalling,evaporation or erosion allows the scale to reform,based on the less stable oxide of the alloy system.

This almost always causes the alloy to enter a behavior regime of rapid surface deterioration from which recovery would be difficult. Indeed, one of the primary aims of the design of high temperature alloys is to suppress or delay the onset of this transition in behavior.

4.MULTIPLE OXIDANT GASES

The situation,where two reactive species are present in the gas phase and where it is possible for more than one product of reaction to be formed,is complex(10,11,12).

In most cases, one of the oxidants will be oxygen or will lead to the formation of an oxide. The other component of the gas will normally be a compound of S,Cl,H or C. Thus the possibility of formation of oxides, sulfides,chlorides etc must be appraised.

The mechanisms available for the formation of an extra phase in the scale , whether or not a physical mixture or duplex scale will form and with what type of distribution are important aspects of this type of attack.

The first step in assessing the multiple oxidant situation is to consider the appropriate phase diagram, in order to decide which phase is in thermodynamic equilibrium with the atmosphere. This phase is then the one which is expected to form initially on exposing the metal to the atmosphere at temperature.

If this phase provides a complete cover of the metal surface then there exists across the thickness of the layer a chemical potential gradient of the metal and of the non-metal from equiibrium with the atmosphere at the gas scale interface to equilibrium with the metal at the metal-scale interface . The effect of this on the behavior of the second oxidant must be taken into account.

The system may respond to the two oxidant situation in several ways(9)

* An oxide scale may form in which the second oxidant is dissolved. This would allow the oxidant to diffuse through the scale to the scale-metal interface where second phase formation can proceed. This can only occur if the activity of the second oxidant in the gas is higher than it is at the scale-metal interface.

* An oxide scale may form in which the second oxidant does not dissolve. In this case the only chance for the second oxidant to attack the metal is to penetrate the scale as a molecular species using physical defects as transport paths.

*It is possible for a scale to form in which the two scale products are intimately mixed to form a duplex or composite scale. In this case the transport and mechanical properties of the scale reflect its structure. The structure or the degree and type of dissemination of the two phases will be influenced strongly by transport and surface energy considerations and may vary from a relatively coarse distribution of particles of one phase in the other to a finely divided but continuous network. In the latter case the structure would take the form of two interpenetrating phase networks and the ionic conductivity would favor that of the phase with the higher conductivity. In the former case of a distribution of discrete particles, the conductivity would reflect that of the matrix phase.

It is also possible that the second phase to be formed may be solid, liquid or gaseous. In the case where both phases are solid the above considerations would hold. Where the second phase is liquid particularly where the system forms a eutectic that includes the metal , severe attack of the metal can be expected particularly at grain boundaries.
Finally the second reaction product may be a gas which introduces the possibility of very rapid and selective attack of the metal. This may give rise to denudation of the reactive metal at the metal surface together with perhaps void formation particularly along grain boundary diffusion paths. This leads to mechanical instability of the surface as well as to enhanced corrosion attack

The special situation at the metal-scale interface with respect to the enhancement of the activity of the second oxidant should be emphasized.

5.EFFECT OF GAS VELOCITY

In general, the oxidation of a metal involves a number of processes occurring in series. These are:

i. Transport of oxidant through the gas phase to the specimen surface.
ii. Reaction at the surface including adsorption, dissociation, and incorporation into a solid reaction product.
iii. Diffusion of anions and/or cations through the reaction product.
iv. Desorption of product gases (if any).
v. Transport of product gases away from the specimen surface.

Gas velocity has a substantial effect on the reaction kinetics when either steps i. or v. are rate controlling. Two types of processes which illustrate these effects are (a) the effect of boundary layer diffusion when the reactive gas is present as a dilute species in a mixture with an inert gas and (b) the effect of one or more reaction product being a gas or vapor.

(a) Dilute Gases

When a metal is reacted with an atmosphere consisting of an active species diluted by an inactive or inert species, the active molecules are rapidly depleted in the gas layers immediately adjacent to the specimen surface. Subsequent reaction can then only proceed as long as molecules of the active species can diffuse through the denuded layer to the metal surface. When the concentration of the active species is low in the atmosphere, this step can become rate controlling. The denuded gas layer next to the metal surface may be regarded as a boundary layer of thickness δ. If the concentrations of the active species at the metal surface and in the bulk gas are expressed, respectively, by the partial pressures p''_i and p'_i then the rate at which active species, i, diffuse across boundary layer may be expressed as

$$j_i = -D_i (p'_i - p''_i)/RT\delta \qquad (5)$$

which represents the maximum possible reaction rate observable under these circumstances. In most cases, at least at the beginning of the reaction, the value of p''_i is very low and can be neglected giving

$$j_i = -D_i\, p'_i / RT\delta \qquad (6)$$

i.e. the reaction rate is directly proportional to p'_i, the partial pressure of the active species in the bulk atmosphere. The sensitivity of the reaction rate to temperature arises solely from the way in which D_i/RT and, to a lesser extent, δ, vary with temperature.

Reaction rate control by this step can be identified most securely by the sensitivity of the reaction rate to the gas flow rate. If the gas flow rate is increased the thickness of the boundary layer, δ, is decreased and, correspondingly, the reaction rate increases according to equation GV1.

It is important to note that the concentration of adsorbed species and, therefore, the rates of surface reactions involving adsorbed species also vary with the partial pressure of the active species in the gas phase. Thus, in order to identify boundary layer transport as the process controlling a constant rate, sensitivity of the rate to both active species partial pressure and gas flow rate must be established.

(b) Volatile Reaction Products

i. Oxidation of Cr and Chromia-forming Alloys

The oxidation of materials which form protective Cr_2O_3 is an important example of how the formation of both a solid and a gaseous reaction product can lead to paralinear oxidation kinetics. The oxidation of Cr is illustrated schematically in Figure 3. The Cr_2O_3 scale grows by solid state diffusion but is thinned by the formation of the volatile species CrO_3 at the scale/gas interface. The evaporation of this species results in more rapid consumption of Cr because of the reduced diffusion distance through the scale. The instantaneous change in scale thickness is the sum of two contributions(13):

$$dx/dt = k_d / x - k_s \qquad (7)$$

where k_d is a constant describing the growth by solid state diffusion and k_s describes the rate of volatilization. Initially, when the diffusion through a thin scale is rapid the effect of volatilization is not significant but as the scale thickens the volatilization becomes comparable and then equal to the rate of diffusive growth. This situation, paralinear oxidation, results in a limiting scale thickness, x_0, for which $dx/dt=0$ which is shown schematically in Figure 4. This leads to the limiting thickness

$$X = k_d / k_s \qquad (8)$$

The constant k_s increases with both oxygen partial pressure and flow rate so that increasing either of these parameters decreases x_0 and, more importantly, increases the rate of Cr consumption. This phenomenon limits the usefulness of chromia-forming alloys in high gas flow rate applications to temperatures of less than 1000°C. It should be pointed out that any process which removes the CrO_3 from the gas surrounding the specimen will have the same effect as increasing the flow rate(14). Figure 5 shows a cross-section of a Ni-30Cr alloy which has been oxidized in contact with a CaO deposit which reacts rapidly with $CrO_3(g)$ to form $CaCrO_4(s)$.

This alloy, which would normally maintain a protective Cr_2O_3 scale, is covered by a scale of NiO because of the excessive loss of Cr even though the gas velocity was very low.

ii. Oxidation of Mo

Molybdenum, unlike Cr which develops a limiting oxide thickness, undergoes complete oxide volatilization at high temperatures in gases with high oxygen partial pressures. The effects of oxide volatility on the oxidation of Mo have been observed at temperatures as low as 475°C(15) and the rate of evaporation above 725°C was such that gas phase diffusion of oxygen to the specimen was rate controlling(16). Under these conditions the rate of Mo consumption increases with increasing flow rate and decreasing specimen size.

iii. Oxidation of Carbon-Carbon Composites

The oxidation of carbon-carbon composites, as well as other forms of carbon, involves the formation of only gaseous reaction products, CO and CO_2. This results in a linear weight loss as a function of time as illustrated in Figure 6. Figure 6 illustrates how this linear rate depends on the oxidation temperature. At temperatures below 600°C these constants obey an Arrhenius relationship with an activation energy of about 43 kcal/mole. In this temperature range the rates are not affected by specimen size or gas flow rate. Clearly the reaction rate is controlled by surface processes. At temperatures above 800°C the rates become essentially independent of temperature. Moreover, the rates are accelerated by increasing gas flow rate and decreasing specimen size. In this temperature range the oxidation rates could be fit to the empirical equation:

$$\text{Oxidation Rate (gm/cm}^2\text{-s)} = 0.001(F/A)^{0.68} \qquad (9)$$

where F is linear flow rate (cm/s) and A is specimen area (cm^2). All of the above data indicate that rate is controlled by gas phase diffusion of oxidant to the specimen in this temperature range.

At temperatures between 600 and 800°C the rates are influenced by flow rate(Fig. 6) and specimen size, but the activation energy is higher than that for gaseous diffusion. In this temperature range the rate is apparently determined by a coupling of the surface process and diffusion in the gas phase.

6. THE PRESENCE OF PARTICULATES IN THE GAS STREAM.

Hot Corrosion

When metals and alloys are used in equipment in which combustion processes occur, deposits consisting of chlorides or sulfates can be formed upon the metallic surfaces. Such deposits may affect transport through the scale by altering the concentrations of atomic defects.(9) When these deposits are liquid, however, they can cause the oxide scale to become non-protective which results in substantially increased corrosion rates(17), Figure 8.

Accelerated oxidation arising from nonprotective oxide scale formation which in turn is caused by a liquid deposit is called hot corrosion. The liquid deposits can cause the oxide scales to become non-protective via a number of processes. Some of these processes have the common characteristic that the normally protective oxide scale reacts with the molten deposits. These reactions can be described as basic or acidic fluxing processes that either make the normally protective oxide thinner or prevent it from becoming continuous through dissolution according to

$$MO + O^{2-} \rightarrow MO_2^{2-} \text{ (basic dissolution)} \qquad (10)$$
$$MO \rightarrow M^{2+} + O^{2-} \text{ (acidic dissolution)} \qquad (11)$$

or cause this scale to become a porous network via dissolution followed by reprecipitation in the molten deposit away from the surface of the metal. Other of these processes establish conditions due to the presence of the melt, whereby additional reactants such as sulfur, chlorine or carbon, begin to affect the corrosion process. The effects produced by these other reactants in addition to oxygen are similar to those described previously for multiple oxidant gases.

Erosion-Corrosion.

The presence of even chemically inactive particles in the scale can compromise the passivity of a normal scale if the particles are flowing at a sufficiently high velocity. This is another deviation from the original assumptions that is commonly met in applications.

It has been shown (18,19), that the interactions between erosion and corrosion can be placed in several distinct regimes, depending on the relative intensities of the two processes. Thus the degradation of the metal under attack can range from erosion dominated to corrosion dominated .

The three primary regimes of interaction are displayed in Figure 9, which shows pure erosion , erosion enhanced oxidation and oxidation affected erosion regimes.

The erosion intensity and the rate at which it removes an oxide layer varies with parameters such as particle velocity,particle loading of the flow,particle type, size and shape and angle of impingement. Similarly the corrosion rate is a function of the gas composition and temperature.

Under conditions where the oxidation rate is very rapid compared with the erosion rate, the situation exists where the erosion affects the scale only to a minor degree. Under these conditions the scale is removed from the surface as if it were a piece of solid oxide with no contact with the metal.

Pure erosion of the metal represents the other extreme where there is no oxidation reaction and is characteristic of a metal at low temperature or in a non oxidizing environment.

The erosion enhanced oxidation regime represents a more equally balanced situation in which the processes of erosion and corrosion proceed at comparable rates.In this case, the erosion holds the protective scale at a thickness that is less than would have been achieved in the absence of erosion and may even be maintained at a constant thickness when a steady state is achieved. In this regime the degradation rate of the metal is increased compared to pure oxidation and the kinetics are modified from a pure parabolic situation to one in which the scale thickness is given as a function of time that involves both the parabolic rate constant and the erosion constant for the oxide.

$$dX/dt = k_p/X - K_{eo} \qquad (12)$$

In the steady state situation, the erosion processes remove scale as quickly as it is being formed and the scale thickness remains constant while metal is being consumed at a constant rate. This process is analogous to the paralinear oxidation of Cr discussed previously. This behavior is shown in Figure 10(20). There have also been reports that,under the influence of erosion, the intrinsic rate of oxidation was increased. Presumably, the erosion process has,in this case,
caused the structure of the scale to be modified to allow more rapid transport. This aspect requires more thorough study.

As the intensity of the erosion is increased, the scale is not capable of reaching a thickness sufficient to contain the erosive impact which is now absorbed partially by the metallic substrate which,in turn,undergos plastic deformation and an uneven surface develops. This is the beginning of another zone of interaction between erosion and corrosion called oxidation affected erosion. This involves particularly severe degradation of the metal under conditions where the scale surface and the metal surface appear to be merged to form a composite surface layer which promotes rapid oxide formation because the oxide never manages to form any degree of passivation. Consequently the oxide is removed rapidly as it forms. There is also substantial rearrangement of the metal surface layer by the erosive impacts that cause the metal to be continually exposed to the atmosphere with consequent enhanced oxidation attack.The details of these surface processes are not fully understood or even documented.

Spalling,which is a problem in the successful retention of a passivating oxide layer on the surface of a metal,is also seen under the influence of erosion to play a role in the mechanism of scale removal. In this case spalling can be provoked to occur under isothermal situations and the evidence that is so far available indicates that spalling occurs preferentially at an interface and when a critical thickness of the relevent layer has been reached. The mechanisms by which the stress leading to separation of the spalling scale from the metal is generated have not been completely described , although several possibilities involving direct stress generation and the interaction of shock waves are being considered(21). This problem is compounded by the general lack of experience in the area of the behaviour of nominally plastic materials under abnormally high rates of application of stress.This is an area that requires further study, however the problem of imposing such high strain rates under experimental conditions remains a limiting factor.

The above observations apply equally well to pure metals as to alloys. In the cases of alloys there are also several additional factors to be considered.

Erosion of an alloy that is simultaneously undergoing oxidation can lead to rapid removal of the surface oxides and thereby impede, or prevent, the formation of a passivating layer of the appropriate oxide. In this way the alloy can be held in a state of extended or permanent transient oxidation. The removal of the surface oxide by erosion will at the very least cause it to reform more quickly and thereby denude the surface layers of the alloy in the metal whose oxide forms the passivating layer.

All of these effects interfere with the usual mechanisms by which alloys oxidize and form passivating scales, the presence of erosion, if slight, will lead at least to more extensive zones of denudation in the alloy surface and,if strong, could lead to such wholesale removal of the surface oxides such that the passivating layer cannot form. This latter behaviour could be characterized by the absence of a zone of enrichment or depletion at the alloy surface.

CONCLUSIONS

A sound theoretical basis exists for the description of metal degradation by high temperature oxidation, and several systems conform quite strictly to the theory. However,most real situations, where metals are exposed to high temperature oxidation, are complex to the extent that it is neccessary to consider two or more distinct,but possibly interacting, phenomena in order to arrive at a valid, or adequate, description of the mechanism(s) that account for the observed, or anticipated behavior.

REFERENCES

1. C. Wagner; Z. Phys. Chem., 21, 25, (1933).
2. D.L. Douglass; Oxidation of Metals and Alloys, ed. D.L. Douglass, ASM, Metals Park, 1971.
3. J. Stringer; J. Corros. Science, 10, 513, (1970).
4. P. Hancock and R.C. Hurst; Advances in Corrosion Science and Technology, ed. R.W. Staehly and M.G. Fontana, Plenum Press, New York 1074.
5. G.D. Oxx, Prod. Eng., 21, 61 (1958).
6. E.J. Felton and F.S. Pettit; Oxid. Metals, 10, 189 (1976).
7. D.P. Whittle and J. Stringer; Phil. Trans. Roy. Soc. 295A, 309 (1980).
8. G.M. Ecer and G.H. Meier; Oxid. Metals; 13, 159, (1979).
9. N. Birks and G.H. Meier; Introduction to High Temperature of Metals, Arnold London, 1983.
10. F.S. Pettit, J.A. Goebel and G.W. Goward; Corros. Sci. 9, 903 (1969).
11. N. Birks; High Temperature Gas Metal Reactions ed. S.A. Jansson, Z.A. Foroulis, AIMME New York, p.322, 1973.
12. A. Rahmel and J.A. Gonzalez; Werkstoffe und Korrosion, 21, 95, (1970) also 22, 283, (1971).
13. C.S. Tedmon; J. Electrochem. Soc., 113, 766 (1966).

14. K.T. Chiang, G.H. Meier, and R.A. Perkins; J. Materials for Energy Systems, 6, 71 (1984).
15. E.A. Gulbransen and W.S. Wysing; TAIME, 175, 628, (1948).
16. E.A. Gulbransen, K.F. Andrew and F.A. Brassart; J. Electroch em. Soc. 110, 952, (1963).
17. F.S. Pettit and C.S. Giggins; "Hot Corrosion" Chapter 12, in Superalloys II, C.T. Sims, N.S. Stoloff and W.C. Hagel eds., John Wiley and Sons, New York, (1987).
18. C.T. Kang, F.S. Pettit and N. Birks; Met. Trans. A, 18, 1785, (1987).
19. S. Hogmark,A. Hammersten and S. Soderberg; Proc. 6th Int. Conf. on Erosion by Liquid and Solid Impact, Cambridge, England, (19- 83), p.37-1.
20. S.L. Chang, F.S. Pettit and N. Birks; Oxid. Metals, 34, 21, (1990).
21. D.L. Rishel,F.S. Pettit and N. Birks; Princpal Mecanisms in Simultaneous Erosion and Corrosion Attack of Metals at High Temperature in Corrosion-Erosion-Wear of Materials at Elevated Temperature, ed. A.V. Levy, Berkeley CA, January 31 - February 2, 1990.

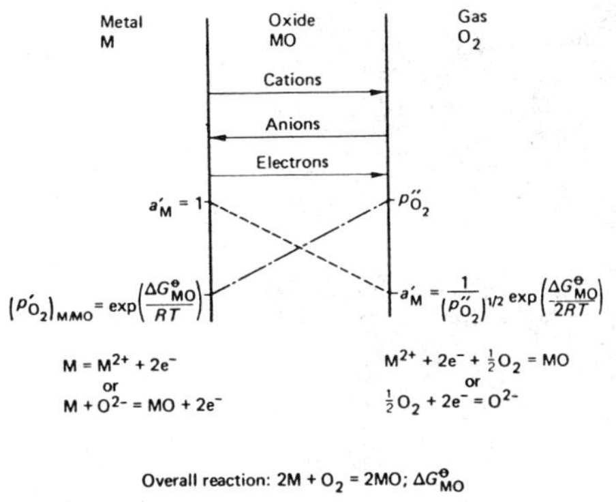

Figure 1: Diagram of scale formation according to Wagner's model (1)

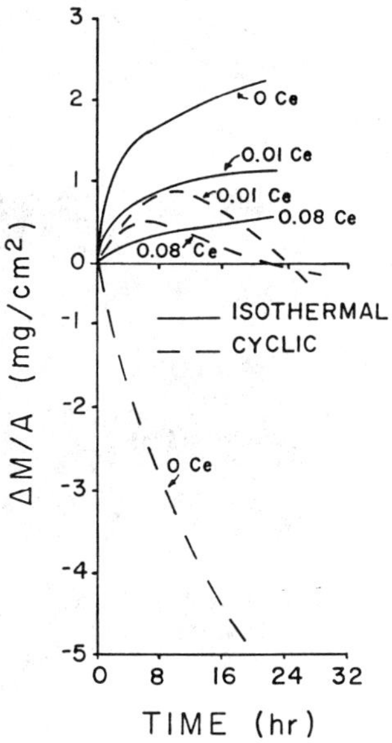

Figure 2: Effect of cerium content on the cyclic oxidation mass loss of Ni-50 Cr.

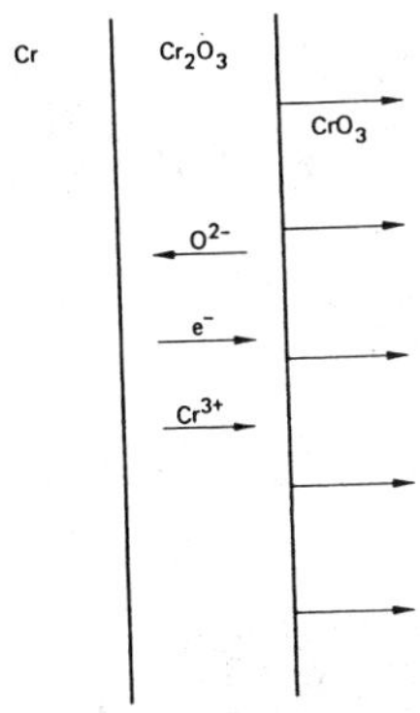

Figure 3: Schematic diagram of combined scale growth and oxide volatilisation from Cr.

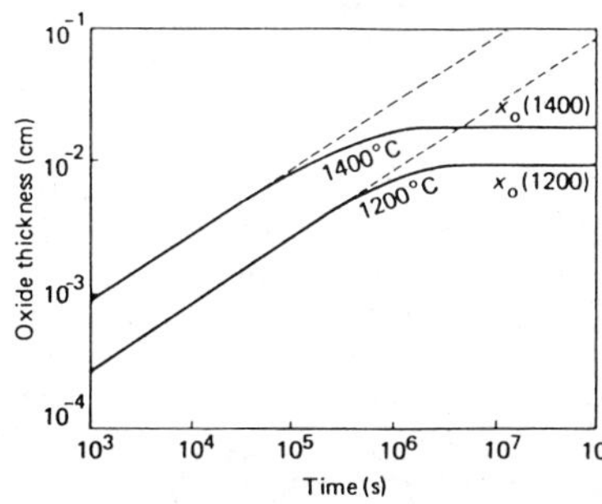

Figure 4: Scale thickness versus time for the oxidation of Cr.

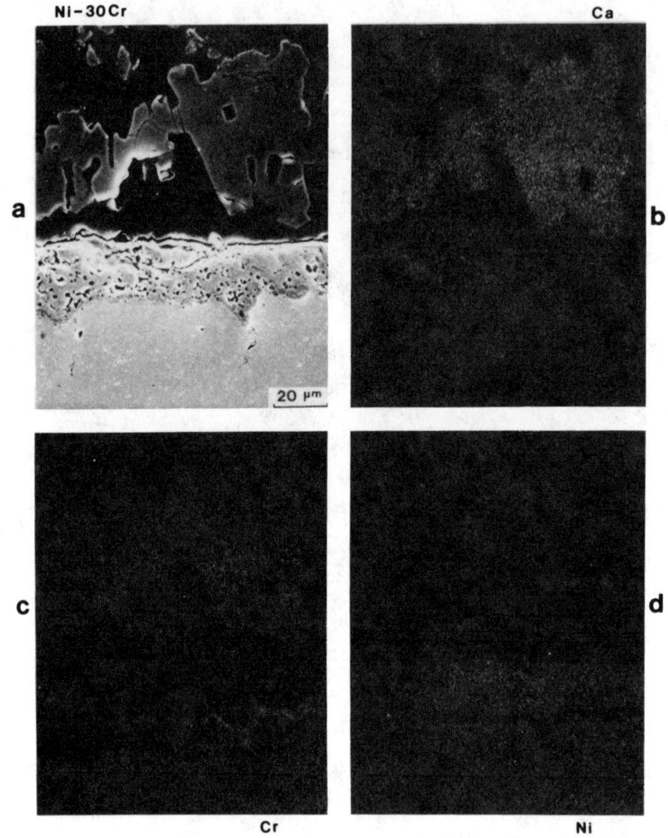

Figure 5: Cross-section of preoxidized Ni-30Cr reoxidized at 900°C with a CaO coating (a) and x-ray maps for Ca, Cr, and Ni (b-d).

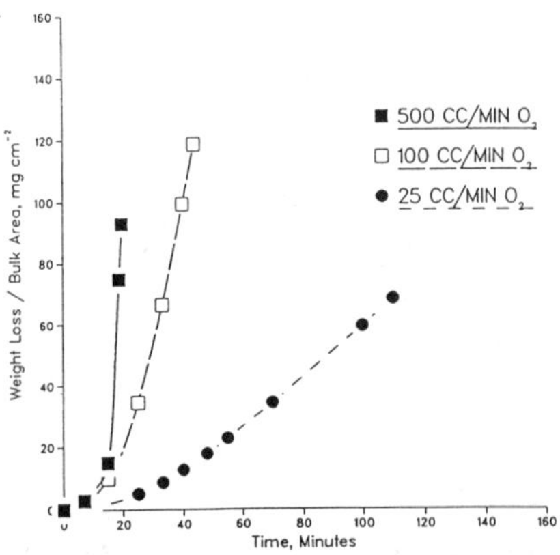

Figure 6: Isothermal oxidation kinetics for uncoated CCC at 700°C shows a dependence of oxidation rate on flow rate. At 700°C the oxidation reaction is under mixed control.

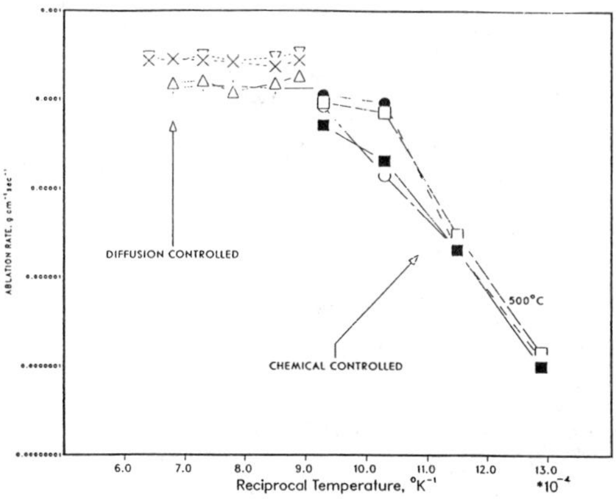

Figure 7: Log oxidation rate (weight loss) vs. 1/T ($^{\circ}K^{-1}$) for samples oxidized in O_2 at 100 cc/min at various temperatures. Diffusion control dominates at high temperatures and surface control dominates at lower temperatures.

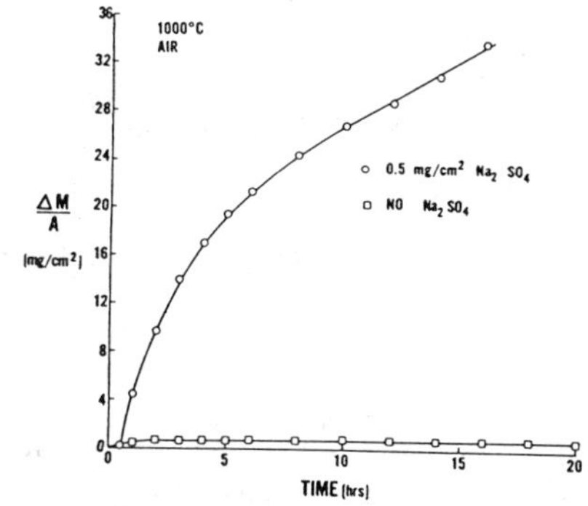

Figure 8: Comparison of the oxidation behavior of the nickel-base super-alloy B-1900 (8Cr, 1 Mo, 4Ta, 10 Co, 0.1C, 0.07Zr, 0.01B , balance nickel) with and without a coating of Na_2SO_4. The accelerated oxidation of the Na_2SO_4 coated alloy appears to self-sustaining.

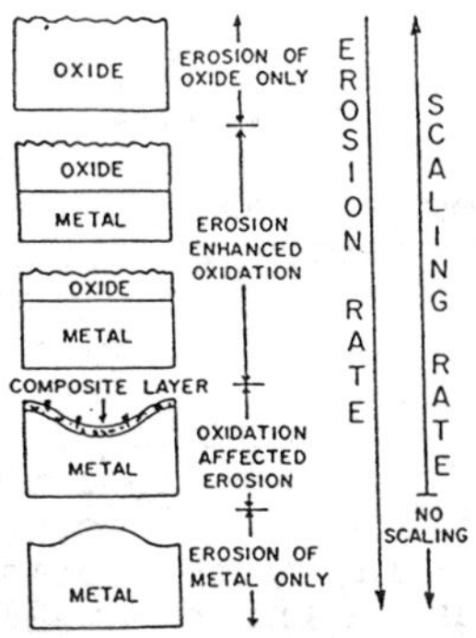

Figure 9: Regimes of interaction between erosion and corrosion indicating the effects of the relative severity of the erosion and oxidation components.

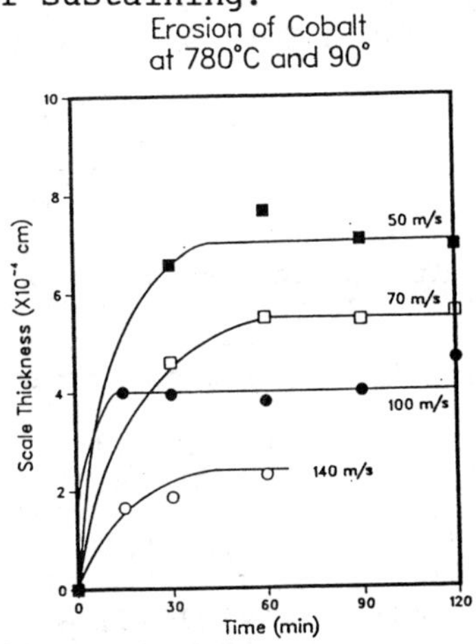

Figure 10: Variation of oxide scale thickness with time for cobalt exposed to erosion-corrosion at 780°C and 90° incidence to airstream.

Evolution of the structure of NiO and CoO scales

J.S. Sheasby
Department of Materials Engineering, The University of Western Ontario, London, Ontario, Canada, N6A 5B9

D.S. Cox
AECL, Station 88, Chalk River, Ontario, Canada, K0J 1J0

B. Gleeson
School of Materials Science and Engineering, University of New South Wales, Kensington, N.S.W. 2033, Australia

H.L. Williams
IBM Canada, North York, Ontario, Canada M3C 1V7

The growth of NiO and CoO scales on pure and impure Ni and Co, and their alloys with Cr and Pt was examined using oxygen tracer techniques. Isotopic imaging was essential as not all the relevant features could be seen by optical microscopy. The complexity of the scales varied considerably between systems and also frequently around single samples. In all cases the outward transport of cations was the basic rare controlling step. On simple columnar scales such cation movement was the only kinetic process, however oxygen movement was essential for the continued growth of more complex scales. The oxygen was predominantly provided by voids that developed in the columnar layer of scales, but could also be provided by a dissociation of the base of the columnar layer or directly from the atmosphere following scale cracking. Such cracks healed after forming. The origin of the major features of scales was accounted for.

Introduction

The growth of CoO and NiO scales at high temperatures on Co and Ni and their alloys has been extensively studied. The systems in broad outline follow classical parabolic behaviour, with the rate of metal diffusion in both oxides far exceeding that of oxygen diffusion. A comparison of the rates of oxidation predicted by theory with those observed in practice has been reviewed by Atkinson (1) with good agreement being found above 1173 K.

This agreement demonstrates quantitatively the dominance of metal transport; however there is continued debate on the role of oxygen transport particularly when the scale structure is complex. Robertson and Manning (2) have developed criteria for the formation of single layer, duplex, and breakaway scales in their "available space" model. They propose that the oxidant has some degree of access to the scale-metal interface in all scales through a dynamic equilibrium in the scale in which old pores seal and new pores open up. Single layers form, with minimal oxygen ingress, when dislocation climb eliminates the space created at the metal-scale interface by outward cation diffusion. Duplex scales form when collapse is inhibited by specimen corners or alloying.

In contradiction, the present authors have shown (3-6) that duplex scales in both systems can grow both with and without the ingress of gaseous oxygen from the atmosphere.

This paper combines the previous work of the authors to develop an understanding of the origin of features in duplex scales.

Methods

Specimens were prepared from commercially pure (nominally 99+%purity) Ni and Co; from Puratronic Ni and Co (Johnson Matthey Ltd); and from Puratronic Ni and Co alloyed with Puratronic Cr and Pt. The Ni and Co were decarburized in wet hydrogen at 1273 K for 24 hours. Samples were prepared to a metallographic finish, electropolished, and then vacuum degassed at 1473 K just prior to oxidation.

Specimens were oxidized at 10 torr oxygen pressure in an apparatus designed for quick complete changes from one oxygen isotope to the other. The specimens were brought to temperature in vacuum followed by the introduction of oxygen.

When changing from $^{16}O_2$ to $^{18}O_2$, the $^{16}O_2$ was pumped out for 45s and 15s later the $^{18}O_2$ introduced. When changing from $^{18}O_2$ to $^{16}O_2$, the $^{18}O_2$ was flushed out by $^{16}O_2$ whilst maintaining the total pressure constant. The intent of the latter procedure was to avoid giving the scale a stoichiometric shock that could cause artifacts such as DIR (7). Trial change-overs with and without evacuation caused no detectable differences, but the "soft" exchange was retained as being the more conservative procedure.

SIMS analysis was performed on polished cross-sections of oxidized specimens using a Cameca IMS 3-f. Images were obtained using 10 KeV Cs primary ions and negative spectrometry. Cs was used because of its enhanced secondary ion yield for electronegative species such as oxygen. The primary ion current was in the range 50-400 nA. This current along with defocussing the primary beam minimized sample charging and gave the best imaging conditions. The field of view was 150 µm. Some of the isotope images shown in this paper have a dark central region due to the age of the viewing screen (channel plate) of the SIMS.

Results

Duplex scales were grown on all the systems studied except for Puratronic Ni. Although leakage of gaseous oxygen to the inner scale occurred frequently, all the duplex scales could form without the ingress of oxygen. The microstructure, but not the thickness, of scales was affected by oxygen leakage. Examination of all the scales that formed without the ingress of oxygen showed one of the two basic patterns of development depicted in Figure 1. The structural consequences of leakage will be described later. The mechanism of the growth of the subcolumnar scales in the two paths, A-B-C-D and A-E-F-G-H, was similar, with the difference being the source of oxygen for the inner scale. The early stages of scale growth on the alloys was not examined. However from observations of mature scales, and from the work of others, it is anticipated that scale development is similar to that shown in Figure 1.

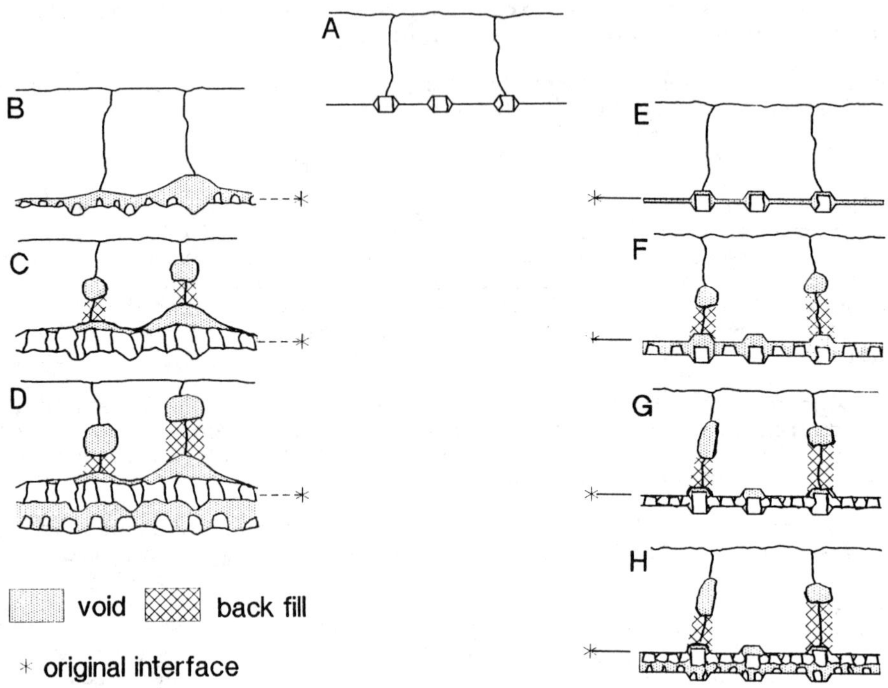

Figure 1. Duplex scale development in the absence of oxygen leakage from the atmosphere. Purer metals and high temperatures tend to follow path A-B-C-D; alloys, impure metal and low temperatures favour path A-E-F-G-H.

Scales were composed initially of a single layer of columnar crystals. Metal-scale interfaces were flat reflecting the initial surface preparation of samples. Tracer oxygen used at the beginning of oxidation remained at the metal-scale interface, and tracer oxygen used at the end of oxidation formed a layer at the scale-gas interface. Interfaces between the isotope types were sharp showing that neither lattice nor grain boundary diffusion of oxygen made a significant contribution to scale growth. The columnar part of scales on all specimens at all stages of growth tended to contain voids. The voids moved across the scale presumably by a CO/CO_2 mechanism from residual carbon in the metal. Such void movement was most clearly seen when they crossed from one isotope layer to the other. When the voids reached the scale-gas interface they vented and filled, as in Figures 2, 9 and 10. Such events were not detectable by optical microscopy.

With further oxidation new oxide crystals and porosity developed at the metal-scale interface as shown in Figures 1A and 3. Thin initial layers of tracer showed that the oxygen for the new crystals came from near the metal-scale interface, and was presumably from the associated porosity. Due to the fine scale of these features the details of their formation and propagation, 1B or 1E, could not be discerned. Pure metals and high temperatures favoured the structures 1A-B-D, and impure metals or alloys, and low temperatures favoured structures 1A-E-H. The determining factor was not a major force as on occasion both structures could be found on single samples.

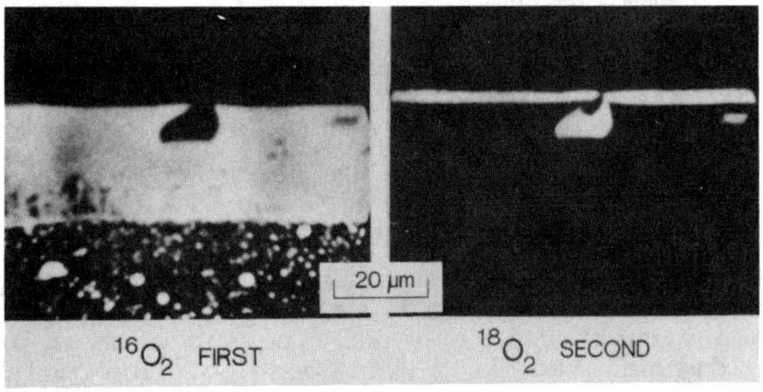

Figure 2. Isotopic images of a scale formed on Puratronic Ni oxidized at 1473 K for 150 min in ^{16}O followed by 30 min in ^{18}O. Scale structure is type 1B. The ^{18}O image shows where two pores floated to the scale-oxygen interface, vented and filled. Such pore filling cannot be detected by optical microscopy.

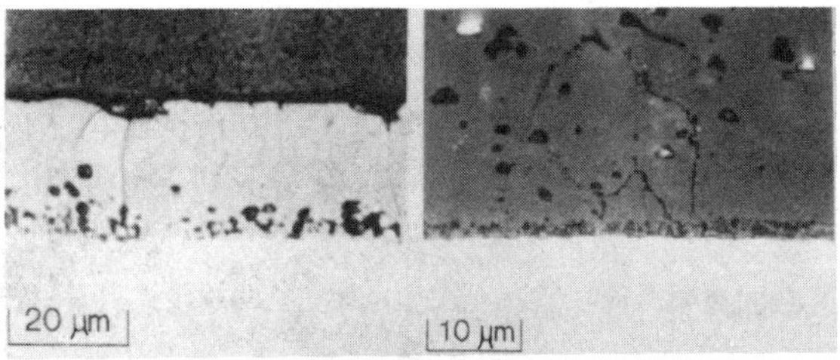

Figure 3. Pore and new grain formation at the columnar oxide - metal interface. a) Puratronic Co oxidized at 1273 K for 30 min. b) commercially pure Co oxidized at 1273 K for 450 min.

Most of the oxygen for the crystals growing on the metal in Figures 1B or 3, came from the dissociation of the base of the columnar scale. This dissociation proceeds unevenly giving it a distinctive waviness that is frequently marked by a line of pores as in Figures 4, 8 and 10. Evidence of the original metal surface is lost by the dissociation, and an initial tracer layer, if present at all, would remain only as fragments at the tips of columnar crystals.

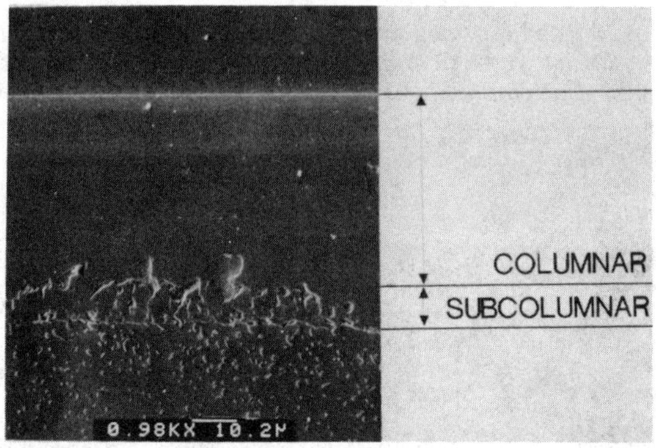

Figure 4. SEM and isotopic images of a scale formed on commercially pure Ni oxidized at 1473 K for 370 min in ^{16}O followed by 60 min in ^{18}O . The scale structure is type 1B. The subcolumnar grains formed from oxygen dissociated from the base of the columnar grains creating a distinctively wavy interface.

The remainder of the oxygen for the new crystals came from a more localized dissociation of the columnar scale creating pores that propagated towards the scale-gas interface. These pores tend to partially fill in, and only rarely are the open channels that fed oxygen to the subcolumnar scale evident, as in Figure 5. Samples that are oxidized in tracer oxygen first show this filling-in as an apparent back diffusion of oxygen, Figure 6.

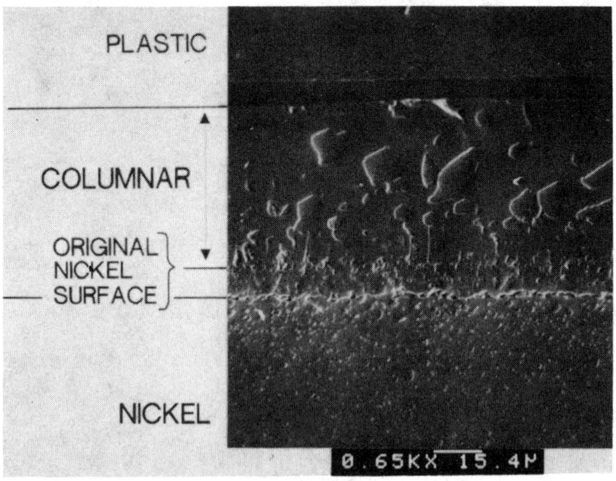

Figure 5. SEM images of a different part of the scale shown in Figure 4. Several channels are present connecting large pores in the columnar oxide to the subcolumnar oxide.

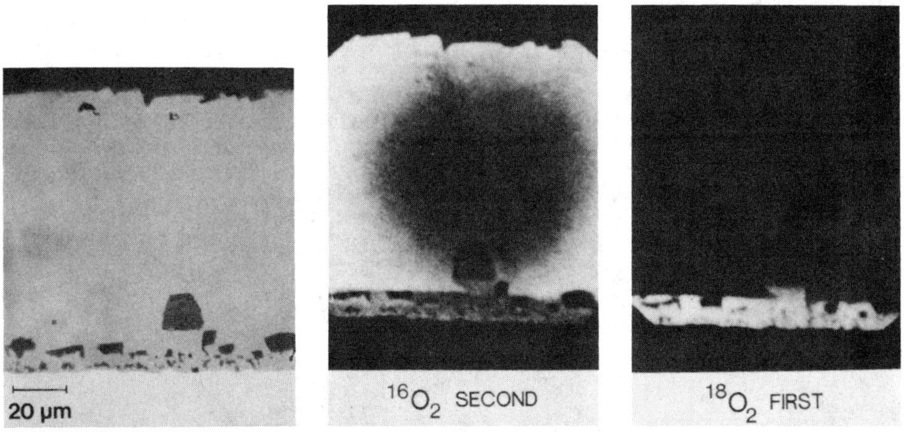

Figure 6. Optical and ^{18}O image of a scale on a Puratronic Co sample oxidized at 1373 K for 45 s in ^{18}O, followed by 300 min in ^{16}O. The large pore feeding oxygen to the subcolumnar oxide has partially filled in.

Scales that follow the Figure 1A-E-H scheme grow similarly except that most of the oxygen required for the new crystals comes from the localized type of dissociation of the columnar scale (pores) leaving the oxide at the original metal interface largely intact. These interfaces frequently form a distinctive linear feature in the scale, such as a grain boundary or line of small pores, Figure 7. As with structures 1 C and D, the larger pores usually back-fill, and so appear to be isolated from the inner oxide.

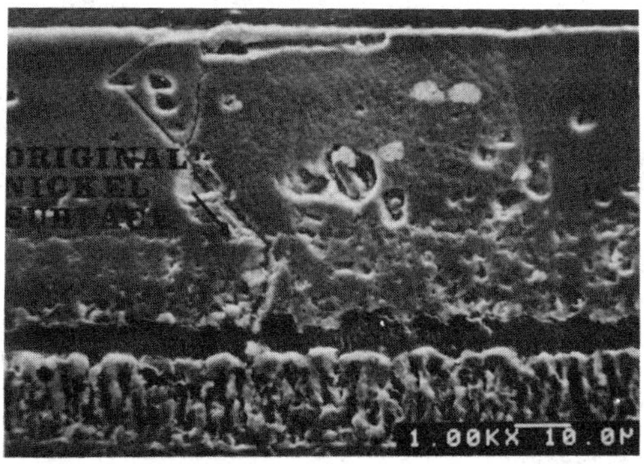

Figure 7. Type 1H scale on a commercially pure Ni sample oxidized at 1373 K for 20.5 hours in ^{16}O, followed by 4.5 hours in ^{18}O. The location of the original Ni surface is defined by the linear grain structure. There was no penetration of ^{18}O into the subcolumnar oxide at this location.

The oxide on the metal surface in structures 1B and 1F grows as isolated crystals that cover about half the surface. The crystals grow out until the separation to the columnar scale is bridged, whereupon the porous inner oxide partially fills in. The layer of inner oxide then separates from the metal presumably because of the enhanced vacancy flux resulting from the cation movement across the full scale. New oxide crystals form on the metal surface, and the cycle repeats to form a crudely layered inner scale, Figures 1D and H. Examples of these scale structures are shown in Figure 8 for 1D, and Figure 7 for 1H.

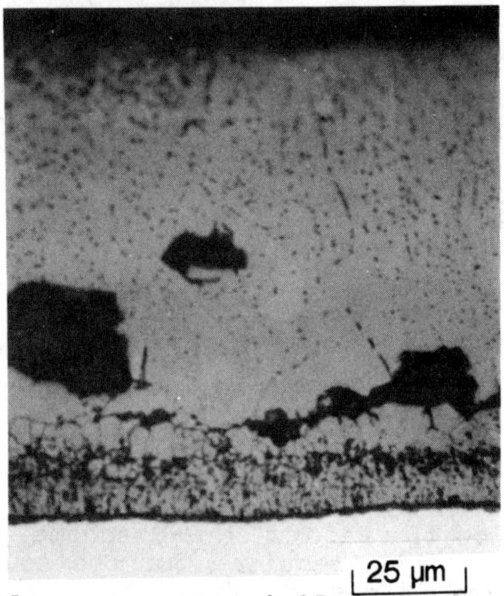

Figure 8. Type 1D scale on a commercially pure Co sample oxidized for 9 hours at 1273 K.

Tracer experiments show that the oxygen for the partial filling of the bridging crystals, and for the formation of new layers of subcolumnar oxide can come entirely from dissociation of the columnar scale. However as the scale thickens there is an increasing probability that oxygen from the atmosphere will gain access to the inner scale by scale cracking or pore venting. This oxygen seals the porous structure from the metal out, both by forming new crystals and by adding to the old. The latter cannot be detected by optical microscopy but form distinctive lateral features in isotope images, Figures 9 and 10. Such oxygen ingress events can occur at any time, or several times, during the formation of a scale, so that scale microstructures can be very complex. Leakage can also be quite local so that the scale structure can change several times down the side of a sample.

Scale structures on the alloys showed features similar to those in Figure 1, and also evidence of oxygen leakage. Leakage was confirmed by tracer experiments. For example, Figure 11 shows a scale grown on a Ni-Cr alloy which is essentially of the type shown in Figure 1D. The greatest difference is the presence of a band of solid oxide (arrowed in the figure) between the wavy base of the columnar oxide and the proper subcolumnar oxide. The band was confirmed to be the result of leakage as this particular specimen was oxidized in tracer oxygen last.

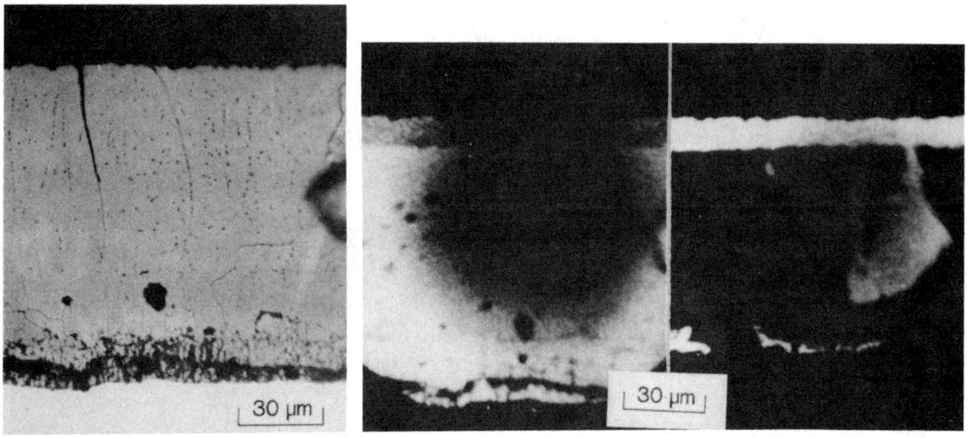

Figure 9. Optical and isotope images of a scale formed on a commercially pure Co oxidized at 1273 K for 7 hours in ^{16}O followed by 2 hours in ^{18}O. Most of the scale growth took place at the scale-gas interface. One large and one small pore vented and filled (not detectable on the optical image). The scale structure was basically type 1D, however the void between the subcolumnar and columnar oxide was partially filled by oxygen from the atmosphere.

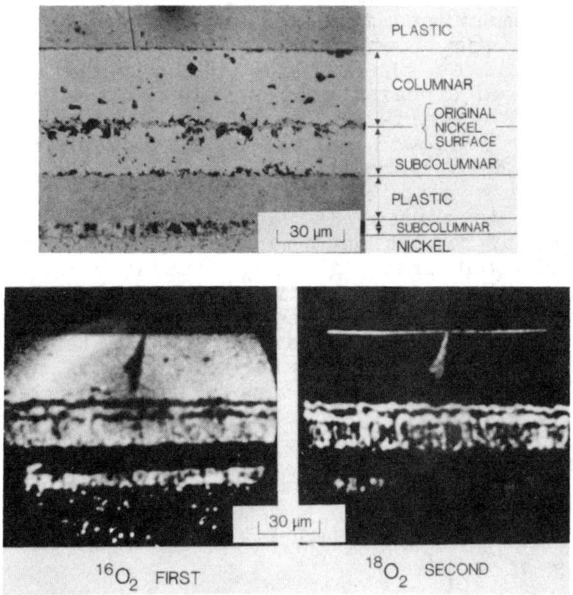

Figure 10. Optical and isotope images of a different location on the scale of the specimen of Figure 7. The scale structure was basically type 1D, however the scale admitted oxygen from the atmosphere on at least one , and possibly more, occasions. The fume-like crystals on the metal are intermixed $^{16}O/^{18}O$.

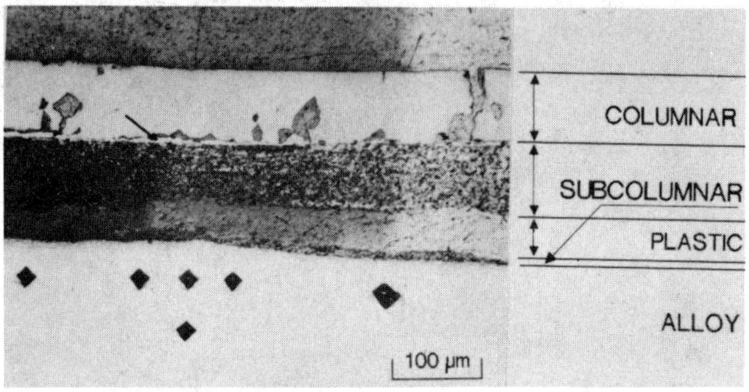

Figure 11. Scale formed on a CO 10 wt% Cr sample oxidized at 1473 K for 67 min in ^{16}O followed by 8 min in ^{18}O. ^{18}O was present at high concentrations in the solid band just beneath the columnar oxide (arrowed), and at the base of the large pores. ^{18}O was also present at lower concentrations throughout the subcolumnar oxide.

The samples used in this study were given a high polish before oxidation and were therefore microscopically flat. In the absence of oxygen movement the metal-scale interface was strictly planar and any deviation from flatness could be attributed to such movement. Specimens that followed the path A-E..H of Figure 1 frequently showed linear features that could be associated with the original metal interface. No such features were present for specimens that followed path A-B..D, as the first formed scale dissociates to form subcolumnar oxide. For this reason tracer first experiments were also of limited use for the location of the original metal interface. It was therefore of interest to observe the behaviour of traditional markers. Pt paint particles, and the Pt in the Pt containing alloys, usually defined a line in the subcolumnar oxide at the location shown in Figure 1 A...D. However there were occasional particles of platinum carried on the surface of pores well into the columnar scale. Similarly Al_2O_3 markers did not stay fixed in space but were carried back by the dissociating columnar scale, Figure 12. A sample marked with Pt paint on only one side gave the structure shown in Figure 13. Subcolumnar oxide only formed on the marked side, with the other side having a simple columnar structure.

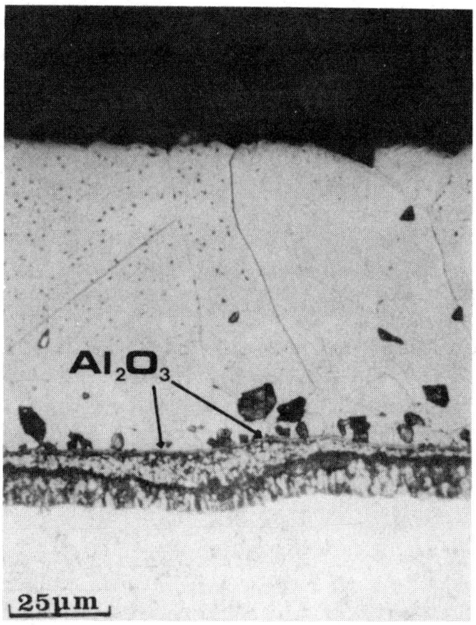

Figure 12. Type 1D structure on a commercially pure Co sample oxidized at 1273 K for 4 hours after marking with Al2O3 powder. The marker has receded with the dissociating base of the columnar scale.

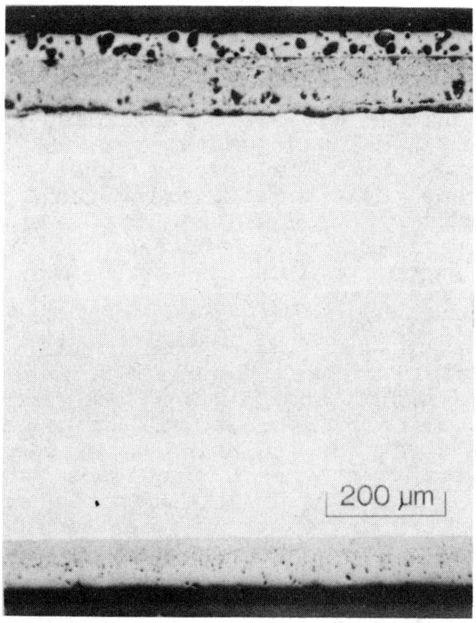

Figure 13. Scale structure across a sample of commercially pure Co oxidized for 5 hours at 1273 K. The top side of the sample was marked with Pt paint and has a 1H type structure. The unmarked side retained a simple columnar structure.

Discussion

This study has established the origin and significance of the major features of NiO and CoO scales. The finer details of exact growth sites, and the causes of nucleation, separation and failure events are beyond the resolution of the techniques used. Scale structures, and the extent of leakage, can vary greatly around a specimen without significantly affecting the total scale thickness. The basic rate controlling process in all scales is therefore the solid state diffusion of cations across the full scale width. Oxygen movement and inner scale growth occurs primarily to allow continued cationic diffusion. The process is better described as " bridging the created space" than filling the available space. Oxygen movement also partially fills in pores in the older inner oxide, and fills behind the large pores in the columnar scale. The latter process occurs by the movement of O_2 from the inner oxide to the columnar part of the scale, indicating that the oxygen potential in the pores is well above the equilibrium value.

The processes of bridging and filling can occur by the rearrangement of oxygen already in the scale. Should the scale crack, or a pore be created, then additional subcolumnar oxide will form until the path is sealed. However contrary to other models (2,3,8,9), such oxygen is not essential for continued scale growth.

In agreement with the "available space" model (2,3), there is no evidence that the new oxide creates its own space. For instance, the columnar scale structure, and the lines marking the original metal surface in 1H type structures, are continuous around specimen corners. However, if the oxygen for the inner oxide comes primarily from the dissociation of the base of the columnar scale, then the thickness of the subcolumnar scale can exceed the maximum allowed by a Pilling-Bedworth calculation (approx 58% for these oxides). It was also observed that the space created by the inability of the scale to recede with the shrinking metal core need not be distributed uniformly around samples (Fig. 13). Vacancy annihilation at a metal scale interface is therefore not easy, allowing vacancy injection into the metal and their availability for void creation at quite distant sites.

The interpretation of scale structures given in this paper would not have been possible without the use of tracer techniques in the imaging mode. Many significant features are localized, and further cannot even be detected by microscopy. The use of relatively thin tracer layers either at the beginning or at the end of runs was a great help. However, even tracer techniques have limitations for investigating the non-steady state processes seen in these scales. For instance, even if tracer oxygen can not be detected in the inner scale of a sample that was oxidized in tracer last, there is the possibility that the scale structure was affected by an earlier failure. Similarly, tracer layers at the beginning of runs are not infinitely thin, and so cannot detect movement of oxygen close to the metal-scale interface that is entirely within the tracer layer.

Conclusions

Simple columnar scales grow by cationic diffusion to form new oxide at the scale-gas interface.
The growth of complex scales (voids, subcolumnar grains, backfilling, and healing) involves processes that cannot be detected by microscopic examination of the grain structure.
Complex scales grow in response to any inability of the columnar scale to follow the retreating metal core.
The oxygen for subcolumnar oxide can be entirely provided by the dissociation of the columnar scale.
This dissociation can be relatively local, resulting in the formation of large pores in the columnar layer (which can partially backfill).
Alternatively, the dissociation can be less localized, resulting in a distinctively wavy base to the columnar layer.
Such subcolumnar oxide is porous, and the porosity increases towards the metal-scale interface.
The scale can crack or develop open porosity at any stage allowing the ingress of oxygen.
This oxygen tends to fill the porosity in the subcolumnar layer, and seals the cracks and pores in the columnar layer.
Small marker particles can move with the surface of the dissociating columnar oxide.

Acknowledgment

This research was funded by the Natural Sciences and Engineering Council of Canada.

References

1. A. Atkinson, Rev. Mod. Phys., 57, 437 (1985).
2. J. Robertson and M.I. Manning, Mat. Sci. Tech., 4, 1064 (1988).
3. D.S. Cox, M.E.Sc. thesis, The University of Western Ontario, 1985.
4. B. Gleeson, M.E.Sc. thesis, The University of Western Ontario, 1987.
5. H.L. Williams, M.E.Sc. thesis, The University of Western Ontario, 1989.
6. J.S. Sheasby, and B. Gleeson, Oxid. Met., 5-6, 379 (1989).
7. T.A. Parthasarathy and P.G. Shewmon, Acta metall. 32, 29 (1984).
8. S. Mrowec, Corrosion Sci., 7, 563 (1967).
9. G. Gibbs and R. Hales, Corros. Sci.,17 487 (1977).

An in situ measurement of stresses generated during scale growth limited by cation/vacancy diffusion

T. Homma
IIS, University of Tokyo, Roppongi, Minato-ju, Tokyo, Japan

An x-ray diffraction technique has been employed for determining stresses in oxide and metal, respectively. A hot-stage capable of "in situ" measurement of elastic strains at elevated temper -atures under controlling atmosphere has been used. Copper specimen has been chosen since a new model was proposed for the case of copper oxidation. The oxidation condition is so selected that the copper is covered with only Cu_2O layer. The mosaic structure in the oxide layer is controlled by changing a fraction of intrinsic misfit grain boundaries. This is realized bycontrolling epitaxial nucleation of the oxide. The strain generation during growth of the oxide on copper shows clear dependence on crystallographic ori -entation of the substrate copper. This characteristic of the strain is described in terms of the intrinsic misfit for a grain boundary in the oxide. No observation of the strains in the metal is compared with the predictions by both the internal oxidation theory and the intrinsic misfit interface dislocation theory.

Introduction

Stresses and strains are experimental facts for growing oxide scales. Quite a variety of interesting and complex behaviors may well be found in such experiments. One example is the stress generation during scale growth limited supposedly by cation vacancy diffusion. The stress generation during growth of NiO scale on pure Ni is a typical case(1,2). Recent evidences on the transport processes in the oxidation of Ni studied using tracers in growing NiO scales showed that, at temperatures below about 1100K, the principal transport mechanism was short-circuit diffusion of Ni in NiO. In this temperature range, inward oxygen transport also contributed to scale growth on Ni and was responsible for growth of the inner layer of duplex scales(3).

The following three possible ways of inward oxygen transport have been pointed out:

1. grain-boundary diffusion
2. grain-boundary fissures produced by dissociation
3. cracks induced by growth stresses.

Although among these possibilities the most probable mechanism by which oxygen gains assess to the inner scale is through cracks, a question of how the stresses to be generated remains(3). The purpose of this paper is to present a new information of generation features of the stresses which obtained by means of in situ x-ray strain measurements during oxidation(4∿6).

Methods

General oxidation characteristics

The oxidation results described here were obtained at temperatures between 573∿923K in pure oxygen at pressures between 25∿110 Pa. Under these conditions only Cu_2O may form on the copper surface(7). In this study no CuO formation was confirmed by using in situ RHEED technique.

Two different types of Cu specimens were employed. The first ones were single-crystal slices($7x5x0.5mm^3$) cut from zone-refined crystal of >99.99% purity grown by the Czochralski method. The flat surfaces were (100) and (111) faces. The second ones were poly crystalline Cu produced by hot-pressed method at 1173K under pressure of $9.8x10^6$Pa for 1800s.

For the epitaxial relationships of Cu_2O on the single crystal surfaces, the most general feature of the orientation relationships is the parallelism of at least one close-packed [110] direction in the metal and the oxide, and the twin-related oxide orientation as shown in Figure 1. These relationships were observed for the oxidation in gaseous oxygen at temperatures from 373 to 1173K (100∿900°C)(8). Since a purpose of our investigation was to obtain clear evidence for the correlation between the stress generation and the grain boundary structure of oxide layer, it was planned to control the structure of oxide, "the mosaic structure", by changing epitaxial behavior. We selected the (100)face for this purpose because the twin related epitaxial behavior could be controlled on the (100)face by changing the misorientation that is the off-angle with in a few degrees from the <100>pole(9,10). The specimens were cut by acid sawing and annealed, and finally electrochemically pol -ished in a phosphoric acid solution, changing their crystallographic orientations. The orientation correction was examined by the Laue back-reflection diffraction method with accuracy of 1/2°.

Stress measurement

The diffractometric residual stress measurement method was employed for determining elastic stress(11). The Bragg angle position of a diffracted beam is much more sensitive to a given change in plane spacing when the angle θ is large than when it is small. In the cubic system, the fractional error in the lattice parameter a caused by a given error in θ approaches zero as 2θ approaches 180°. We therefore used particular characteristic lines of x-rays for each diffraction measurements, respectively. The experimental conditions for the diffraction measurement have been shown in Table 1.

A proper scanning speed (0.5°/s) was employed to minimize the influence of change in strain generation feature during the measurement. When the diffraction intensity was weak, as in the case of oxide, large time constant (10s) and small scanning speed (0.2s/min) were conbined, in order to result in a relatively large averaging-of-counts effect for recording a proper profile.

TABLE 1

Experimental conditions for the diffractometric residual stress measurement method,"$\sin^2\psi$-method",using parallel beam of area $2x4mm^2$.

Diffraction Planes	Diffraction Angles(2θ)	Remarks
Cu_2O (222)	136.56	Cr $K\alpha_1$
Cu_2O (311)	125.75	Cr $K\alpha_1$
Cu (400)	163.50	Co $K\alpha_1$
Cu (222)	135.53	Fe $K\alpha_1$

In a general case for the high temperature oxidation, the stress generated in the oxide layer is usually biaxial one because the stress normal to the surface may be zero at a free surface. In order to determine the stress parallel to the specimen surface, we measure the strains along two mutually orthogonal directions.

The fundamental equations on which practical procedures for the stress analyses are based will be derived and explained by using the ellipsoid of strain as shown in Figure 2. The details of that are shown in Appendix.

Practical measurements were conducted by using "the $\sin^2\Psi$-meth-od" which is one of the side-inclination methods". Since this method is simple to operate and uses a parallel x-ray beam in this study, it has a following advantage. That is, a reduction in error due to the geometrical misalignment which can be realized by a rotation of the angle Ψ by $\sim$35°. In these measurements Ψ-values were chosen as follows: Two specific Ψ anles(0° and 29.5°,refered to Figure 1.b) for the epitxially grown Cu_2O on the single crystal coppers and several angles(0°,15°,25°, and 35°) for the Cu_2O grown on the poly-crystalline copper specimens.

A peak position of diffracted beam measured by a fixed time counting method as the angular setting was determined by numerical treatments. The least squares method was used for determination of the lattice strain as a function of $\sin^2\Psi$

A hot stage capable of "in situ" measurement of elastic strain at elevated temperatures up to about 1200K under controlling atmos-phere has been used. A schematic detail is given in Figure 3, and a detailed description has been reported elsewhere(12).

X-ray $\sin^2\Psi$-integral method is also applied to analyze the residual strain distribution in the specimen which has a stress gradient in the depth direction and a non-linear relation of lattice strains measured by x-ray diffraction to $\sin^2\Psi$.

These methods were applied to obtain an evidence of the strain generation in comparison with a prediction by the internal oxidation theory(13) and the intrinsic misfit interface dislocation theory(14).

Details of the $\sin^2\Psi$-integral method are shown in Appendix and have been reported elsewhere(15).

Results

Strain generation during oxidation

For polycrystalline specimens, the lattice strains as function of $\sin^2\psi$ were measured. Figure 4 shows plots of a versus $\sin^2\psi$ for one set of experimental measurements of the(111) planes of the cu-rous oxide formed on a polycrystalline copper surface, where a is the lattice spacing of (100) plane derived from the spacing of concerned diffraction plane, (111). It is evident that the two plots are approximately linear and that compressive stresses are generated in the oxide at 573K and 773, respectively. Vertical bars (at ψ=0° and ψ=30°) show a spread of the measured points.

Though a corresponding measurement was performed for the sub-strate metal, no evidence of stress generation was obtained in the metal even at 573K.

Similar stress generation behavior was observed almost contin-uously after an initial oxidation period.

According to the elastic anisotropy effect (see Appendix), the correction factor was obtained by measuring of polycrystalline specimen.

Figure 5 shows the effect on the plots of a versus $\sin^2\psi$. These plots were obtained for two set of experimental measurements of the same oxide,as that of Figure 4 (corresponding to the oxida-tion run at 773K), after cooling to room temperature. It is clear that these linear lines are derived from two set of the experimen-tal measurements of the(111) and the(311) planes of the same oxide formed on a polycrystalline copper surface.

It is worthwhile to note that the strains evaluated from dif-ferent diffraction planes in the same polycrystalline-specimen are different. This fact may arise from an anisotropy in deforma-tion which implies the model of a homogeneous stress distribution to be reasonable instead of a homogeneous strain distribution.

The elastic anisotropy is responsible for this difference.

In case of Cu_2O grown epitaxially on the single crystal copper surfaces, the(222) and the(311) Bragg reflections of the oxide occur from the atomic planes in the oxide parallel to the surface (ψ=0°) and inclined to the surface (ψ=29.5°),as shown in Figure 1 (b), respectively. A change in the interatomic distance of these atomic planes due to the anisotropy-effect, can be corrected with a correction factor which was obtained by means of the $\sin^2\psi$-meth-od for the corerelated hkl-diffractions, using polycrystalline copper specimen at a given temperature. The correction factor was evaluated as $\varepsilon_{111}/\varepsilon_{311} = \sim 0.87$. This figure is well corresponding to the value derived from the elastic compliance coefficients of a cubic crystal, assuming a homogeneous stress distribution model.

For single crystal specimens, stresses generated during epi-taxial growth of Cu_2O scale on the (100) and the (111) copper surfaces are identified as compressive mode. Figure 6 shows depend-ence on temperature and crystal-orientation.

It is evident that the oxide exhibits a pseudo elastic plastic transformation at elevated temperature. This is in agree-ment with the fact that an increase in temperature increases dis-location mobility thus facilitating plastic flow at lower stress-es.

The investigation of mechanical properties of Cu_2O scales by Menzies and Aldred(16) shows that the oxide exhibited a pseudo brittle-ductile transformation at 360°C(633K) when slow deformation rates were used.

Though the surface cracks were found in the Cu_2O specimens, during compressional experiments at <400°C(673K), there is no clear information for higher temperatures.

It is found that crack nucleated at local stress concentrations where slip bands crossed and also where they met grain boundaries.

It might be expected that,in the transformation temperature range,the plastic deformation in the scale,in combination with that in the metal, due to active operation of slip systems, occurs conspicuously depending on the crystallographic orientation.

Figure 7 shows the strain generation behavior in the oxide and its dependence on the temperature and the metal surface orientation.

Qualitatively, the strain generation behaviors are almost similar for the single crystal specimens.

SEM-observation revealed a formation of micro-voids in the scale. Some scales showed flaking off from the metal surface during cooling. The exposed metal surface revealed the formation of regular facetted voids at the metal-oxide interface. The (111) surface showed more conspicuously when the oxidation ran at lower temperature(i.e. 573K).

Quantitatively, the strains were generated differently. The stress values are evaluated quantitatively assuming that the stress normal to the surface is zero and therefore the scale is subjected to principal stresses $\sigma_1$and σ_2 in a biaxial mode (i.e. $\sigma_1=\sigma_2$) in its surface plane, and providing that the elastic con- is $2.8x10^{10}N/m^2$(16) and Poisson's ratio is 0.4 for Cu_2O. Table 2 shows the results.

TABLE 2

Summary of the results of the stress measurements

Specimens	Temperature/K		
	573	773	973
Cu(100)	6∿13 (15∿23)	0∿2 (10∿17)	∿0 (13∿16)
Cu(111)	3 ∿ 7 (13∿18)	0∿1 (10∿13)	———
Polycrystal	———	∿1 (15)	———

Stress,$\sigma x10^{-7}N/m^2$, $\Delta\sigma$:$\pm 2x10^7N/m^2$, figures in a parenthesis are values measured at room temperature due to athermal effect.

Effect of a mosaic structure

The lattice strains were generated a little differently on the (100)faces of copper with a slight misorientation within a few degrees. Figure 8 shows a corresponding change in the lattice strains in the cuprous oxide with a change in the mosaic structure controlled by the epitaxial behavior on copper surface.

The histogram indicates the relative amounts of the twin related orientations in the oxide formed on Cu(100) face.

(A) and (B) are the results for the specimens with 1°-misorientation that is an off angle from Cu(100) surface, (C) 3° to [011]azimuth, and (D) 4° to [011]azimuth.

Oxidation was done at 573K in pure oxygen at 0.5 Torr for 2000s.

The smaller strains are observed in the oxides which have a simple preferrential orientation like a single crystal with a small lattice misfit or a coherent grain boudary(C and D in Figure 8).

Strain generation in metal

Since the internal oxidation model predicted a possibility of the strain generation due to oxygen gradient into the metal, the $\sin^2\psi$-integral method was used to obtain an evidence for the strain generation near the metal/oxide interface. No detectable strain was measured. This is probably due to the low oxygen solubility and the low Yield strengh of copper metal.

Stress generation mecnanism

First, the experimental evidences obtained are surveyed here briefly.

An evidence obtained previously by Atkinson(3),that single layer scale growing by lattice diffusion of Ni and no contribution from O-transport were observed at 1273K(1000°C) on the (100) face of single crystal Ni, gives the following suggestion. At the higher temperatures above $\sim$0.5Tm where Tm is the melting temperature of the oxide, 2230K for NiO, a plastic deformation may proceed more easily. This alteration of mechanical properties may reduce the micro-crack-generation induced by growth stresses.

There is the similar expectation that such features do exist in the Cu_2O scale formation. In the case of Cu_2O, above $\sim$0.5Tm (Tm,1503K for Cu_2O) the micro-crack-generation may be reduced under loading of the low level stresses.

It is possible to predict that cation transport occurs via grain boundaries in cases of short-circuit diffusion and through lattice for the ideal case.

A question,"how are the stresses generated", arises now. There is considerable controversy regarding whether formation of new oxide can occur within the bulk scale. Conclusive evidence has not been presented(17).

The model in which the stress may arise from contraction of the metal core and the nucleation of crack where vacancies have condenced at the metal/oxide interface, is most probable, for accepting the evidences obtained in this work.

The mosaic structure effect observed on the (100)Cu crystals in this work may be explained by a vacancy annihilation model proposed by Pieraggi and Rapp(14).

Summary and conclusions

Experimental techniques have been develped for determining

stresses in oxide and metal during high temperature oxidation.

The strain generation during growth of the cuprous oxide on copper specimens shows a dependence on crystallographic orientation of the substrate copper. The surface orientation dependence in the strain generation is as follows,

$$\varepsilon(100) > \varepsilon(\text{polycrystalline}) > \varepsilon(111).$$

It is possible to explain the observed temperature dependence of the strain generation. At temperatures above $\sim 0.5Tm$ where Tm is a melting temperature of the oxide, that is 1053K for Cu_2O, plastic-deformation occurs in the oxide relieving the elastic strain.

The oxide exhibited a pseudo elastic-plastic transformation in a temperature range including 0.5Tm.

No detectable strain was measured in the metal in comparison with the strain generation in the oxide though a detectable strain was detected in the oxide below the transformation temperature. This indicates that there are no mechanical equilibrium for the Cu_2O/Cu system during high temperature oxidation.

The most plausible stress generation model for the oxide is that of the contraction of the metal, which is consistent with the many experimental facts.

No observation of strain in the metal may call for further reexamination.

The mosaic structure effect on the stress generation needs to reinforce with a vacancy injection model.

Acknowledgments

The author gratefully acknowledge my indebtness to Professor W.W.Smeltzer and Dr. J.V.Cathcart for receiving suggestions in the beginning of this study. This work was supported by a grant from the Ministry of Education, Science and Culture,53-247056.

References

1. F.N.Rhines and J.S.Wolf,Metallurgical Trans.,1,1707(1970)
2. F.N.Rhines and R.G.Connell Jr.,Stress Effects and the Oxidation of Metals,ed.by J.V.Cathcart,P.94,Metallurgical Soc.AIME NY(1975)
3. A.Atkinson,R.I.Taylor, and P.D.Goode, Oxid.Metals, 13,519 (1979)
4. Y.Pyun,T.Homma, and K.Takakuwa, J.Japan Inst.Metals,47,663 (1983)
5. T.Homma and Y.Pyun, Proc.JIMIS-3 (1983),High Temperature Corrosion, Trans Japan Inst. Metals,Supplement, 161
6. T.Homma and Y.Pyun,Intern. Conf. Residual Stresses,ed.by G.Beck,S.Denies, and A.Simon,Elsevier Applied Science,London (1989)P.279
7. T.Bretheau,B.Pellissier,and B.Sieber, Acta Met.,29,1617(1981)
8. T.Homma and T.Yoneoka,Jpn.J.Appl.Phys.11,1231(1972)
9. T.Homma and T.Yoneoka,J.Appl.Phys.,46,1459(1975)
10. A.T.Gwathmey and K.R.Lawless,The Surface Chemistry of Metals and Semiconductor,ed. by H.C.Gatos,John Wiley and Sons,Inc. NY(1960)P.483
11. C.S.Barrett and T.B. Masalski,Struture of Metals,McGraw-Hill, Inc.,NY(1966)P.466
12. K.Kataoka,T.Yamazawa,Y.Pyun,and T.Homma,Trans.ISIJ,24,365

(1984)
13. J.B.Condon,S.S.Cristy, and J.R.Kirkpatrick, Oak Ridge Y-12 Plant,Y-2394(1988)
14. B.Pieraggi and R.A.Rapp, Acta Metall.,36,1281(1988)
15. T.Bretheau,B.Pellissier, and B.Sieber, Acta Metall.,29,1617 (1981)
16. I.A.Menjies and P.Aldred, J.Electrochem.Soc.,116,1414(1969)
17. F.H.Stott and M.I.Manning, ed.by P.Kofstad,A.Rahmel,R.A.Rapp and D.I.Douglass, Oxid.Metals,32,125(1989)

Appendix

$\sin^2\Psi$-method

The principal stresses are σ_1, σ_2, and σ_3 and the principal strains, $\varepsilon_1, \varepsilon_2$, and ε_3 in isotropic bodies.

σ_3 is directed normal to the surface of a polycrystalline specimen. There is no stress possible at a free surface in the absence of an external load, that is $\sigma_3=0$. The strain normal to the sample surface is

$$\varepsilon_3 = -\frac{\nu}{E}(\sigma_1+\sigma_2). \qquad (1)$$

The stress parallel to the surface at angle ϕ from axis number 1 in igure 2 is

$$\sigma_\phi = \sigma_1\cos^2\phi + \sigma_2\sin^2\phi. \qquad (2)$$

The following approximate form may be reduced,

$$\sigma_\phi = \frac{d\psi-d\perp}{d\perp}\cdot\frac{E}{1+\nu}\cdot\frac{1}{\sin^2\psi} \qquad (3)$$

where $d\perp$ be the spacing in the stressed perpendicular to the surface, $d\psi$ is the spacing in the direction specified by ψ,ϕ, E is Young's modulus, and ν is Poisson's ratio.

Elastic anisotropy effect

In the case of stress measurement of the epitaxially-grown Cu_2O, it is necessary to consider the anisotropy-effect of the elastic constants. In cubic crystals, the relation between a strain ε and a stress σ can be given as

$$\varepsilon = [s_{11}-(2s_{11}-2s_{12}-s_{44})\Omega]\sigma \qquad (4)$$

where s_{11}, s_{12}, and s_{44} are the elastic compliance coefficients, and Ω is given as a function of direction cosine, l,m, and n as follows(16)

$$\Omega = l^2m^2 + m^2n^2 + n^2l^2. \qquad (5)$$

Using values of the elastic compliances for MgO which has a cubic structure (s_{11}= 297.1GN/m^2, s_{12}= 95.4GN/m^2, and s_{44}= 156.1 GN/m^2), a difference due to the anisotropy is estimated to be within 15% for strains along [111] and [311] directions.

For Cu_2O, the correction was done by using $\Omega_{111}/\Omega_{311}$-ratio which was obtained from the measurement of poly-crystalline specimen.

The $\sin^2\psi$-integral method

In this method, a strain at point 0,see Figure 2(b), is defined as follows, assuming that the strain changes linearly in X, Y, and Z direction,respectively,

$$\varepsilon_{\phi\psi} \equiv A(\psi)+B(\psi)\cdot X+C(\psi)\cdot Y+D(\psi)\cdot Z. \qquad (6)$$

Considering the strain as weighted function in the depth direction, the above equation can be rewritten.

Integration can be performed numerically from a series of polynomial using the Simpson's rules.

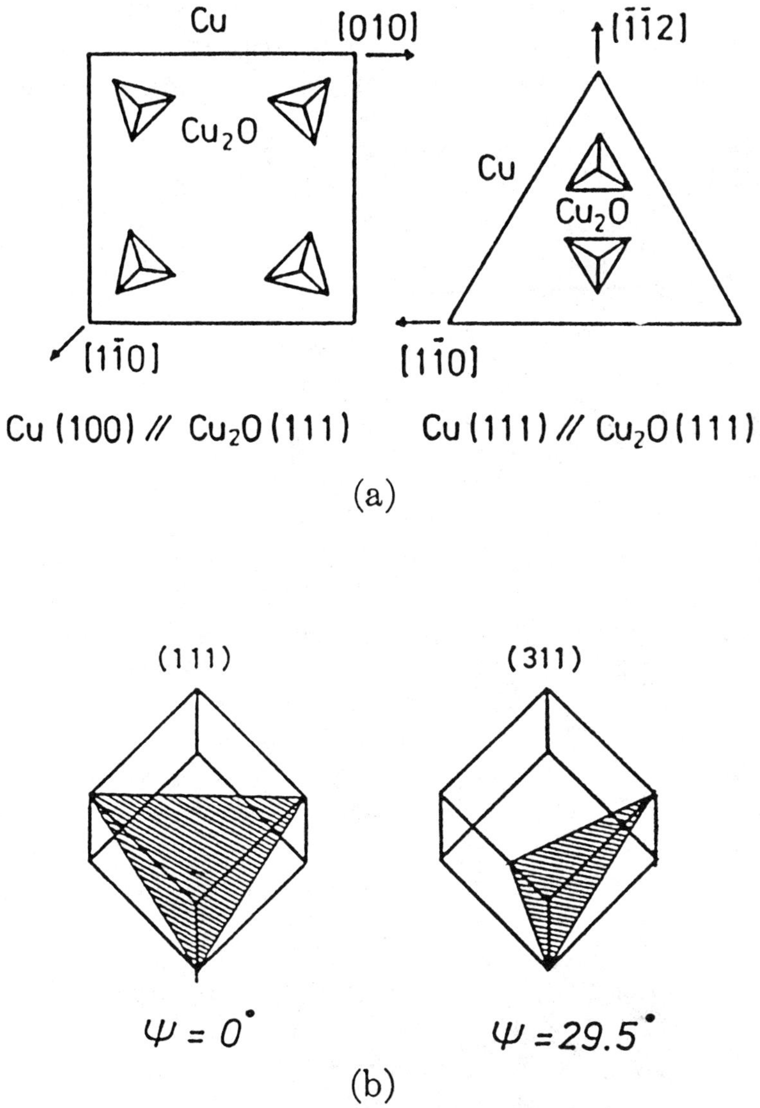

FIG. 1. Schematic diagram showing that (a) epitaxial relation for the Cu_2O formed on the (100) and the (111) faces of copper, and the twin-related oxide orientation. (b) Miller indices of lattice planes of Cu_2O lattice. Specific values, Ψ, indicate angles between the lattice planes of the form (hkl) and the (111).

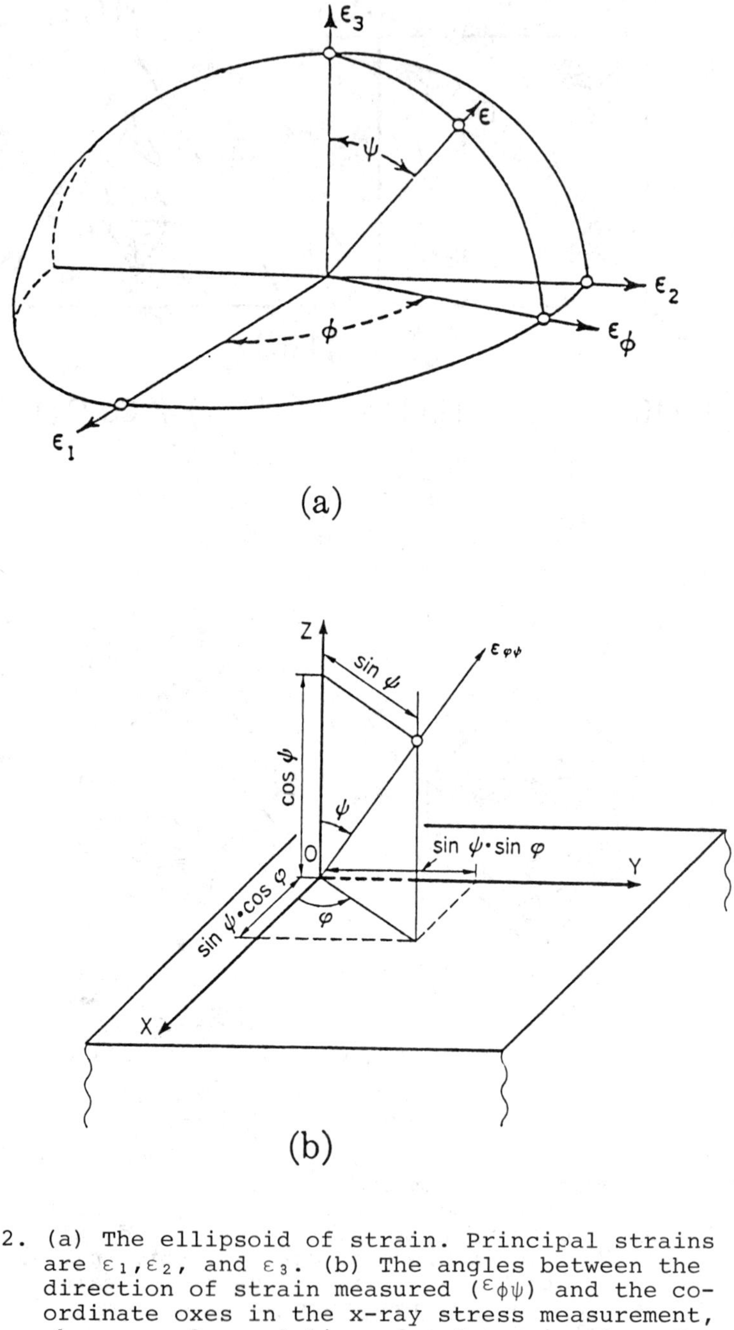

FIG. 2. (a) The ellipsoid of strain. Principal strains are $\varepsilon_1, \varepsilon_2$, and ε_3. (b) The angles between the direction of strain measured ($\varepsilon_{\phi\psi}$) and the coordinate oxes in the x-ray stress measurement, where angular relations for determination of the strain component in the ϕ direction.

HOT STAGE

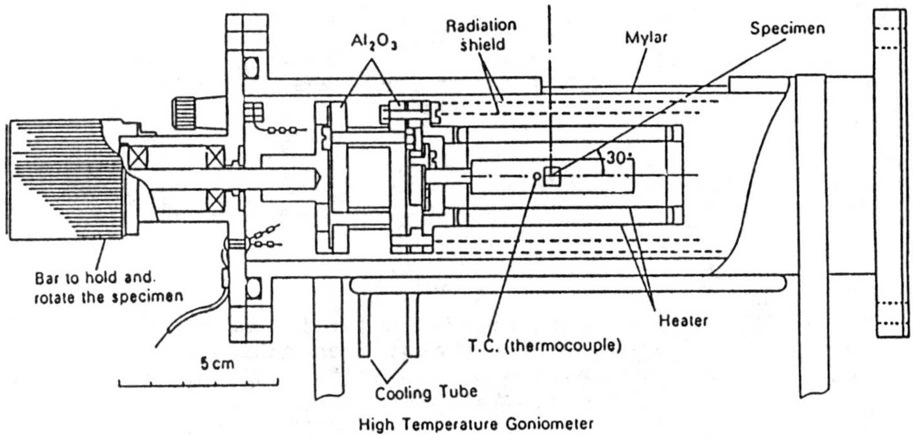

Measurable angle---$2\theta = 120 \sim 165^{\circ}$

Atmosphere ....Vacuum 10^{-5} Torr

FIG. 3. A high temperature -heating apparatus for the measurement of oxidation stresses generated in the metal/oxide scale system, by x-ray diffraction method.

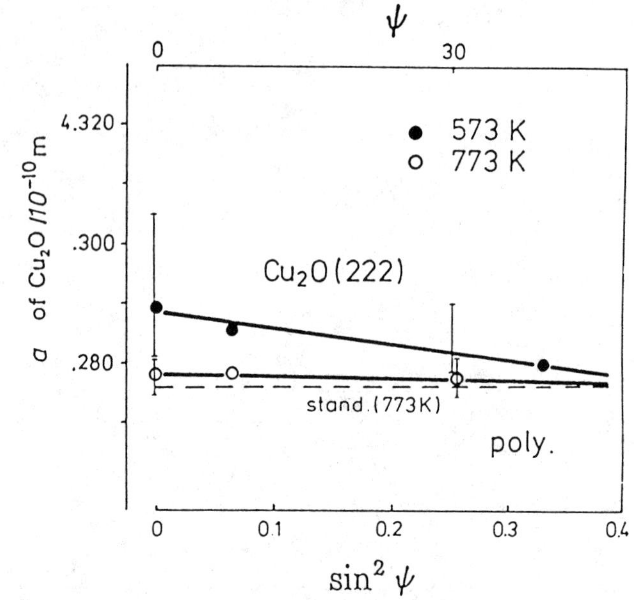

FIG. 4. ψ-dependence of the lattice spacing, indicating a certain generation of compressive stress in the Cu_2O grown on a polycrystalline copper at 573K, and a slight generation at 773K.

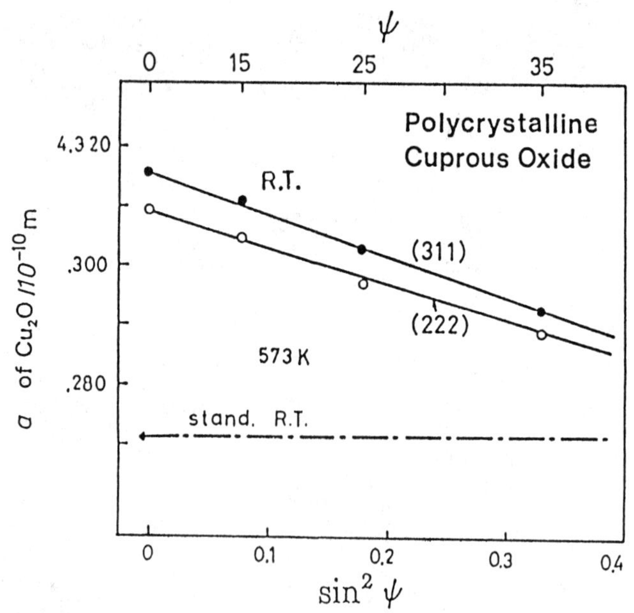

FIg. 5. Lattice strains in the same oxide formed at 773K as that of FIG.4 after cooling to room temperature, showing elastic-anisotropy effect.

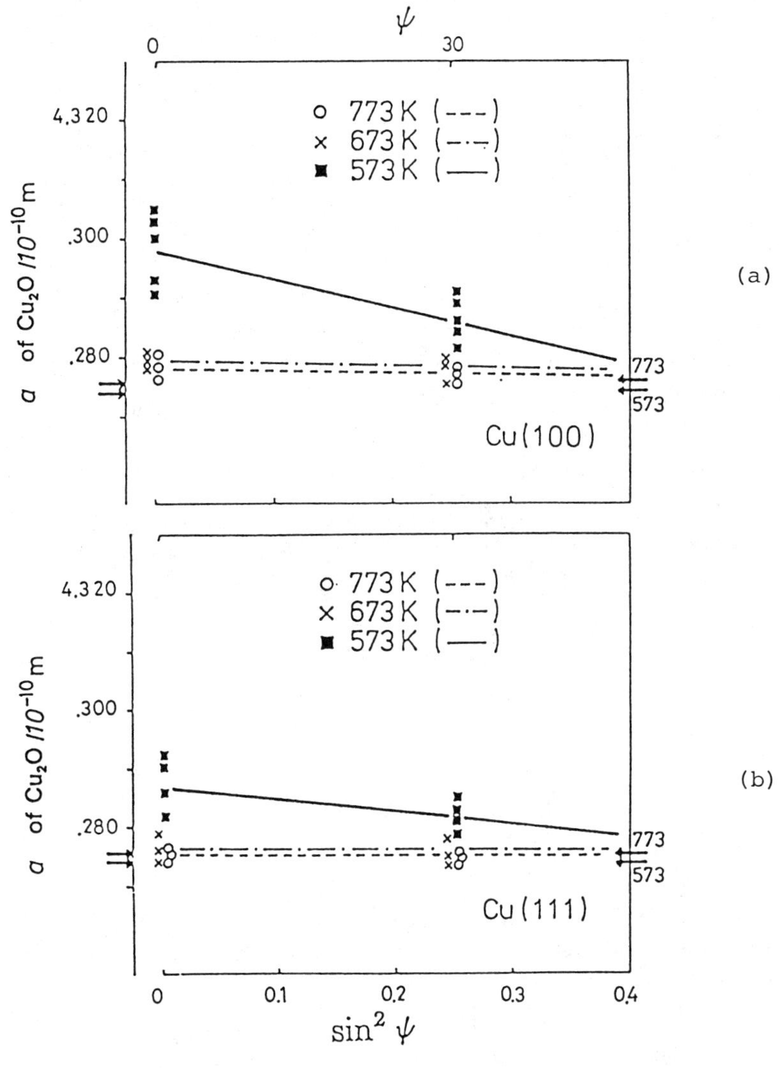

FIg. 6. ψ-dependence of the lattice spacing. (a) indicating lattice strain in the oxide grown epitaxially on Cu(100). Solid line represents compressive strain, broken and chained lines no strain observed at higher temperatures. (b) lattice strain in the oxide grown epitaxially on Cu(111).

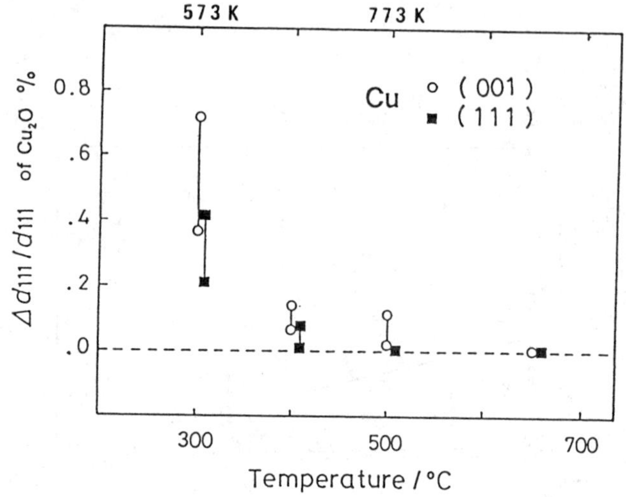

FIG. 7. Variation of the lattice strains in the cuprous oxide scale as a function of oxidation temperature for the different metal surface orientation, showing development of plastic deformation with increasing temperature.

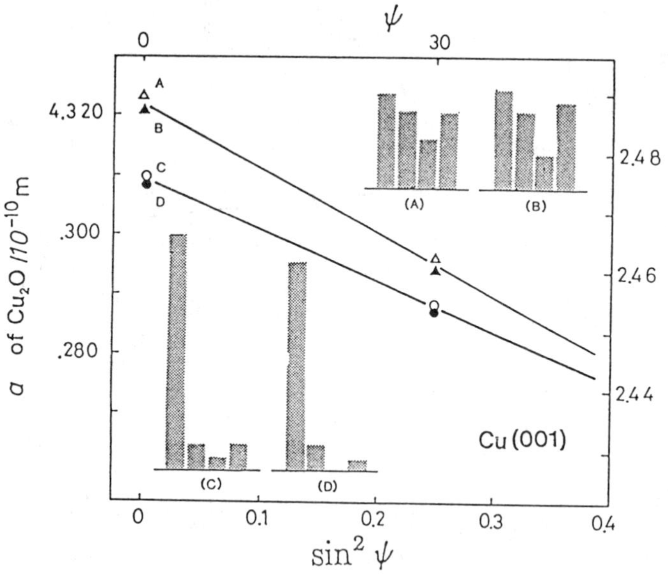

FIG. 8. Comparison of lattice strains generated in the cuprous oxides with different mosaic structure. Histogram indicates the relative amounts of two the twin-related orientations in the oxide on Cu(100) face. (A) and (B), misorientation ∿1°, (C) 3° [011], and (D) 4°[011].

Isothermal oxidation properties of pure copper

C. Zhou
Department of Materials Science and Engineering, Massachusetts Institute of Technology, Cambridge, Massachusetts 02137, U.S.A.

W.W. Smeltzer
Institute for Materials Research and Department of Materials Science and Engineering, McMaster University, Hamilton, Ontario, Canada, L8S 4M1

Abstract

The isothermal oxidation kinetics of pure copper and the oxide scale structures were studied from 800 K to 1200 K with $P_{O_2} \geq P_{O_2}(Cu_2O/CuO)$. The growth rates of the duplex scales containing layers of Cu_2O and CuO were examined based on experiments and physical modelling. Good agreement is obtained for the parabolic oxidation rates determined from experiment and by modelling computation using independently obtained diffusion coefficients for copper and oxygen. Growth of the Cu_2O layer is attributed to singly charged and neutral copper vacancies with minor concentration of oxygen interstitials. CuO behaves as an n-type oxide; copper interstitials and oxygen vacancies may be the mobile defects during growth of this oxide layer.

Key words

Cu oxidation, Cu_2O, CuO.

Introduction

The oxidation properties of copper have been extensively reviewed (1-5). Oxidation kinetics over a wide temperature range from 73K to 1273K reported by different investigators have been interpreted on the basis of the five simple laws of metal oxidation: linear, parabolic, cubic, direct and inverse logarithmic rate equations (2). Utilizing a model for the growth of a duplex oxide scale on the metal core (6-8), Garnaud (9) interpreted the mechanism for parabolic growth of the duplex Cu_2O/CuO scale. In this contribution, the isothermal oxidation kinetics are examined of copper exposed in various atmospheres of $P_{O_2} \geq P_{O_2}(Cu_2O/CuO)$ over a large temperature range from 800 to 1200K. A computation is carried out to evaluate the parabolic oxidation rate constants by Wagner's oxidation theory using recently published tracer diffusivity data for copper and oxygen in Cu_2O.

Experimental

Copper disc specimens for oxidation experiments, 9.5 mm diameter and 5 to 10 mm thickness, were prepared from 99.998 wt% pure rods which contained to 1μm diamond paste using kerosene as lubricants, degreased with acetone and then with methanol and ultrasonically cleaned in these liquids before each oxidation run.

Most copper specimens were oxidized in flowing oxygen at 1 atm pressure for the reaction temperatures selected; some were oxidized in static oxygen at 0.33 atm and others at the equilibrium Cu_2O/CuO pressure by heating the metal specimen buried in a compacted powder of these oxides corresponding to a Rhines pack. The oxidation kinetics were measured thermogravimetrically using an Ainsworth type RV, semi-automatic balance and a Cahn RG electronic continuously recording microbalance using assemblies as previously described (10,11). Oxide surface morphologies and cross-sections were examined using optical and scanning electron (SEM) microscopies; phase identification of oxides were performed by X-ray powder diffraction using Cu-Kα radiation.

Results

Morphological Structures of Oxide Layers

A duplex Cu_2O/CuO scale developed on copper exposed to oxygen at 1 and 0.33 atm. The inner Cu_2O layer grew rapidly whilst the outer CuO layer grew slowly. SEM images of the surface morphology are shown in Fig.1. The surface exhibits a lace pattern due to preferential thickening of oxide at the boundaries of the fine polycrystalline oxide in the outer CuO layer. This morphological pattern persisted up to complete oxidation of a copper specimen as shown in Fig.1b. On the other hand, when a copper specimen was annealed and oxidized in the Cu_2O/CuO powder compact, only Cu_2O was grown as large well defined grains as shown in Fig.2. The average size of these oxide grains 100 ~ 200μm corresponded closely to the grain size of the metal substrate. Grooves decorated the oxide grain boundaries and they were confined to the external surface region of the scale as shown by complementary examination of transverse sections. This boundary grooving of CuO possibly arises from the low equilibrium oxygen pressure of the oxides in the pack.

A fractured cross-section of a specimen exposed for 32 h in flowing oxygen and exhibiting the Cu_2O/CuO duplex scale is shown in Fig.3. The oxide scale has developed uniformly. The Cu_2O layer is 600-700 μm thick but the CuO layer is too thin for determination of its thickness. Micrographs of the highly magnified CuO layer are shown in Fig.4; the Cu_2O/CuO interface is of a crystallographic faceted appearance. For all examined specimens oxidized at P_{O_2} = 1 and 0.33 atm for exposures ranging from 5 to 32 h at 1173K, the CuO layer was 8 to 10 μm and negligibly thin in comparison to the rapidly growing Cu_2O layer.

Growth Rates of the Oxide Layers and Scales

Figure 5 shows the thickness measurements of the scales formed at 1173K and P_{O_2} = 1, 0.33 atm and the Cu_2O/CuO equilibrium oxygen pressure as determined by optical and SEM microscopies using polished cross-sections. The thin CuO layer in practical terms was negligible in comparison to the thick Cu_2O layer. The different oxygen pressures give identical results within the experimental reproducibility of measurements. Thus, the influence of oxygen pressure on the oxidation kinetics is not significant at pressures larger than that for Cu_2O/CuO equilibrium. In light of the previous observations on the compact nature of the CuO layer (Fig.4), the oxygen activity for the Cu_2O/CuO equilibrium provided the true reaction environment for growth of the Cu_2O layer.

Several specimens were oxidized for extended time at 1173K following entire consumption of the copper substrate, and the slow growth of the CuO layer followed by means of thermogravimetry and thickness measurements from cross-sections. Figure 6 illustrates the thickening of CuO on Cu_2O. The parabolic type plot demonstrates that the growth kinetics for the CuO layer may be expressed as

$$\ell^2 - \ell_o^2 = k_p'(t-t_o) \quad (1)$$

where $k_p' \approx 2 \times 10^{-10} cm^2/s$ by least squares method. In this expression, t_o is the time when the metal substrate was oxidized completely as determined by thermogravimetry, $\ell(t)$ and ℓ_o are the thickness of CuO at time t and prior to complete oxidation of the copper substrate, respectively.

The oxidation kinetics of copper as measured gravimetrically at several temperatures ranging from 826 to 1189K are presented in Fig.7. The curve band for the reproducibility of oxidation at 1173K was determined by experiments at P_{O_2} = 1 and 0.33 atm on six different specimens. Growth rate of the duplex scale was independent of the oxygen pressure. The oxidation curves at other temperatures were only obtained at P_{O_2} = 1 atm. As illustrated by the plots of $(\Delta W/A)^2$ vs. time in Fig.7b, the oxidation kinetics were parabolic. The corresponding linear correlation factors are 0.997 ~ 1.0. As shown by the Arrhenius plot of the parabolic oxidation rate constants in Fig.8, log k_p vs 1/T, the curve is not linear becoming of smaller slope at lower temperature.

Discussion and Conclusions

The experimental results clearly demonstrate that the magnitude of the parabolic oxidation rate constants for growth of a Cu_2O scale or a Cu_2O/CuO duplex scale are largely determined by the predominant growth of Cu_2O. When P_{O_2} > Cu_2O/CuO equilibrium pressure, a compact but thin CuO layer always formed on the Cu_2O layer. The growth of CuO was observed in the initial exposure stage up to 900s, but its thickness did not increase significantly during the rest of the reaction period.

The slow parabolic growth of the CuO layer independent of oxygen pressure suggests that CuO behaves like a n-type semiconductor. This implies that either oxygen vacancies or/and copper interstitials are the dominant lattice defects. The concentration of these types of point defects would vary as $P_{O_2}^{-1/6}$ and the self-diffusion coefficient would be of the form

$$D^s(CuO) = A(T)\, P_{O_2}^{-1/6} \quad (2)$$

where A(T) is a constant of Arrhenius form. The parabolic oxidation rate constant for CuO growth then can be described as

$$k_p(CuO) = \int_{P_2}^{P} A(T)P_{O_2}^{-1/6} d\,ln\,P_{O_2} = 6A(T)[P_2^{-1/6} - P^{-1/6}] \quad (3)$$

where P is the atmospheric pressure and $P_2 = P_{O_2}(Cu_2O/CuO)$. When $P >> P_2$,

$$k_p(CuO) = 6A(T)P_{O_2}^{-1/6}(Cu_2O/CuO) \quad (4)$$

and hence independent of atmospheric oxygen pressure. With $\Delta G^o(1173K) = -20,700$ J/Mol for formation of CuO from Cu_2O(12) and the oxidation results from Eqn.(1), $k_p(CuO) = 2 \times 10^{-10} cm^2.s$, $A(T) = 7 \times 10^{-11} cm^2/s$ at 1173K. This type of computation with sufficient experimental results on CuO growth could provide a convenient and valid way to determine the diffusivity of transporting species in CuO.

The innermost Cu_2O layer of a duplex scale accounted predominantely for the amout of oxide which developed by parabolic kinetics. Thus the flux of reactants across this layer and the parabolic oxidation rate constant remained independent of atmospheric oxygen pressure because the oxygen activity driving the reaction was established by local equilibrium at the Cu_2O/CuO interface. In a relatively recent investigation dealing with diffusion in Cu_2O, Peterson and Wiley (13) have advanced a method to obtain accurate assessment of the relationship between point defect migration and oxidation kinetics. They advance a defect model involving neutral and singly charged copper vacancies, electron holes and singly charged oxygen interstitials. They noted also that the activation energy for oxidation corresponded to this energy for copper vacancy diffusion in Cu_2O. Using the results for copper and oxygen diffusion (13,14) in conjunction with the Wagner theory of oxidation, the parabolic oxidation rate constant for Cu_2O growth can be expressed as

$$k_p(Cu_2O) = \int_{P_1}^{P_2} (1/2D_{Cu} + D_O) d\,ln\,P_{O_2} \quad (5)$$

where P_1, the oxygen pressure of Cu/Cu_2O equilibrium, and P_{O_2}, the oxygen pressure for Cu_2O/CuO equilibrium, are given by (12)

$$P_1 = exp - \Delta G_1^o/2RT; \; \Delta G_i^o = 40500 + 3.92T\,log\,T - 29.5T \text{ cal/mol}$$

$$P_2 = exp - \Delta G_2^o/RT; \; \Delta G_2^o = 34950 + 6.1T\,log\,T - 44.3T \text{ cal/mol}$$

Upon employing the tracer-diffusion coefficients of species in Cu_2O reported in literature as

$$D_{Cu}^* = D_1^o P_{O_2}^{1/8} exp\left(-\frac{Q_1}{RT}\right) + D_2^o P_{O_2}^{1/4} exp\,(-Q_2/RT)$$

where the first and last terms refer to singly charged and neutral copper vacancies, respectively

$$D_1^o = (3.7 \pm 1.1) \times 10^{-4} cm^2/s; \; D_2^o = (4.1 \pm 2.2) \times 10^{-1} cm^2/s$$

$$Q_1 = 101.5 \pm 2.7 \text{ kJ/mol}; \; Q_2 = 169.8 \pm 5.8 \text{ kJ/mol}.$$

The oxygen tracer diffusion coefficient can be expressed as

$$D_O^* = (3 \times 10^{-3}) P_{O_2}^{0.4} \exp\left(-\frac{149300 J/mol}{RT}\right)$$

Parabolic oxidation constants, k_p', k_p'' and k_p''' contributed from each type of point defect, V_{Cu}', V_{Cu}^x and O_i' can therefore be assigned and computed. The parabolic oxidation rate constant for Cu_2O growth then becomes a summation of the above constants: $k_p(Cu_2O) = k_p' + k_p'' + k_p'''$. The relationship beteween the tracer and self diffusivity is described by a correlation factor f. Because the copper cations lie in a FCC sublattice in Cu_2O and the dominant atomic defects in the anion sublattice are assumed to be oxygen interstitials, $D_{Cu} = D_{Cu}^*/f_{Cu}$ and $D_O = D_O^*/f_O$ where $f_{Cu} = 0.78$ and $f_O = 1.0$. Consequently, the expressions of k_p for each type of defect are as follows:

$$k_p' = 1.9 \times 10^{-3} \exp\left(-\frac{101500 J/mol}{RT}\right) (P_2^{1/8} - P_1^{1/8})$$

$$k_p'' = 1.05 \times 10^{-3} \exp\left(-\frac{169800 J/mol}{RT}\right) (P_2^{1/4} - P_1^{1/4})$$

$$k_p''' = 7.5 \times 10^{-3} \exp\left(-\frac{149300 J/mol}{RT}\right) (P_2^{0.4} - P_1^{0.4})$$

The computed values for these individual rate constants and the overall $k_p(Cu_2O)$ are plotted in Fig.9 as log k_p vs 1/T. The experimental results for $k_p(Cu_2O)$ (Figs.7 and 8) are also given in this figure. One notes the good agreement in the high temperature regime between the computed and experimental $k_p(Cu_2O)$. The computed values are definitely lower than those determined from the experimental results in the lower temperature range between 826 and 900K which is attributed to a component by short-circuit diffusion in the thinner polycrystalline scales. Evaluations of the individual terms in the parabolic oxidation constant at 1173K and P_{O_2} = 1 atm determined from diffusion of each type of defect and a comparison of the calculated to the experimental k_p (Cu_2O) are given in Table I.

One concludes that the calculated and experimental values for the parabolic oxidation rate constant are in agreement within the accuracy of the diffusion measurements, that copper transport is predominant as the ratio of metal to oxygen migration is ~ 10^2 and that less than 30% of metal transport occurs through neutral copper vacancies. It appears, notwithstanding, that a definite conclusion cannot be drawn of the relative role of neutral and singly charged copper vacancies until it is proven that the variable oxygen pressure dependence on defect concentrations, electrical conductivity and copper diffusion is not caused by electrical screening effect of positive holes on single charged cation vacancies.

For the case of $P_{O_2}(Cu/Cu_2O) < P_{O_2} \leq P_{O_2}(Cu_2O/CuO)$, Cu_2O is the only product of oxidation for which several investigations are found for $T \geq 1173K$(15-17). The growth kinetics of this oxide layer were considered using the preceding computation method. That part of the computation for

TABLE I

Calculated and Measured Terms of Parabolic Oxidation Rate Constant for Copper to Cuprous Oxide at 1173K and P_{O_2} = 1 atm.

Rate Constant	Calculated (cm^2/s)	Measured (cm^2/s)
$k_p''(V_{Cu}'')$	9.7×10^{-9}	
$k_p'(V_{Cu}')$	2.5×10^{-8}	
$k_p(O_i')$	3.2×10^{-10}	
$k_p(Cu_2O)$	$3.6 \pm 1.1 \times 10^{-8}$	$3.9 \pm 0.2 \times 10^{-8}$

$k_p(Cu_2O)$ at 1173, 1223 and 1273K and comparison of these calculated values to those determined experimentally (17) are given in Fig.10. The scatter of the results based upon experiment around the computed curves is again in the error range of the previously presented tracer-diffusion coefficients. There is accordingly good agreement at $T \geq 1273K$ between the present experimental results, published work (15,17) and the simulation computational model based upon independent measurements of copper and oxygen diffusivities for $k_p(Cu_2O)$.

Acknowledgements

The investigators are indebted to the Natural Sciences and Engineering Research Council of Canada for financial support of this research.

References

1. R.F. Tylecote, J. Inst. Metals, 78, 25 (1950-1951).
2. A. Ronnquist and H. Fischmerister, J. Inst. Metals, 89, 65 (1960).
3. K. Hauffe, Oxidation of Metals, Plenum Press, New York, (1965).
4. H. Leidheiser, The Corrosion of Copper, Tin and their alloys, John Wiley and Sons, Inc., P28 (1971).
5. P. Kofstad, High Temperature Corrosion, Elsvier Publishers (1989).
6. G.J. Yurek, J.P. Hirth and R.A. Rapp, Oxid. Met., 8 265 (1974).
7. G. Valensi, in Pittsburgh International Conference on Surface Reactions (Corrosion Publ. Co., Pittsburgh (1948), P.146.
8. R.D. Shaw and R. Rolls, Corros. Sci. 14, 443 (1974).
9. G. Garnaud, Oxid. Met. 11, 127 (1977).
10. P. Mayer and W.W. Smeltzer, J. Electrochem. Soc. 119, 626 (1972).
11. A.G. Goursat and W.W. Smeltzer, ibid., 120, 390 (1973).
12. O. Kubaschewski, Metallurgical Thermochemistry, 4th ed., Pergamon Press (1967).
13. N.L. Peterson and C.L. Wiley, J. Phys. Chem. Solids, 42, 697 (1981).
14. F. Perinet, S. Barbezat and C. Monty, J. Physique Collq. C6, 315 (1980).
15. C. Wagner and K. Grunewald, Z. Physik. Chem. B40, 455 (1938).

16. W.J. Moore and B. Selikson, J. Chem. Phys., 19, 1539 (1951).
17. J.P. Baur, D.W. Bridges and W.M. Fassell, J. Electrochem. Soc. 103, 273 (1956).

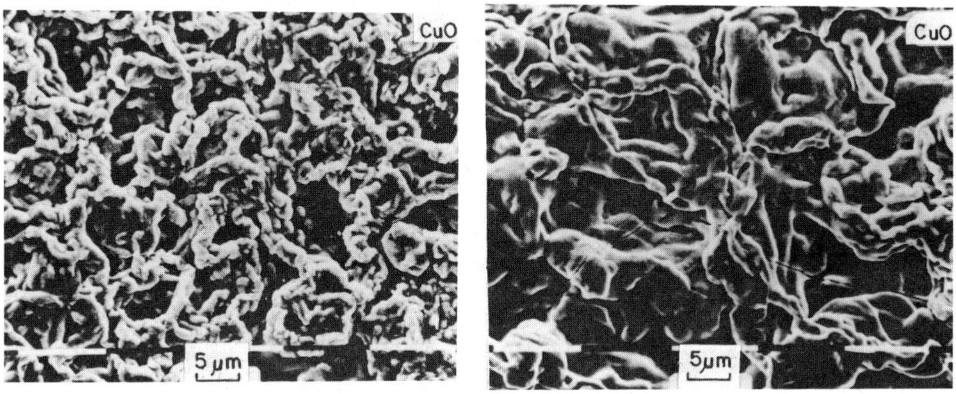

(a) (b)

FIGURE 1: The surface morphology of copper oxidized at 1173K: (a) oxidized for 32 h in flowing oxygen at P_{O_2} = 1 atm; (b) oxidized for 2 h after 22 h giving complete oxidation of comparatively thin copper specimen.

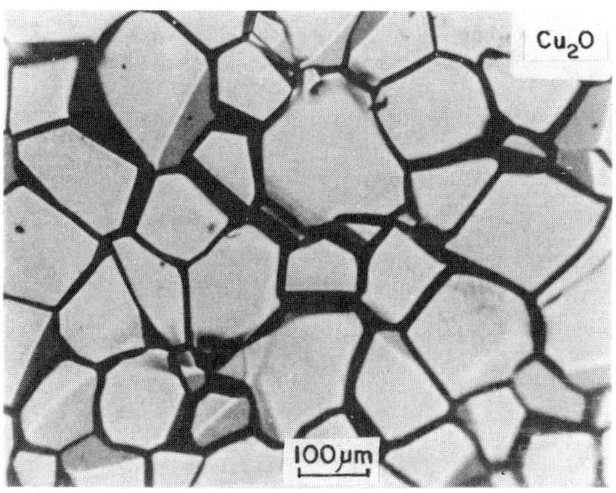

FIGURE 2: The surface morphology of copper oxidized at 1173 K for 5 h in CuO/Cu_2O pack.

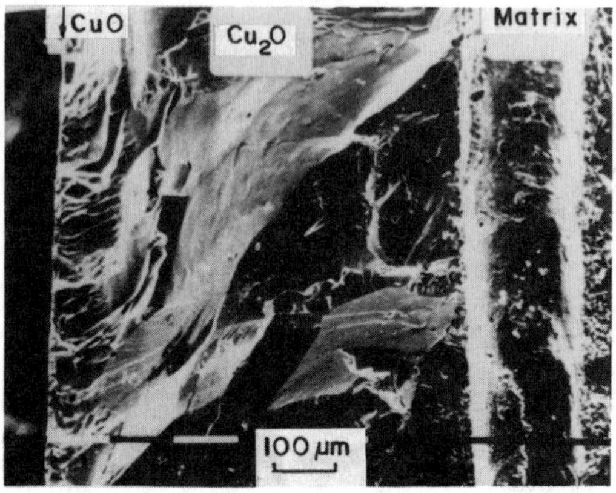

FIGURE 3: A fractured cross-section of duplex Cu_2O/CuO scale formed on copper at 1173 K after 32 h in flowing oxygen at P_{O_2} = 1 atm.

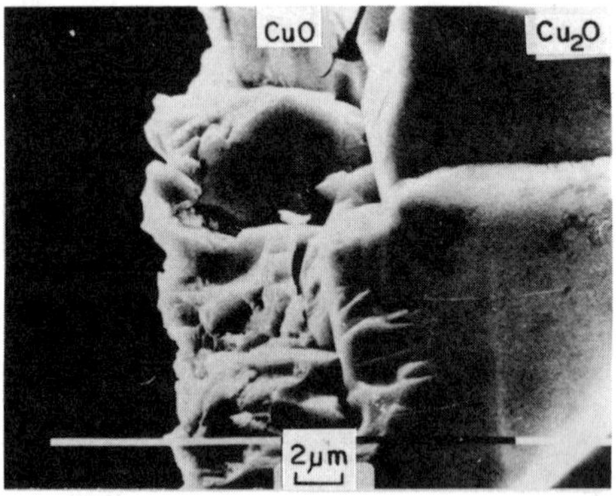

(a)

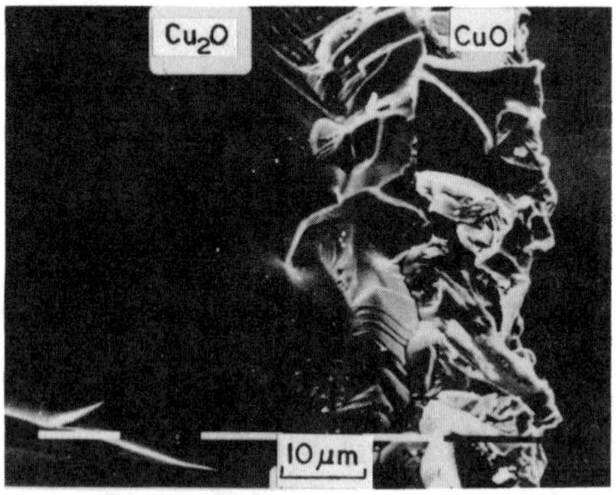

(b)

FIGURE 4: Section of duplex Cu_2O/CuO scale exhibiting CuO layer at high magnification: (a) scale of Fig.3; (b) Cu layer formed on Cu_2O after 2 h oxidation following complete oxidation of copper specimen.

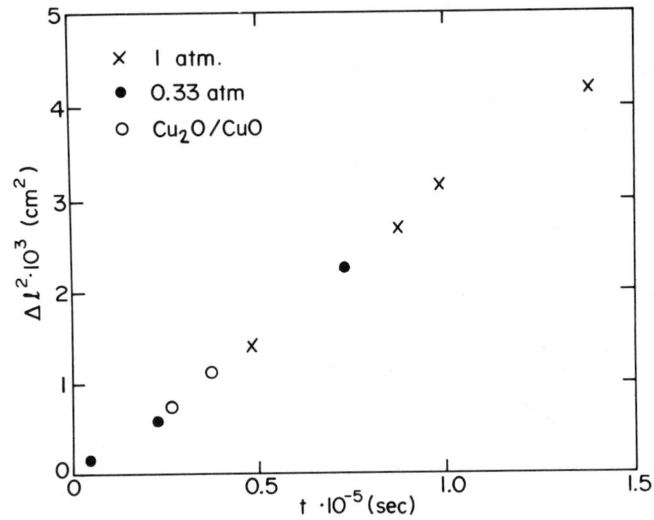

FIGURE 5: Parabolic plot of the thickness measurements of the oxide scales formed at 1173 K and P_{O_2} = 1 atm, 033 atm and the equilibrium Cu_2O/CuO oxygen pressure.

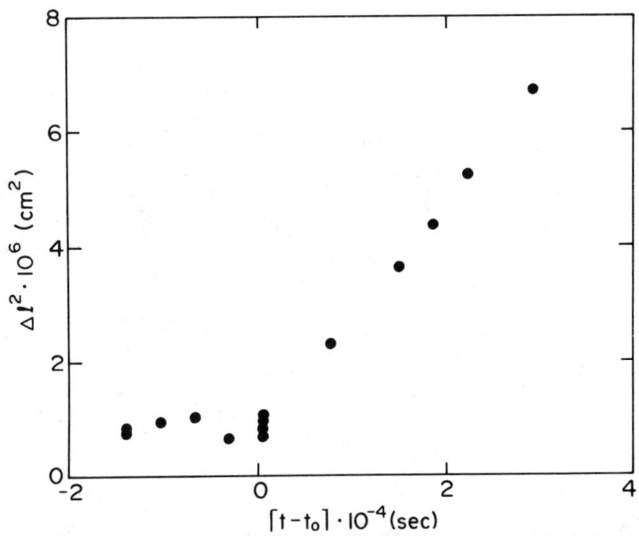

FIGURE 6: Growth of CuO layer at 1173 K after entire consumption of copper substrate up to time $t = t_o$.

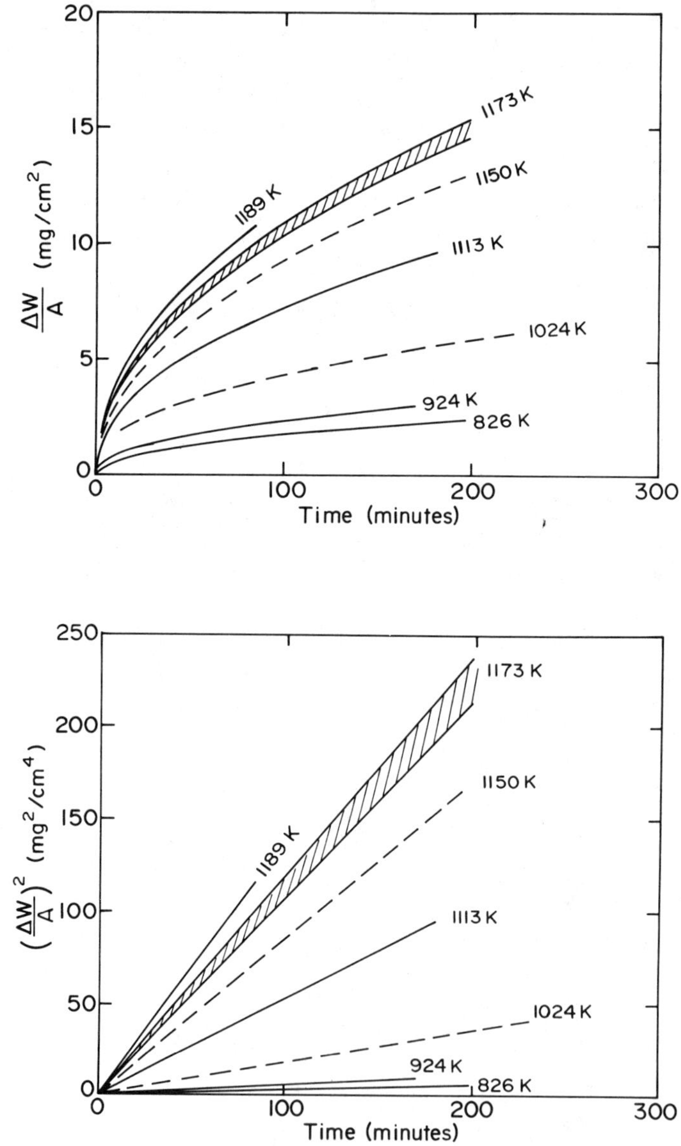

FIGURE 7: Thermogravimetric results for growth of Cu_2O/CuO duplex scales on copper at $P_{O_2} > P_{O_2}(Cu_2O/CuO)$ at temperatures 826 to 1189 K: (upper) linear plotting; (lower) parabolic plotting.

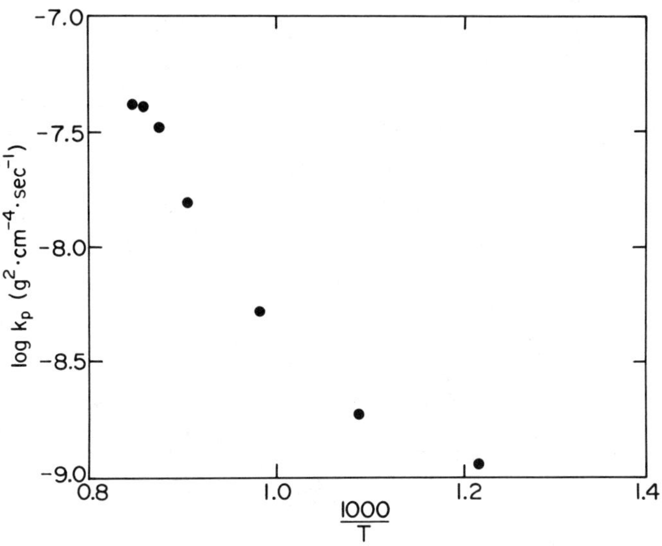

FIGURE 8: Arrhenius plot of log k_p vs 1/T from the oxidation results given in Fig.7.

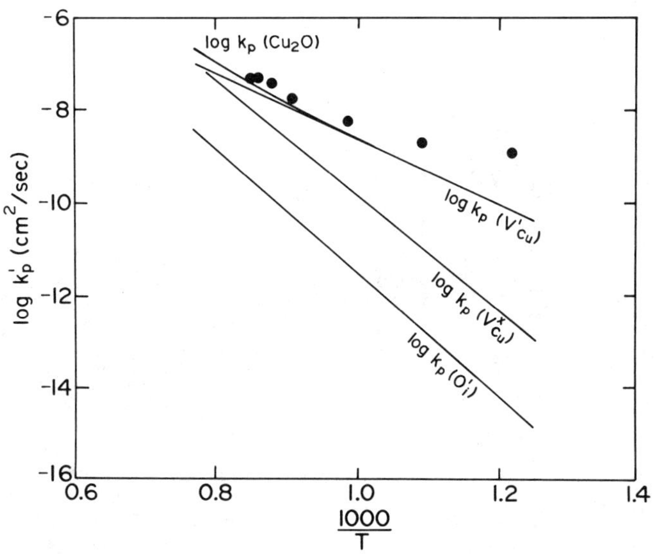

FIGURE 9: Present experimental results and computed result for log k_p vs 1/T from tracer diffusivities of $V_{Cu}^{'}$, V_{Cu}^{x} and $O_i^{'}$ at $P_{O_2} = P_{O_2}$ (Cu_2O/CuO).

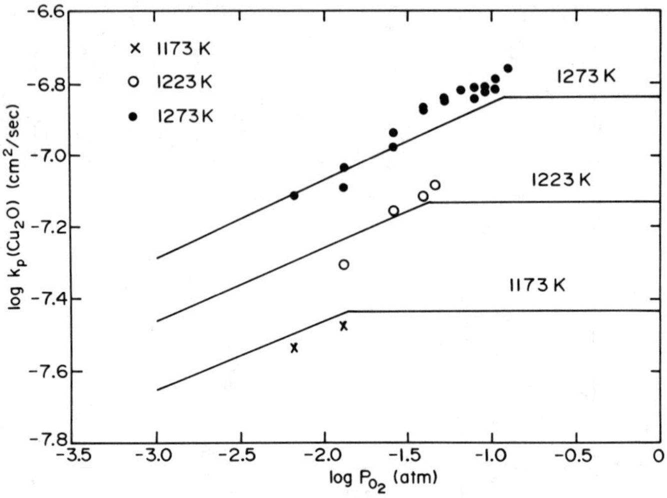

FIGURE 10: A comparison of the parabolic oxidation constants for growth of the Cu_2O scale formed on copper at several oxygen pressures at 1173, 1223 and 1273 K determined experimentally (15,17) and those calculated using tracer-diffusivities for V'_{Cu}, V^x_{Cu} and O'_i.

Diffusion in quasi-steady-state irreversible systems; epitaxy of nickel telluride from halide vapour phase

Y.K. Rao and Y. Do

Department of Materials Science and Engineering, RB-10, University of Washington, Seattle, Washington 98195, U.S.A.

Abstract

The epitaxy of single crystal non-stoichiometric nickel telluride $NiTe_y$ in closed chemical vapor transport ampoule reactor from vapor phase composed of $NiCl_2$, NiCl, Te_2, Te_5, $TeCl_2$, $TeCl_4$, and Cl_2, is an archetype of a class of irreversible phenomena that are driven by chemical potential gradients. The tellurium-deficient $NiTe_x$ solid constitutes the source, which at the higher temperatures of the source-zone, reacts with the chlorine solvent producing $NiCl_2$, $TeCl_2$, Te_2, and so forth. The substrate is a tellurium-rich $NiTe_y$ solid, held at a lower temperature in the deposition zone, on which the epitaxial growth occurs by reactions between the nickel chlorides, tellurium, and chlorides of tellurium. The quasi-steady-state growth rate of the telluride is computed by assuming virtual equilibria in the end zones with diffusive mass transport of vapor species between them constituting the irreversible phenomenon that mitigates the chemical potential (or partial pressure) differential.

Introduction

The nickel-tellurium system exhibits three solid-state compounds: β-phase (Ni_3Te_2) with composition from $NiTe_{0.667}$ to $NiTe_{0.692}$; γ-phase with a relatively narrow range from $NiTe_{0.77}$ to $NiTe_{0.775}$; and δ-phase which has a large range of homogeneity from $NiTe_{1.09}$ to $NiTe_{2.00}$ (1). Brennan and co-workers (2) have synthesized nickel telluride, NiTe, using organometallic reagents bis (cyclooctadiene) nickel, $Ni(COD)_2$, and trialkyl-phosphine telluride, $(C_2H_5)_3PTe$. Tellurides such as CdTe(s) have successfully been applied in solar energy conversion (3). The "diluted magnetic semi-conductor" cadmium manganese telluride $Cd_xMn_{1-x}Te(s)$ has generated considerable interest due to its unusual magneto-optical properties which are attributed to the presence of the transition element Mn (4,5).

Single crystals of tellurides can be grown from the melt by the Bridgman technique (4). The growth of single crystals, in bulk or as thin films, of binary compounds is carried out most advantageously by vapor phase epitaxy (VPE) in a closed ampoule reactor. Schäfer (6) had given an elegant review of the potential applications of closed chemical vapor transport reaction systems in crystal growth. A useful listing of the experimental and theoretical developments in chemical vapor transport (CVT) was furnished by Mercier (7). The direction of vapor transport between the source and seed, and the feasibility of forming a deposit on the seed can be predicted on the basis of equilibrium calculations. The experimental investigation of Wiedemeier and Csillag (8) on the vapor transport of $SnS_2(s)$ using $SnI_4(g)$ or $I_2(g)$ as the initial transport agent revealed excellent qualitative agreement between the observed growth or etching mode and the equilibrium pressure-difference, for the SnS_2 species, between the two zones.

In a closed ampoule reactor with an arbitrary temperature profile, the mass-transport of gaseous species from the source to the seed (or vice-versa) occurs by the following mechanisms: ordinary diffusion due to concentration-gradient, laminar (or Stefan) flow induced by the molar volume changes accompanying the chemical reactions, thermal (or Soret) diffusion, and natural convection due to gravity. The modeling of chemical vapor transport in closed reactors received considerable attention over the last three decades (9-18). An apriori calculation of the rate of deposition or etching can be attempted using the three-zone model which visualizes the reactor as consisting of a source-zone at T_s, a connecting-zone with a temperature-gradient, and a deposition-zone at T_d; equilibrium is assumed between the gas and condensed phases in each of the two end-zones, with the transport of species between the zones occurring by the diffusion mechanisms (6).

The principal author and co-workers (19,20) have applied the three-zone model to compute vapor transport rates of Ni(s), NiO(s), and $YBa_2Cu_3O_{7-y}(s)$ in the closed reaction systems Ni-Cl, Ni-Cl-O, and Y-Ba-Cu-Cl-O, respectively. The present work focuses on the vapor deposition, in a horizontal ampoule reactor, of a tellurium-rich solid solution $NiTe_y(s)$ using the tellurium-deficient solution $NiTe_x(s)$ as the source, with $Cl_2(g)$

as the initial transport agent. The solid solution compositions selected are within the δ-phase region of the Ni-Te system (1). The deposition rates are computed for the condition of virtual gas-solid equilibrium in the two end-zones and a quasi-steady-state mass-transport of gas species, through the connecting-zone, by diffusion and Stefan flow. The model employed by the authors (19,20) differs from the earlier attempts (6-18) in some important aspects: instead of considering each of the end-zones in isolation in order to calculate the equilibrium state, full coupling is maintained between the three zones for any specific gas species; and phase compositions are determined for a given $T_s \rightarrow T_d$ set at increasing initial amounts of the halogen in the reactor. The diffusion-limited transport rates are calculated using the Stefan-Maxwell formalism.

Westrum and co-workers (21,22) determined the heat capacities, dissociation pressures, and phase-relations for non-stoichiometric nickel tellurides at a number of compositions in the tellurium-rich region. Mills (23) critically reviewed the published data and recommended values for standard enthalpy of formation, standard entropy, and vapor pressures of $NiTe_{1.1}(s)$ and $NiTe_{2.0}(s)$ compounds. The composition $NiTe_{2.0}$ changes from a single-phase to a two-phase region at temperatures greater than 450^oC. The solid solution $NiTe_{1.9}(s)$ is stable at considerably higher temperatures, with the second-phase, a liquid containing 86.5 atom% Te, appearing at 712.5^oC. The composition $NiTe_{1.7}(s)$ remains a single-phase solution at 800^oC and higher. The crystal structure changes from hexagonal NiAs-type at $NiTe_{1.1}(s)$ to hexagonal $Cd(OH)_2$-type for $NiTe_{2.0}(s)$. In the unit cell of the idealized hexagonal NiTe(s) structure, two nickel atoms occupy positions 0,0,0; 0,0,1/2 while the two tellurium atoms are in the positions 1/3,2/3,3/4; 2/3,1/3,1/4; as the tellurium-content of the solution is increased to $NiTe_2(s)$, one of the nickel positions - probably 0,0,1/2 site - remains vacant while the second position at 0,0,0 is occupied by nickel (1).

Chemical Vapor Transport

The vapor deposition of $NiTe_y(s)$ on a suitable substrate at temperature T_d can be brought about using Cl_2, $NiCl_2$, or $TeCl_2$ as the solvent (or transport agent), with the source $NiTe_x(s)$ held at a higher temperature of T_s; in the present work, the stoichiometric parameters x=1.7 and y=1.9, respectively. Fig. 1 shows the schematic diagram of the closed ampoule reactor: source-zone at T_s, deposition-zone at T_d, and connecting-zone with a constant temperature-gradient w. The system is prepared by placing $NiTe_x(s)$ in the source-zone, evacuating the reactor, and then filling it with a predetermined amount of the gaseous transport agent, Cl_2. In practice, a specially wire-wound furnace is used to produce the desired temperature distribution along the reactor length.

The chlorination of nickel telluride in the source-zone produces gaseous species $NiCl_2$, $TeCl_2$, NiCl, $TeCl_4$, Te_2, and Te_5 which diffuse to the substrate at T_d in the deposition-zone where the reverse reaction occurs forming a tellurium-rich nickel telluride crystal. The condensed phase $NiTe_{1.7}(s)$ is in

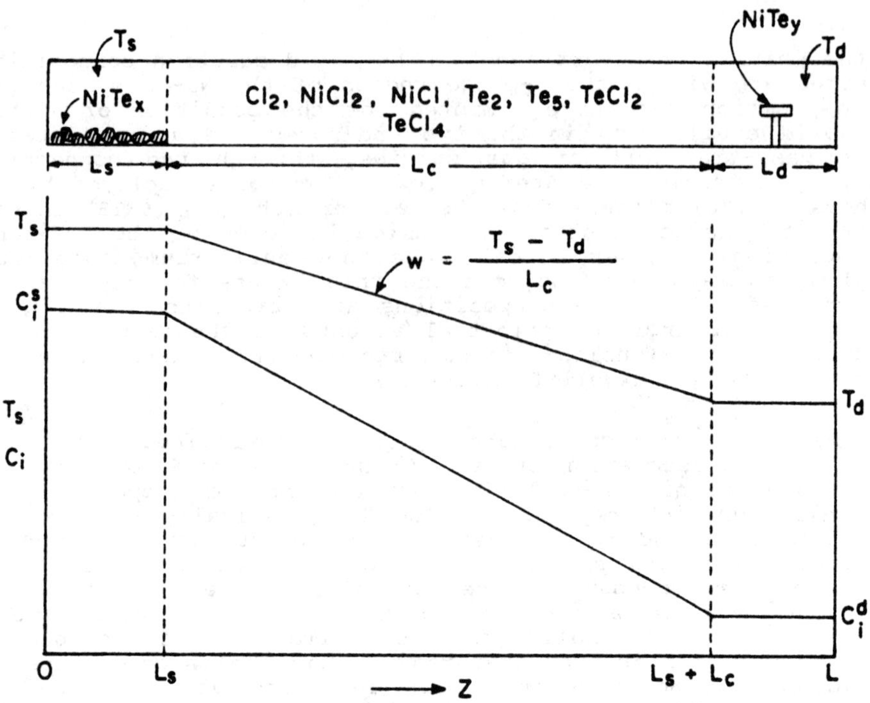

Figure 1. Closed chemical vapor transport (CVT) reactor

a state of virtual (or near) equilibrium with vapor species Cl_2, $NiCl_2$, NiCl, Te_2, $TeCl_2$, Te_5, and $TeCl_4$ in the source-zone. A similar state of virtual equilibrium exists between the condensed-phase $NiTe_{1.9}$(s) and vapor species in the deposition-zone. The transport of various species occurs by diffusion which acts as a small perturbing force on the virtual equilibrium state.

In view of the fact that the vapor species containing nickel or tellurium are derived from the condensed phase, it is reasonable to expect the gas composition in the immediate vicinity of the nickel telluride to conform to the stoichiometric constraint. Thus, for the source-zone of the reactor,

$$\frac{n^s_{Te}}{n^s_{Ni}} = \frac{n^s_{TeCl_2} + n^s_{TeCl_4} + 2n^s_{Te_2} + 5n^s_{Te_5}}{n^s_{NiCl_2} + n^s_{NiCl}} = 1.7 \quad (A.1)$$

where n^s_i denotes the number of moles of the i-th species in the source-zone. The superscripts 's', 'd', and 'c' are used to denote the quantities in the source, deposition, and connecting zones, respectively.

Source Zone

The number of independent chemical reactions in this zone can be found from the Gibbs stoichiometric rule (26). The atom matrix for the eight species, $NiTe_{1.7}(s)$, $NiCl_2$, $TeCl_2$, $TeCl_4$, NiCl, Te_2, Te_5, and Cl_2, present in this zone has a rank of three, same as the number of elements (Ni, Te, Cl); consequently, there can be at the most five linearly independent chemical equilibria. These are chosen as follows:

$$NiTe_{1.7}(s) + 2.7\ Cl_2 = NiCl_2 + 1.7\ TeCl_2 \qquad (1s);\epsilon_1^s$$

$$NiCl_2 = NiCl + 1/2\ Cl_2 \qquad (2s);\epsilon_2^s$$

$$1/2\ Te_2 + Cl_2 = TeCl_2 \qquad (3s);\epsilon_3^s$$

$$1/2\ Te_2 + 2\ Cl_2 = TeCl_4 \qquad (4s);\epsilon_4^s$$

$$2\text{-}1/2\ Te_2 = Te_5 \qquad (5s);\epsilon_5^s$$

where ϵ_1^s, ϵ_2^s, ϵ_3^s, ϵ_4^s, and ϵ_5^s are the extents of reactions for (1s), (2s), (3s), (4s), and (5s), respectively. The gaseous species are transported through the connecting-zone to the deposition-zone and quasi-steady-state conditions are established. The number of moles of each species in the source-zone can be expressed as follows:

$$n^s_{NiCl_2} = \epsilon_1^s - \epsilon_2^s - n^{od}_{NiCl_2} - n^c_{NiCl_2} \qquad (S.1)$$

$$n^s_{NiCl} = \epsilon_2^s - n^{od}_{NiCl} - n^c_{NiCl} \qquad (S.2)$$

$$n^s_{TeCl_2} = 1.7\epsilon_1^s + \epsilon_3^s - n^{od}_{TeCl_2} - n^c_{TeCl_2} \qquad (S.3)$$

$$n^s_{Te_2} = -0.5\epsilon_3^s - 0.5\epsilon_4^s - 2.5\epsilon_5^s - n^{od}_{Te_2} - n^c_{Te_2} \qquad (S.4)$$

$$n^s_{Te_5} = \epsilon_5^s - n^{od}_{Te_5} - n^c_{Te_5} \qquad (S.5)$$

$$n^s_{TeCl_4} = \epsilon_4^s - n^{od}_{TeCl_4} - n^c_{TeCl_4} \qquad (S.6)$$

$$n^s_{Cl_2} = n^o_{Cl_2} - 2.7\epsilon_1^s + 0.5\epsilon_2^s - \epsilon_3^s - 2\epsilon_4^s - n^{od}_{Cl_2} - n^c_{Cl_2} \qquad (S.7)$$

$$n^s_{1.7} = n^o_{1.7} - \epsilon_1^s \ :\ NiTe_{1.7}(s) \qquad (S.8)$$

where n_i^c is the number of moles of the i-th species in the connecting-zone and n_i^{od} represents the number of moles of the i-th species transferred to the deposition-zone prior to the onset of the deposition of the $NiTe_{1.9}(s)$ phase on the substrate. The extent of reaction ϵ_3^s is negative with the result that a significant amount of $Te_2(g)$ exists in the source-zone gas-phase. It will be noted that the initial amounts of $Cl_2(g)$ and $NiTe_{1.7}(s)$ are $n^o_{Cl_2}$ and $n^o_{1.7}$, respectively.

Deposition Zone

Under quasi-steady-state conditions, the gas species that diffuse to the deposition-zone are consumed by chemical reactions. The tellurium-rich $NiTe_{1.9}(s)$ solid solution is deposited at a rate governed by the mass-transport processes. The nucleation and interfacial reactions are considered to occur very rapidly and the gas-solid system in this zone is only slightly displaced from equilibrium. The species present include the seven gas species identified earlier and $NiTe_{1.9}(s)$. There are five independent chemical equilibria which are selected as follows:

$$NiCl_2 + 1.9\ TeCl_2 = NiTe_{1.9}(s) + 2.9\ Cl_2 \qquad (1d);\epsilon^d_1$$

$$NiCl_2 = NiCl + 1/2\ Cl_2 \qquad (2d);\epsilon^d_2$$

$$1/2\ Te_2 + Cl_2 = TeCl_2 \qquad (3d);\epsilon^d_3$$

$$1/2\ Te_2 + 2\ Cl_2 = TeCl_4 \qquad (4d);\epsilon^d_4$$

$$2\text{-}1/2\ Te_2 = Te_5 \qquad (5d);\epsilon^d_5$$

where ϵ^d_1, ϵ^d_2, and so forth, are the extents of reactions. As deposition occurs, ϵ^d_1 increases with time. The amounts of the species in the deposition-zone are written in terms of n^{od}_i and the extents of reactions.

$$n^d_{NiCl_2} = n^{od}_{NiCl_2} - \epsilon^d_1 - \epsilon^d_2 \qquad (D.1)$$

$$n^d_{NiCl} = n^{od}_{NiCl} + \epsilon^d_2 \qquad (D.2)$$

$$n^d_{TeCl_2} = n^{od}_{TeCl_2} - 1.9\ \epsilon^d_1 + \epsilon^d_3 \qquad (D.3)$$

$$n^d_{Te_2} = n^{od}_{Te_2} - 0.5\ \epsilon^d_3 - 0.5\ \epsilon^d_4 - 2.5\ \epsilon^d_5 \qquad (D.4)$$

$$n^d_{Te_5} = n^{od}_{Te_5} + \epsilon^d_5\ ;\ n^d_{TeCl_4} = n^{od}_{TeCl_4} + \epsilon^d_4 \qquad (D.5,D.6)$$

$$n^d_{Cl_2} = n^{od}_{Cl_2} + 2.9\epsilon^d_1 + 0.5\epsilon^d_2 - \epsilon^d_3 - 2\epsilon^d_4 \qquad (D.7)$$

$$n_{1.9} = n^o_{1.9} + \epsilon^d_1 \qquad (D.8)$$

On adding together equations (S.1), (S.2), (D.1), and (D.2), it is found that

$$\epsilon^s_1 - \epsilon^d_1 = n^s_{NiCl_2} + n^d_{NiCl_2} + n^c_{NiCl_2} + n^s_{NiCl} + n^d_{NiCl} + n^c_{NiCl} \qquad (A.2)$$

Similarly, the summation(S.3)+(S.6)+(D.3)+(D.6)+2(S.4)+2(D.4) +5(S.5)+5(D.5) yields

$$1.9(\epsilon_1^s-\epsilon_1^d)-0.2\epsilon_1^s=n_{TeCl_2}^s+n_{TeCl_2}^d+n_{TeCl_2}^c+n_{TeCl_4}^s+n_{TeCl_4}^d$$

$$+n_{TeCl_4}^c+2(n_{Te_2}^s+n_{Te_2}^d+n_{Te_2}^c)+5(n_{Te_5}^s+n_{Te_5}^d+n_{Te_5}^c) \quad (A.3)$$

The chlorine atom balance provides the following relationship:

$$n_{Cl_2}^o=n_{Cl_2}^s+n_{Cl_2}^d+n_{Cl_2}^c+n_{NiCl_2}^s+n_{NiCl_2}^d+n_{NiCl_2}^c+n_{TeCl_2}^s$$

$$+ n_{TeCl_2}^d+ n_{TeCl_2}^c+ 2(n_{TeCl_4}^s+ n_{TeCl_4}^d+ n_{TeCl_4}^c)$$

$$+ 0.5(n_{NiCl}^s + n_{NiCl}^d + n_{NiCl}^c) \quad (A.4)$$

These relations are simplified by assuming that in the connecting-zone the concentration of each gas species varies linearly with distance Z. The error in the equilibrium total pressure in the closed ampoule reactor introduced by such an assumption was found to be about 0.1% for the $NiCl_2$-$NiCl$-Cl_2 system (19). Let V_s, V_c, and V_d denote the volumes of the source, connecting, and deposition zones, respectively. For a uniform-sized tubular reactor of cross-section S_c,

$$V_s = S_cL_s;\ V_c = S_cL_c;\ V_d = S_cL_d$$

It can be easily demonstrated that

$$n_i^c = 0.5\ V_c\ (C_i^s + C_i^d) \quad (A.5)$$

For the particular case when $V_s = V_d$ and $V_c = rV_s$, equation (A.5) becomes

$$n_i^c = 0.5rV_s(C_i^s+C_i^d) = 0.5r(n_i^s+n_i^d)$$

On combining this result with equations (A.1), (A.2), and (A.3), for $0.2\epsilon_1^s(1+0.5r)^{-1}\to 0$, the following stoichiometric constraint is obtained for the deposition zone.

$$1.9(n_{NiCl_2}^d+n_{NiCl}^d) = n_{TeCl_2}^d+n_{TeCl_4}^d+2n_{Te_2}^d+5n_{Te_5}^d \quad (A.6)$$

Assuming ideal gas behavior, for $V_s = V_d$, equations (A.1), (A.4), and (A.6) can be transformed into relations between partial pressures.

$$P_{TeCl_2}^s+P_{TeCl_4}^s+2P_{Te_2}^s+5P_{Te_5}^s= 1.7(P_{NiCl_2}^s+P_{NiCl}^s) \quad (A.7)$$

$$P_{TeCl_2}^d+P_{TeCl_4}^d+2P_{Te_2}^d+5P_{Te_5}^d = 1.9(P_{NiCl_2}^d+P_{NiCl}^d) \quad (A.8)$$

$$P_{Cl_2}^s+P_{NiCl_2}^s+P_{TeCl_2}^s+2P_{TeCl_4}^s+0.5P_{NiCl}^s+ \frac{T_s}{T_d}(P_{Cl_2}^d+P_{NiCl_2}^d$$

$$+P_{TeCl_2}^d+2P_{TeCl_4}^d+0.5P_{NiCl}^d) = n_{Cl_2}^o\cdot\frac{RT_s}{V_s}(1+0.5r)^{-1} \quad (A.9)$$

Under quasi-steady-state conditions, the total pressure in the reactor remains same in all three zones.

$$P^{s}_{Cl_2}+P^{s}_{NiCl_2}+P^{s}_{NiCl}+P^{s}_{TeCl_2}+P^{s}_{TeCl_4}+P^{s}_{Te_2}+P^{s}_{Te_5}$$
$$= P^{d}_{Cl_2}+P^{d}_{NiCl_2}+P^{d}_{NiCl}+P^{d}_{TeCl_2}+P^{d}_{TeCl_4}+P^{d}_{Te_2}+P^{d}_{Te_5} \quad (A.10)$$

Application of the extended form of the Gibbs phase rule (27) reveals that the Ni-Cl-Te system in the closed ampoule reactor possesses two degrees of freedom which are satisfied by specifying the temperatures of the source (T_s) and deposition (T_d) zones, respectively. The equilibrium constants for the ten reactions given earlier are expressed in terms of partial pressures.

$$K_{1s} = \frac{P^{s}_{NiCl_2}(P^{s}_{TeCl_2})^{1.7}}{(P^{s}_{Cl_2})^{2.7}} \quad ; \quad K_{2s} = \frac{P^{s}_{NiCl}(P^{s}_{Cl_2})^{0.5}}{P^{s}_{NiCl_2}} \quad (A.11,12)$$

$$K_{3s} = \frac{P^{s}_{TeCl_2}}{(P^{s}_{Te_2})^{0.5}\,P^{s}_{Cl_2}} \quad ; \quad K_{4s} = \frac{P^{s}_{TeCl_4}}{(P^{s}_{Te_2})^{0.5}\,(P^{s}_{Cl_2})^{2}} \quad (A.13,14)$$

$$K_{5s} = \frac{P^{s}_{Te_5}}{(P^{s}_{Te_2})^{2.5}} \quad ; \quad K_{1d} = \frac{(P^{d}_{Cl_2})^{2.9}}{P^{d}_{NiCl_2}(P^{d}_{TeCl_2})^{1.9}} \quad (A.15,16)$$

$$K_{2d} = \frac{P^{d}_{NiCl}\,(P^{d}_{Cl_2})^{0.5}}{P^{d}_{NiCl_2}} \quad ; \quad K_{3d} = \frac{P^{d}_{TeCl_2}}{(P^{d}_{Te_2})^{0.5}\,P^{d}_{Cl_2}} \quad (A.17,18)$$

$$K_{4d} = \frac{P^{d}_{TeCl_4}}{(P^{d}_{Te_2})^{0.5}\,(P^{d}_{Cl_2})^{2}} \quad ; \quad K_{5d} = \frac{P^{d}_{Te_5}}{(P^{d}_{Te_2})^{2.5}} \quad (A.19,20)$$

For a given $T_s \leftrightarrow T_d$ set, the equilibrium constants are evaluated from the corresponding standard free energy changes for the various reactions; these data drawn from published compilations and other sources (22-25,28) are summarized in Table 1. The fourteen relations, that is, equations (A.7-A.20) are sufficient to determine the equilibrium partial pressures of the seven species in each of the two end zones. The iterative equilibrium constant method (29) is used to obtain the solution for specified reactor size and other experimental conditions.

Growth Rates for $(Te/Ni)^s$:1.7 and $(Te/Ni)^d$:1.9

The dimensions of the closed ampoule reactor are chosen as follows:

$$L_s = L_d = 4 \text{ cm};\ L_c = 80 \text{ cm};\ S_c = 24.544 \text{ cm}^2$$

$$V_s = V_d = 98.175 \text{ cm}^3;\ V_c = 20\ V_s = 20\ V_d\ ;\ r = 20$$

The gas-phase compositions were computed for three $T_s \leftrightarrow T_d$ sets with the initial amount of the chlorine gas in the reactor varied in the range 0.00001-0.00012 mole. The results for 0.2×10^{-4} mole $Cl_2(g)$ are listed in Table 2. It is seen that at 1054.4K (T_s)/ 970.2K (T_d), the mass-transport of species $NiCl_2$, NiCl, Cl_2, and Te_2 occurs from left to right (Figure 1) while that of the species Te_5, $TeCl_2$, and $TeCl_4$ is in the opposite direction. A similar behavior is noted for the 1020K/939.6K and 1054.4K/926K sets. Thus, deposition of $NiTe_{1.9}(s)$ is favored for the three $T_s \leftrightarrow T_d$ combinations (Table 2); this is true at other initial chlorine concentrations. The formation of $NiCl_2(s)$ from the gas-phase, by condensation $NiCl_2(g) \rightarrow NiCl_2$, is not favored under these conditions.

Mass-Transport

The transport of species between the two end-zones is brought about largely by diffusion and Stefan flow. Convection effects are not significant in horizontal ampoule reactors, especially at low pressures. The contribution of thermal diffusion to the overall mass-transfer is small; using the data on thermal diffusion factors for Ne-CF_4 system published by Dunlop et al. (30), it can be shown that the migration of the heavier species (Te_2) to the colder zone (T_d) produces a decrease of 2-3% in the source zone mole fraction of Te_2.

At quasi-steady-state, the molar fluxes of the species must satisfy the following requirements: (1) no net flux of chlorine between the two end-zones (2) the fluxes of tellurium and nickel are in the proportion of 1.9:1.0. In other words

$$N_{Cl_2} + N_{NiCl_2} + N_{TeCl_2} + 2N_{TeCl_4} + 0.5N_{NiCl} = 0 \qquad (A.21)$$

$$1.9(N_{NiCl_2} + N_{NiCl}) = 2N_{Te_2} + N_{TeCl_2} + 5N_{Te_5} + N_{TeCl_4} \qquad (A.22)$$

Inspection of the partial pressures in Table 2 reveals that the species Cl_2, Te_5, and $TeCl_4$ are negligible. After simplification, the foregoing relations together yield

$$N_{Te_2} = 1.45\ N_{NiCl_2} + 1.2\ N_{NiCl} \qquad (A.23)$$

$$N_{TeCl_2} = -N_{NiCl_2} - 0.5\ N_{NiCl} \qquad (A.24)$$

Table 1
Standard free energy changes for different reactions
$\Delta G_i^o = A + B.T \ln T + CT$, cal

Reaction(s)	A	B	C
(1s)	-44,086	7.758	-84.25
(2s),(2d)	57,275	2.277	-40.12
(3s),(3d)	-45,926	-0.151	12.13
(4s),(4d)	-70,489	-3.646	65.48
(5s),(5d)	-51,850	-1.316	61.90
(1d)	50,030	-8.557	93.10

Table 2
Diffusion-limited deposition rates of $NiTe_y(s)$ in a closed CVT reactor containing 0.00002 mole $Cl_2(g)$ with the stoichiometric constraints (A.7) and (A.8)

(atm)	$T_s \rightarrow T_d$ 781.4°C/697.2°C	747°C/666.6°C	781.4°C/653°C
$P^s_{NiCl_2}$	6.6986×10^{-4}	5.6866×10^{-4}	5.8364×10^{-4}
P^s_{NiCl}	1.1842×10^{-5}	4.0890×10^{-6}	1.1660×10^{-5}
$P^s_{Cl_2}$	2.3470×10^{-10}	2.4216×10^{-10}	1.8376×10^{-10}
$P^s_{Te_2}$	5.4516×10^{-4}	4.2172×10^{-4}	4.8072×10^{-4}
$P^s_{Te_5}$	1.1521×10^{-9}	1.3666×10^{-9}	8.4115×10^{-10}
$P^s_{TeCl_2}$	6.8665×10^{-5}	1.3017×10^{-4}	5.0482×10^{-5}
$P^s_{TeCl_4}$	9.0308×10^{-16}	2.4743×10^{-15}	5.1982×10^{-16}
$P^d_{NiCl_2}$	6.1914×10^{-4}	5.0521×10^{-4}	4.8506×10^{-4}
P^d_{NiCl}	1.8105×10^{-6}	6.0843×10^{-7}	3.7839×10^{-7}
$P^d_{Cl_2}$	9.0289×10^{-11}	8.2736×10^{-11}	8.2808×10^{-11}
$P^d_{Te_2}$	5.0519×10^{-4}	3.4213×10^{-4}	2.8118×10^{-4}
$P^d_{Te_5}$	7.7190×10^{-9}	6.8484×10^{-9}	6.2446×10^{-9}
$P^d_{TeCl_2}$	1.6932×10^{-4}	2.7669×10^{-4}	3.5990×10^{-4}
$P^d_{TeCl_4}$	2.0469×10^{-15}	4.3869×10^{-15}	6.7529×10^{-15}
P	1.2955×10^{-3}	1.1246×10^{-3}	1.1265×10^{-3}
N_{NiCl_2} *	7.0045×10^{-9}	8.9018×10^{-9}	13.8837×10^{-9}
N_{Te_2}*	2.5236×10^{-9}	5.0798×10^{-9}	12.8609×10^{-9}
N_{TeCl_2} *	-3.3452×10^{-9}	-4.9150×10^{-9}	-10.4979×10^{-9}
N_{Te}/N_{Ni}	0.243	0.589	1.097

*mole·cm^{-2}·s^{-1}

where N_i is the molar flux (in mole·cm^{-2}·s^{-1}) of the i-th species. For the purposes of detailed calculations, the letters A,B,C,D,E,F, and G are used to denote the species $NiCl_2$, Te_2, $TeCl_2$, NiCl, Te_5, Cl_2, and $TeCl_4$, respectively. According to Stefan-Maxwell formalism, the change in the partial pressure of species A over a diffusion distance of dZ is given by

$$-dP_A = RT \sum_i [D_{Ai}^{-1}(N_A X_i - N_i X_A)] \cdot dZ; \; i=B,C, \ldots G \qquad (A.25)$$

where D_{AB}, D_{AC}, ... are ordinary diffusion coefficients for the pairs A-B, A-C, ..., respectively and X_i is the mole fraction for the i-th species at Z=Z in the connecting-zone. The Fuller-Schettler-Giddings equation (31) for binary diffusivity can be expressed as

$$D_{AB} = D_{AB}^{o} \cdot P^{-1} \cdot T^{1.75} \qquad (A.26)$$

The binary coefficient D_{AB}^{o} can be calculated from the molecular weights and diffusion volumes for the two species. The latter are 37.7 for Cl_2 (31), 51.6 for $NiCl_2$, 32.1 for NiCl, 46.4 for Te_2, 62.2 for $TeCl_2$, 101.2 for $TeCl_4$, and 116.0 for Te_5, respectively. The molar fluxes of species B,C, ..., are expressed relative to that of A using

$$\gamma_A^B = N_B/N_A, \quad \gamma_A^C = N_C/N_A, \quad \ldots, \quad \gamma_A^i = N_i/N_A \qquad (A.27)$$

which together with equation (A.25) provide

$$N_A = \frac{-1}{RT} [\Sigma(X_i/D_{Ai}) - X_A \Sigma(\gamma_A^i/D_{Ai})]^{-1} \cdot \frac{dP_A}{dZ} \qquad (A.28)$$

Introducing $u=T^{0.25}$ and noting that the temperature-profile is linear with a gradient of w (in $^oC \cdot cm^{-1}$), equation (A.28) is transformed to

$$N_A = \frac{w}{4R} [\Sigma(X_i/D_{Ai}^{o}) - X_A \Sigma(\gamma_A^i/D_{Ai}^{o})]^{-1} \cdot \frac{dX_A}{du} \qquad (A.29)$$

A binary equivalent diffusivity D_{AV} is defined by (32)

$$N_A = -\frac{D_{AV}}{RT} \cdot \frac{dP_A}{dZ} + X_A N_A \psi_A \qquad (A.30)$$

where

$$\psi_A = 1 + \gamma_A^B + \gamma_A^C + \gamma_A^D + \gamma_A^E + \gamma_A^F + \gamma_A^G \qquad (A.31)$$

By combining the last three equations and treating ψ_A as essentially independent of u (or Z), it is found that

$$N_A = \frac{-D_{AV}^{o} w}{4R \psi_A} \cdot \frac{d \ln (1 - X_A \psi_A)}{du} \qquad (A.32)$$

At steady-state, N_A is constant over the entire length of the reactor, with the result equation (A.32) can be integrated with

(i) $u = u_s$(or $T=T_s$), $X_A = X_A^s = P_A^s/P$

(ii) $u = u_d$ (or $T=T_d$), $X_A = X_A^d = P_A^d/P$

to obtain the following relation for the molar flux of A.

$$N_A = \frac{D^o_{AV}\ w}{4R\ \psi_A(u_s-u_d)}[\ln(1-\psi_A X^d_A)-\ln(1-\psi_A X^s_A)] \qquad (A.33)$$

An estimate of ψ_A for $NiCl_2$ is made using equations (A.23), (A.24), and (A.31).

$$\psi_A = 1.45 + 1.7\ (N_D/N_A) \qquad (A.34)$$

The binary equivalent coefficient D^o_{AV} is given by

$$D^o_{AV} = (1-X_A\ \psi_A)[\Sigma(X_i/D^o_{Ai}) - X_A\beta_A]^{-1} \qquad (A.35)$$

$$\beta_A = 0.1966 + 0.6858\ (N_D/N_A) \qquad (A.36)$$

Relations similar to equations (A.33)-(A.36) can be deduced for species B, C, and D. Furthermore

$$\psi_B = 1.0 + 0.5\ (N_D/N_B) \qquad (A.37)$$

$$\psi_C = -1.45 + 0.975\ (N_D/N_C) \qquad (A.38)$$

$$\psi_D = 0.975\ \psi_C/(1.45 + \psi_C) = 0.5 + (N_B/N_D) \qquad (A.39)$$

$$\beta_B = -0.0737 + 0.167\ (N_D/N_B) \qquad (A.40)$$

$$\beta_C = -1.3976 + 0.4325\ (N_D/N_C) \qquad (A.41)$$

$$\beta_D = -0.1577 + 0.3418\ (N_B/N_D) \qquad (A.42)$$

The data on partial pressures listed in Table 2 are used in combination with the above equations to calculate the mass-transfer rates of $NiCl_2$, Te_2, and so forth. For each species, the mole fractions in the source and deposition zones are found and the average $\overline{X}_i$ evaluated therefrom. Using these averages together with β_A=0.1966 and ψ_A=1.45, which implies that the molar flux N_D of NiCl is negligible, D^o_{AV} is calculated. For the set 1054.4K/970.2K (Table 2), it follows that

$$D^o_{AV} = 1.8179\times10^{-6}\ cm^2\cdot atm\cdot s^{-1}\cdot K^{-1.75}\ ;\quad w = 1.0525\ K\cdot cm^{-1}$$

$$u_s = 5.6984\ ;\quad u_d = 5.5810\ ;\quad u_s-u_d = 0.117337\ K^{0.25}$$

$$X^s_A = 0.517067\ ;\quad X^d_A = 0.477916\ ;\quad \overline{X}_A = 0.497491$$

Substitutions provide

$$N_{NiCl_2} = 7.0045\times10^{-9}\ mole\cdot cm^{-2}\cdot s^{-1}$$

The molar fluxes of Te_2 and $TeCl_2$ were calculated in a similar fashion (Table 2). The ratio N_{Te}/N_{Ni} of tellurium to nickel molar fluxes is seen to range from 0.243 (at 0.00002 mole Cl_2) to 0.748 (at 0.00012 mole Cl_2), substantially less than that required for the stoichiometry $NiTe_{1.9}$(s). This is attributed to the imposition of the stoichiometric constraints (A.7) and (A.8) in calculating the gas-phase partial pressures. The three-zone model is modified as described in the next section.

Growth Rates for $N_{Te}/N_{Ni} = 1.9$ and $N_{Cl}=0$

The gas-solid virtual equilibrium is considered to prevail in the end-zones with the quasi-steady-state diffusive mass transport between them occurring in such a manner that the relations (A.23) and (A.24), involving the molar fluxes of Te_2, $NiCl_2$, NiCl, and $TeCl_2$, are satisfied. That the gas-phase in the immediate vicinity of $NiTe_x(s)$ or $NiTe_y(s)$ have the same Te/Ni ratio as the condensed phase is now abandoned. By combining equation (A.24) with (A.38) and (A.23) with (A.39), the following constraints are obtained:

$$0.975\, N_A = -(0.5\, \psi_C + 1.7)N_C \qquad (A.43)$$

$$(0.725 - 0.475\, \psi_C)N_A = (1.7+0.5\, \psi_C)N_B \qquad (A.44)$$

Using appropriate analogs of equation (A.33) for the molar fluxes of $NiCl_2(A)$, $Te_2(B)$, and $TeCl_2(C)$ in the above constraints, and making the simplification,

$$\ln(1- \psi_i X_i^d) - \ln(1- \psi_i X_i^s) = \psi_i (X_i^s - X_i^d)/(1-\psi_i X_i^s)$$

together with cross-multiplication and substitution of $X_i=P_i/P$, the following relations are obtained:

$$PP^s_{NiCl_2} - C_3 PP^s_{TeCl_2} - C_3 \psi_A P^s_{NiCl_2} P^d_{TeCl_2} + (C_3\psi_A - \psi_C) P^s_{NiCl_2} P^s_{TeCl_2}$$

$$= PP^d_{NiCl_2} - \psi_C P^d_{NiCl_2} P^s_{TeCl_2} - C_3 PP^d_{TeCl_2} \qquad (A.45)$$

$$PP^s_{NiCl_2} + \psi_B P^d_{NiCl_2} P^s_{Te_2} + (C_4\, \psi_A - \psi_B) P^s_{NiCl_2} P^s_{Te_2}$$

$$+ C_4 PP^d_{Te_2} = PP^d_{NiCl_2} + C_4 PP^s_{Te_2} + C_4 \psi_A P^s_{NiCl_2} P^d_{Te_2} \qquad (A.46)$$

where

$$C_3 = -(D^o_{CV}/D^o_{AV})(0.5\, \psi_C + 1.7)/0.975 \qquad (A.47)$$

$$C_4 = (D^o_{BV}/D^o_{AV})(0.5\, \psi_C + 1.7)/(0.725 - 0.475\, \psi_C) \qquad (A.48)$$

Of the four ψ_i's, only one is independent and it is chosen as ψ_C. In these equations, the input values of D^o_{iV} and ψ_C are found from the gas-phase compositions of the previous section, a portion of which is shown in Table 2. The virtual equilibrium partial pressures of the seven species in the source and deposition zones were calculated as in the earlier section except that the stoichiometric constraints (A.7) and (A.8) were replaced by the equations (A.45) and (A.46). These data were used to compute the molar fluxes of the four species $NiCl_2$, Te_2, $TeCl_2$, and NiCl; and the results are given in Table 3. The partial pressures computed with the modified constraints differed only slightly from those in Table 2; the difference $P_i^s - P_i^d$, however, changed significantly resulting in the desired net fluxes of Ni and Te.

Table 3
Molar Fluxes of Gas Species ($mole \cdot cm^{-2} \cdot s^{-1}$)
for different end-zone temperatures

$n^{o}_{Cl_2}$, mole	$NiCl_2$	Te_2	$TeCl_2$	$NiCl$
		Set 1: 1054.4K/970.2K		
0.00001	0.974×10^{-9}	2.779×10^{-9}	-1.958×10^{-9}	0.921×10^{-9}
0.00002	2.700×10^{-9}	4.717×10^{-9}	-3.125×10^{-9}	0.600×10^{-9}
0.00004	4.266×10^{-9}	6.614×10^{-9}	-4.441×10^{-9}	0.351×10^{-9}
0.00006	4.700×10^{-9}	7.170×10^{-9}	-4.875×10^{-9}	0.260×10^{-9}
0.00008	4.850×10^{-9}	7.298×10^{-9}	-4.952×10^{-9}	0.209×10^{-9}
0.00010	4.810×10^{-9}	7.203×10^{-9}	-4.898×10^{-9}	0.177×10^{-9}
0.00012	4.714×10^{-9}	7.026×10^{-9}	-4.792×10^{-9}	0.156×10^{-9}
		Set 2: 1020K/939.6K		
0.00002	4.400×10^{-9}	6.555×10^{-9}	-4.100×10^{-9}	0.250×10^{-9}
0.00004	4.700×10^{-9}	6.970×10^{-9}	-4.590×10^{-9}	0.143×10^{-9}
		Set 3: 1054.4K/926.0K		
0.00002	9.600×10^{-9}	1.453×10^{-8}	-9.193×10^{-9}	0.826×10^{-9}

The deposition rate R_d (in $mg \cdot hr^{-1}$) is found from the molar fluxes using

$$R_d = (N_{NiCl_2} + N_{NiCl} + 2N_{Te_2} + N_{TeCl_2}) S_c (313.89)(3600) \qquad (A.49)$$

For set 1, even though the toal pressure incresed monotonically from 7.3227×10^{-4} atm at 10^{-5} mole $Cl_2(g)$ to 6.0584×10^{-3} atm at 1.2×10^{-4} mole $Cl_2(g)$, the deposition rate passes through a maximum of 0.408 $mg \cdot hr^{-1}$ at 8×10^{-5} mole $Cl_2(g)$ (at 4.2731×10^{-3} atm); similar pressure dependence was noted in the Sn-S-I system (8). In set 2, with comparable ΔT between end zones, but at lower temperatures, R_d is larger, especially at the low pressure of 1.0686×10^{-3} atm for 2×10^{-5} mole $Cl_2(g)$. With a 50% greater ΔT (set 3), the deposition rate of 0.84 $mg \cdot hr^{-1}$ at a total pressure of 1.0816×10^{-3} atm compares very favorably to 0.267 $mg \cdot hr^{-1}$ of set 1 at 1.2809×10^{-3} atm.

References

1. J. Barstad, F. Gronvold, E. Rost, and E. Vestersjo, Acta Chem. Scand. 20, 2865 (1966).
2. J.G. Brennan, T. Siegrist, S.M. Stuczynski, and M.L. Steigerwald, J. Amer. Chem. Soc. 111, 9240 (1989).
3. A.L. Fahrenbruch, J. Cryst. Growth 39, 73 (1977).
4. J. Stankiewicz and A. Aray, J. Appl. Phys. 53, 3117 (1982).
5. J. Misiewicz, P. Becla, E.D. Isaacs, P.A. Wolff, D. Heiman, L.R. Ram-Mohan, and J.M. Wrobel, J. Appl. Phys. 63, 2396 (1988).
6. H. Schäfer, Chemical Transport Reactions, Academic Press, New York (1964).
7. J. Mercier, J. Cryst. Growth 56, 235 (1982).
8. H. Wiedemeier and F.J. Csillag, J. Cryst. Growth 46, 189 (1979).
9. R.F. Lever, J. Chem. Phys. 37, 1078; 1174 (1962).
10. G. Mandel, J. Chem. Phys. 37, 1177 (1962); ibid. 40, 683 (1964).
11. G. Mandel, J. Phys. Chem. Solids 23, 587 (1962).
12. R.F. Lever and G. Mandel, J. Phys. Chem. Solids 23, 599 (1962).

13. H. Watanabe, T. Nishinaga, and T. Arizumi, J. Cryst. Growth 17, 183 (1972).
14. M.M. Faktor, I. Garrett, and R. Heckingbottom, J. Cryst. Growth 9, 3 (1971).
15. M.M. Faktor and I. Garrett, J. Cryst. Growth 9, 12 (1971).
16. B.I. Nölang and M.W. Richardson, J. Cryst. Growth 34, 198 (1976); ibid. 34, 205 (1976).
17. M.W. Richardson and B.I. Nölang, J. Cryst. Growth 42, 90 (1977).
18. C. Paorici and L. Zecchina, J. Cryst. Growth 83, 453(1987).
19. Y.K. Rao, Y. Do, and S.S. Kim, in Proceedings of the 4th National Seminar on Crystl Growth: Current Trends in Crystal Growth and Characterization, ed. K. Byrappa, University of Mysore, Mysore, India (1989).
20. Y.K. Rao and Yang Do, High Temp. Sci. (in press), (1990).
21. E.F. Westrum, Jr., C. Chou, R.E. Machol, and F. Gronvold, J. Chem. Phys. 28, 497 (1958).
22. E.F. Westrum, Jr. and R.E. Machol, J. Chem. Phys. 29, 824 (1958).
23. K.C. Mills, Thermodynamic Data for Inorganic Sulphides, Selenides, and Tellurides, p. 95 and p. 472, Butterworths, London, U.K. (1974).
24. I. Barin and O. Knacke, Thermochemical Properties of Inorganic Substances, Springer-Verlag, Berlin (1973).
25. I. Barin, O. Knacke, and O. Kubaschewski, Thermochemical Properties of Inorganic Substances. Supplement, Springer-Verlag, Berlin (1977).
26. The Collected Works of J. Willard Gibbs, Vol. I, Thermodynamics, p. 62, Yale University Press, (1957).
27. Y.K. Rao, Chem. Eng. Educ. 19, 40 (1985).
28. M.W. Chase, Jr., C.A. Davies, J.R. Downey, Jr., D.J. Frurip, R.A. McDonald, and A.N. Syverud, JANAF Thermochemical Tables, 3rd edn., American Chemical Society, Washington, D.C. (1986).
29. Y.K. Rao, Stoichiometry and Thermodynamics of Metallurgical Processes, Cambridge University Press (1985); Y.K. Rao, M. Donley, and H.G. Lee, J. Materials Educ. 5, 309 (1983).
30. P.J. Dunlop, H.L. Robjohns, and C.M. Bignell, Ber. Bunsenges. Phys. Chem. 91, 156 (1987).
31. E.N. Fuller, P.D. Schettler, and J.C. Giddings, Indust. Engng. Chem. 58 (5), 18 (1966).
32. H.-W. Hsu and R.B. Bird, A.I.Ch.E., Journal 6, 516 (1960).

High-temperature sulphidation of iron-based alloys

T. Narita
Department of Metallurgical Engineering, Faculty of Engineering, Hokkaido University, Kita 13-jo, Nishi 8-chome, Kita-Ku, Sapporo 060, Japan

Introduction

The sulfidation of metals at high temperatures has become a matter of increasing concern in industrial applications, particularly in energy conversion systems using fossil fuels. Practical and theoretical aspects of the sulfidation mechanism of commercial and artificial alloys have been established (1,2,3), and the results suggest that sulfidation rates are several orders of magnitude higher than oxidation rates. This is mainly because sulfides have high diffusivities because of the considerable nonstoichiometry and the formation of eutectic melts with metals at relatively low temperatures.

This degradation accompanied with internal sulfidation has been often observed in nickel-containing alloys in aggressive environments with multi-components of oxidants. Here a liquid phase with a nickelous sulfide-nickel eutectic melt forms at the relatively low temperature of 918 K. While, due to the higher melting temperature of the eutectic melt in an Fe-S system iron-based alloys would be expected to have better anti-sulfidation properties than nickel-based alloys. A series of investigations of sulfidation of iron-based binary and ternary alloys was carried out by the present author and co-workers(4,5,6,7), and there are reports of the sulfidation of iron-based alloys by other investigators.

In the present paper the sulfidation behavior of iron-based alloys is summarized, and diffusional analysis is applied to the Fe-Cr alloy sulfidation at intermediate and low sulfur pressures to get a better understanding of this sulfidation behavior. The calculated sulfidation rates and concentration profiles were compared with the experimental observations, and the effects of sulfur partial pressures and alloying elements on sulfidation resistance of iron-based alloys are evaluated.

Sulfidation Behavior of Fe-Cr Alloys

Narita and Smeltzer (8,9) sulfidized an Fe-26.6at%Cr alloy at 973, 1073, and 1173K at sulfur pressures from 10^4 to 10^{-6} Pa. From microstructual observations the scale structures could be classified into three groups(8), as exemplified in Fig.1; A triplex $(Cr,Fe)S_x$-$(Cr,Fe)_3S_4$-$(Fe,Cr)_{1-x}S$ scale forms at high sulfur pressure (range 1), a duplex $(Cr,Fe)S_x$-$(Fe,Cr)_{1-x}S$ scale at intermediate pressures (range 2), and a single-phase $(Cr,Fe)S_x$ at low sulfur pressures (range 3). At 973K these pressure ranges were as follows: range 1 $P(S_2)>10^{-2}$Pa, range 2 $10^{-2}Pa<P(S_2)<10^{-5}$Pa, and range 3 $P(S_2)<10^{-5}$Pa.

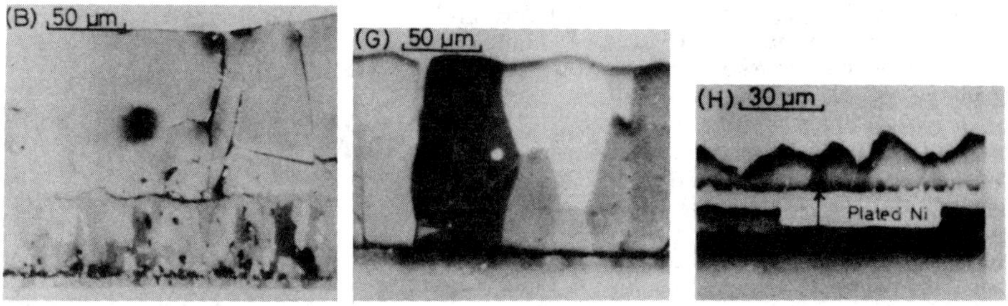

FIG.1. Optical micrographs (polarized light) of the scales formed on the Fe-26.6Cr alloy at 1073K under various sulfur pressures: 10^2Pa(B), 10^{-4}Pa(G), and 10^{-5}Pa(H).

The critical sulfur pressures, 10^{-2} and 10^{-5}Pa at 973K, became higher at higher temperatures, 1 and 10^{-4}Pa at 1073K, and to 10^2 and 10^{-2}Pa at 1173K. The lower sulfur pressures of the two critical pressures are close to the dissociation pressures of ferrous sulfide at each temperature.

Figure 2 shows the sulfur pressure dependence of sulfidation rates obtained for an Fe-26.6at%Cr alloys at 973, 1073, and 1173 K. The sulfidation rates at each temperature display a slow increase with decreasing sulfur pressure in the range 1, and after reaching a maximum in the range 2 the sulfidation rates decreased rapidly at sulfur pressures in the range 3. It could be concluded that changes in sulfidation rates with sulfur pressures appears to arise because of changes in sulfides and their scale structure.

Maximum chromium and sulfur concentration within the sulfide phases of the

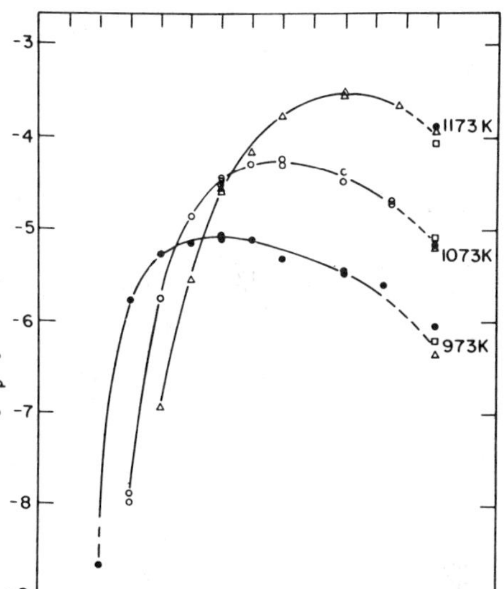

FIG.2. Plots of log k_p vs log Ps_2 for the Fe-26.6Cr alloy.

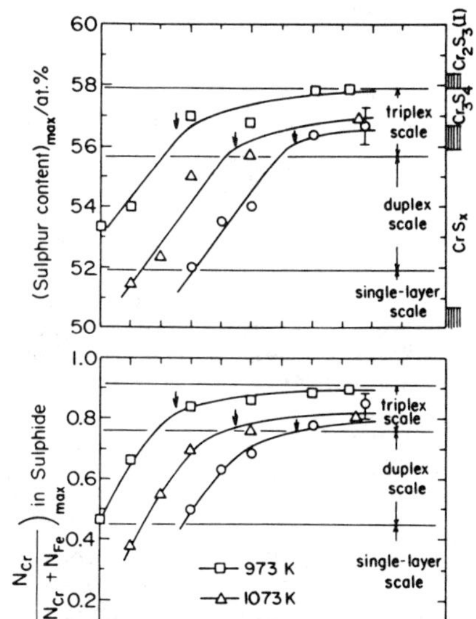

FIG.3. Maximum S and Cr compositions of the sulfides formed on the Fe-26.6Cr alloy.

different types of scales were determined from the concentration profiles, and the dependence of these values on sulfur pressures at each temperature are shown in Fig.3(9). Error limits of these concentrations are 0.5at%S, and the arrow designations in each graph in Fig.3 show the transitions from growth of a triplex to duplex scale.

According to these composition consideration and the phase diagram for the Fe-Cr-S system(10) as shown in Fig.4, spinel $FeCr_2S_4$ can be expected to form between the intermediate layer of monoclinic $(Cr,Fe)_3S_4$ and the outer layer of hexagonal $(Fe,Cr)_{1-x}S$ at high sulfur pressures. It was experimentally demonstrated that this spinel $FeCr_2S_4$ layer could not be identified in sulfides with a triplex layer structure formed at high sulfur pressures, region 1.

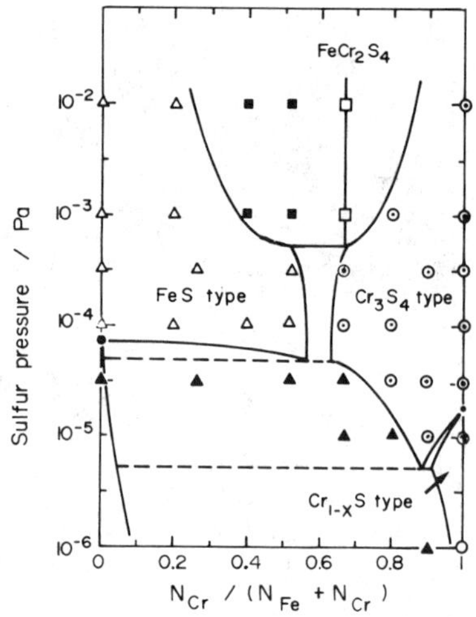

FIG.4. Phase relations in the Fe-Cr-S system at 1073K.

This conflict may be explained by assuming that the thickness of the $FeCr_2S_4$ layer is such that it remained undetected by the conventional optical microscopy, because of low diffusivity (11), in contrast to the high diffusivity of the $(Fe,Cr)_{1-x}S$ phase. As in Fig.4, the phase diagram of the Fe-Cr-S system at 1073K shows that with decreasing sulfur pressures the spinel $FeCr_2S_4$ decomposed into monoclinic $(Cr,Fe)_3S_4$ and hexagonal $(Fe,Cr)_{1-x}S$. The critical sulfur pressure for the phase transition is between 10^{-3} and $10^{-3.5}$Pa at 1073K. Accordingly, it is possible that the inner and outer layers are the $(Cr,Fe)_3S_4$ and $(Fe,Cr)_{1-x}S$ of the duplex scale formed at intermediate sulfur pressures. Experimental results indicate that the inner layer is composed of hexagonal $(Cr,Fe)S_x$ but not monoclinic $(Cr,Fe)_3S_4$. At low sulfur pressures in the range 3 hexagonal $(Cr,Fe)S_x$ was identified in the sulfidation product of the single layer structure, instead of the monoclinic $(Cr,Fe)_3S_4$ deduced from the Fe-Cr-S phase diagram. This sulfide phase is not finnally established and further investigation is necessary.

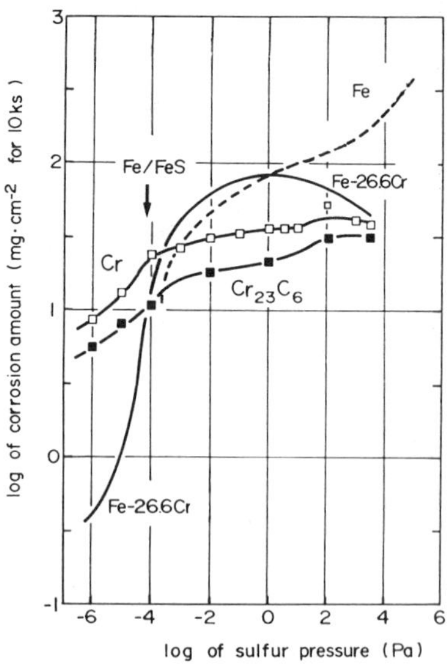

Figure 5 shows corrosion amounts per 10 ks at 1073 K for the Fe-26.6at%Cr alloy. Here amounts of corrosion for chromium and chromium carbide (12) are shown for comparison purpose. It was demonstrated that sulfidation rates of the alloy become higher than that of pure iron at intermediate sulfur pressures, in the range 2, and this seems to be due to a change in phase from monoclinic Cr_3S_4 to hexagonal CrS_x with decreasing sulfur pressure and also due to a doping effect of Cr^{+3} into the ferrous sulfide resulting in higher diffusivities.

FIG.5. Corrosion amounts per 10 ks at 1073K for Fe and Fe-26.6Cr alloy.

At low sulfur pressures the sulfidation rates were several orders of magnitude lower than those at intermediate and high sulfur pressures. This may be due to the fact that the sulfidation rates are determined by chromium diffusion from the bulk of the alloy towards the alloy/scale interface, and theoretical interpretation will be given in the later section.

At low sulfur pressures, several kinds of internal degradation modes were found in iron-based alloys (4): usual internal sulfidation, preferential attack along grain boundaries named grain boundary sulfidation, and intermediates between these. As internal degradation may be expected to cause the loss of mechanical strength in the alloy matrix, the width of the internal degradation zone was measured for the Fe-Cr alloys sulfidized at 1073K for various corrosion times at each sulfur pressure.

Figure 6 shows the parabolic rate constants determined from measurements of mass gain, scale thickness, and alloy decrescence plotted against the chromium content for Fe-Cr alloys at 1073K and at 10^{-5}Pa sulfur pressure. At this low sulfur pressures the sulfidation rates obtained from mass gain and scale thickness increased significantly with increasing chromium content. This composition dependence of sulfidation rates is mainly due to the rapid sulfidation of chromium at low sulfur pressures, as shown in Fig.5, because chromium sulfides formed at low sulfur pressures is the very diffusive hexagonal CrS_x.

Figure 6 also demonstrates that the corrosion rates determined from the metal decresecence due to both general corrosion and grain boudnary sulfidation are several orders of magnitude larger than those obtained from the mass-gain and scale thickness measurements, even when the alloys were sulfidized below the dissociation pressures of ferrous sulfides. This shows that internal sulfidation and its prevention will become an increasing concern in the development of anti-sulfidation alloys.

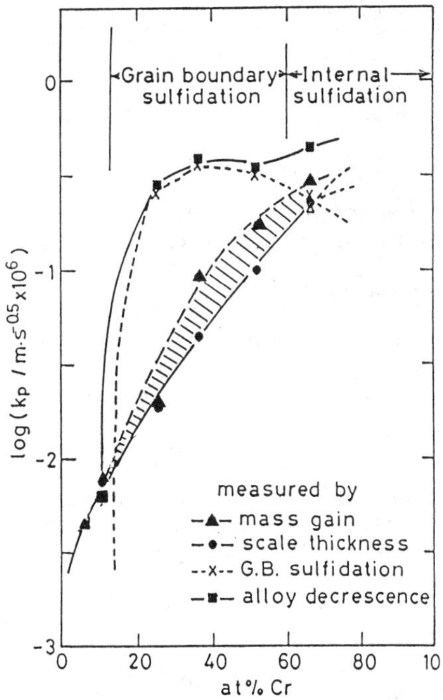

FIG.6. Parabolic rate constants obtained by various methods at 1073 K and 10^{-5}Pa sulfur pressure for Fe-Cr alloys.

Diffusional Analysis

General Equations

The concentration profile of chromium in the alloy phase from the alloy/scale interface toward the alloy bulk is given by the following relationship, which was derived by solving Fick's second law under a moving boundary condition (the prime superscript is the values at the alloy/scale interface).

$$N_{Cr} = N'_{Cr} + (N^o_{Cr}-N'_{Cr}) \cdot [erf(0.5r/(D_{all})^{1/2})-erf((0.5a)^{1/2})]/erfc[(0.5a)^{1/2}] \quad [1]$$

where D_{all} is the chemical diffusivity of the Fe-Cr alloy and N^o_{Cr} and N'_{Cr} are bulk and interface chromium compositions; $r(x/(t)^{1/2})$ is a normalized rate constant and $a(k/D_{all})$ is a normalized alloy distance.

The average sulfide composition C^o_{CrS} in the $(Fe,Cr)_{1-x}S$ scale for the alloy side is,

$$C^o_{CrS} = N'_{Cr} + 2D_{all}(V_{eq}/V_{all})(2/k)^{1/2}(dN'_{Cr}/dr) \quad [2]$$

where N'_{Cr} and dN'_{Cr}/dr are the alloy composition and its gradient at the alloy/scale interface; V_{eq}/V_{all} is the ratio of equivalent volume of sulfide to the alloy molar volume, and k is a parabolic rate constant defined by eq.[10].

Combining eq.[1] and [2] and rearranging, the average sulfide composition is,

$$C^o_{CrS}=(N^o_{Cr}- N'_{Cr})/F(u) + N'_{Cr} \quad [3]$$

where $F(u)= u/ \exp(u^2)[1-erf(u)]$, $u=(V_{all}/V_m)(0.5k/D_{all})^{1/2}$, and V_m is the molar volume of the solid solution sulfide.

For the scale side it is

$$C^o_{CrS} = [C'_{CrS}-(D'_{Fe}/k)(dC'_{CrS}/dy)]/W' \quad [4]$$

where $W'= p + (p-1)C'_{CrS}$ and $p\ (D_{Fe}/D_{Cr})$ is the ratio of self-diffusivities of iron to chromium in $(Fe,Cr)_{1-x}S$. The value of C^o_{CrS} can be obtained by integrating the sulfide composition across the surface scale.

Assuming the alloy and sulfide phases to be thermodynamically ideal, the composition and sulfur activty at the alloy/scale interface are given by the following equations;

$$N'_{Cr} = C'_{CrS}/[R+(1-R)C'_{CrS}] \quad [5]$$

and

$$a'_s = [R-(R-1)C'_{CrS}]\exp(dG_{CrS}/RT) \quad [6]$$

where $R = \exp[(dG_{CrS}-dG_{FeS})/RT]$, and dG_{CrS} and dG_{FeS} are free energy changes for the formation of chromium and ferrous sulfides.

According to the ternary diffusion theory proposed by Wagner (13), Fick's first law for the growth of the solid solution scale is given by the following relation:

$$D_{Fe}(1-C_{CrS})[(-d\ln a_{FeS}/dC_{CrS})(dC_{CrS}/dy)+d\ln a_s/dy] + D_{Cr}C_{CrS}[(-d\ln a_{CrS}/dC_{CrS})(dC_{CrS}/dy)+d\ln a_s/dy] = k \quad [7]$$

By assuming that $dC_{CrS}/dy=0$ in the scale, the following relation is obtained:

$$d\ln a_s/dy = pk/[D_{Fe}(p-(p-1)C^o_{CrS})] \quad [8]$$

where the iron self-diffusivity in $(Fe,Cr)_{1-x}S$, D_{Fe}, must be a function of sulfide composition and sulfur activity. Values for D_{Fe} are not available in the literatures, and analogous to the equation adopted for ferrous-cobaltous sulfide (14), the iron self-diffusivity was assumed to be given by the following relation:

$$D_{Fe} = D^{o}_{Fe}[\ 1+(B-1)C^{o}_{CrS}\ ]^{1/3}(a_s)^{2/n}\ ;\ n=5 \qquad [9]$$

where D^{o}_{Fe} is the iron self-diffusivity in ferrous sulfide (15) and B is the ratio of the vacancy concentration of ferrous sulfide (16) to chromium sulfide (17).

Substituting eq.[9] into eq.[8] and integrating with y between 0 and 1, and with a_s between a''_s and a'_s, the parabolic rate constant is given by the following relation:

$$k = (n/2)\cdot D^{o}_{Fe}\cdot[\ 1+(B-1)C^{o}_{CrS}\ ]^{1/3}\cdot[1-(1-1/p)C^{o}_{CrS}]\cdot[(a''_s)^{2/n}-(a'_s)^{2/n}] \qquad [10]$$

Computer Simulation

The initial input data are temperature T, alloy composition N^{o}_{Cr}, ambient sulfur pressure $P''(S_2)$ $[(a''_s)^2]$, and corrosion time t. The iron/chromium self-diffusivity ratio was assumed to be 1.8. With the parabolic rate constant (k), concentration profiles can be calculated in the alloy substrate. Iron self-diffusivity, D^{o}_{Fe}, in ferrous sulfide was obtained from the results after Flyt et al.(15). The alloy diffusivity in eq.[10] is given as functions of temperature and bulk alloy composition N^{o}_{Cr}(18):

$$D_{all} = \exp[1.2-6.1\times N^{o}_{Cr}-53{,}000(\mathrm{cal/mol})/RT] \qquad [11]$$

Here it should be mentioned that no corrosion data was used for the present calculation.

By solving eqs.[3],[4],[5],[6], and [10] simultaneously, the k, N'_{Cr}, C'_{CrS}, a'_s as well as C^{o}_{CrS} parameters are obtained.

At sulfur pressures below the dissociation pressure of ferrous sulfide the sulfide composition is confined in the area of chromium rich sulfide phase. The compositional limitation was introduced in the calculations by assuming the alloy and sulfide phases to be thermodynamically ideal in eq.[5]. This consideration was disregarded by Bastow et al.(19) in their calculation, resulting in significant error for oxidation of binary alloys at low oxygen pressures.

Figure 7 shows the results calculated for the Fe-26.6 at.%Cr alloy at 1073 K as a function of ambient sulfur partial pressures. It demonstrates that when the alloy was

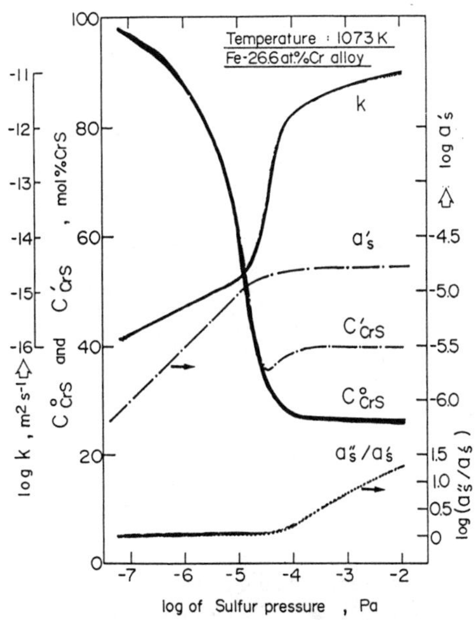

FIG.7. Corrosion data calculated and their changes with sulfur pressures for the Fe-26.6Cr alloy at 1073 K.

sulfidized at sulfur pressures above 10^{-4} Pa the rate constants k decreased gradually with decreasing sulfur pressures due to both the decreases in cation diffusivities and in the difference between the sulfur activity of a_s'' and a_s' across the scale, the values of a_s', C_{CrS}', and C_{CrS}^o are nearly constant. This is due to the simultaneous sulfidation of iron and chromium to form the $(Fe,Cr)_{1-x}S$ scale, and the cation diffusion through the scale is the rate determining process.

With further decreases in the sulfur pressures all the parameters changed rapidly between 10^{-4} and $10^{-4.5}$Pa sulfur pressure; the rate constant and sulfur activity at the alloy/scale interface decreased by several orders of magnitude and the average sulfide composition increased rapidly from 26.6 to more than 95 mol.%CrS. A similar result was obtained for the sulfide composition at the interface C_{CrS}'. This may be a result of the selective sulfidation of chromium and the fact that the sulfidation rate is determined by the alloy diffusion. This calculated critical sulfur pressure is close to the dissociation pressure of ferrous sulfide, $10^{-4.2}$ Pa, at 1073 K.

Sulfidation Rates and Concentration Profiles

Figure 8 shows the concentration profiles across the alloy substrate and sulfide scale formed on the Fe-37 at.%Cr alloy sulfidized at 1073 K under different sulfur pressures.

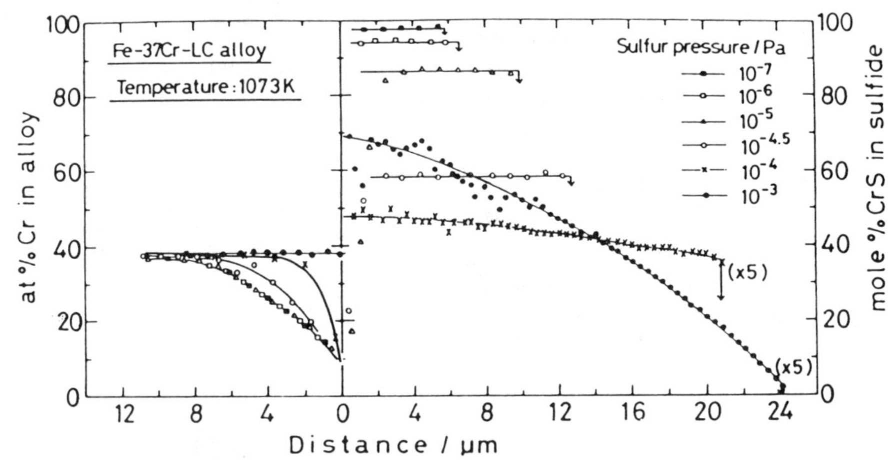

FIG.8. Concentration profiles across the alloy and surface scale of Fe-37Cr alloy sulfidized at 1073 K under various sulfur pressures. The calculated profiles in the alloy were given by solid lines.

At a 10^{-3} Pa sulfur pressure, the chromium concentration profile is virtually unchanged, whereas chromium in the sulfide scale enriched in the vicinity of the alloy scale interface and decreased toward the scale/gas interface. Diffusion process in the surface sulfide scale is a rate determining step and the slow diffusing species is chromium. With decreasing sulfur partial pressures, a chromium depleted zone forms near the alloy surface and there is chromium enrichment in the sulfide. The sulfide concentration profile becomes almost flat due to the very small sulfur activity difference across the scale irrespective of the cation diffusivities in the sulfide.

Figure 9 shows the observed and calculated concentration profiles across the alloy substrate and sulfide scale when the Fe-Cr alloys were sulfidized at

1073 K and at a 10^{-5} Pa sulfur pressure. Here the alloy grains far from grain boundaries were selected for measurement with EPMA, and the measured profiles are shown by symbols and the calculated profiles are given by solid lines.

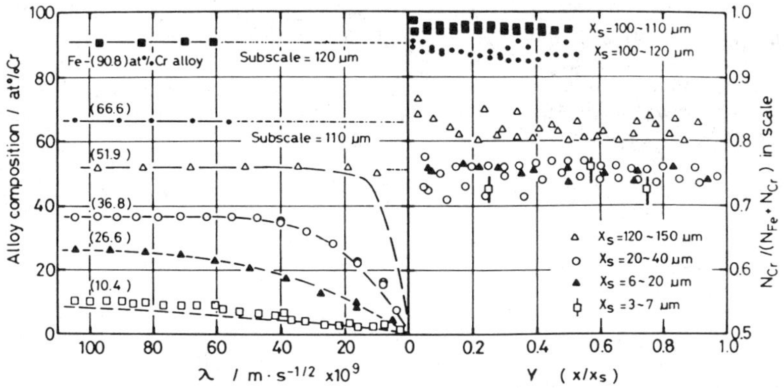

FIG.9. Concentration profiles across the alloy and surface sulfide scales of Fe–Cr alloys sulfidized at 1073 K and 10^{-5}Pa sulfur pressures.

It was demonstrated that there are good agreement between the calculated and observed cocentration profiles across the alloy substrate and surface scale. The average sulfide composition is about 75 mole%CrS for alloys containing up to 52 at.%Cr, and increased to 97 mole%CrS for the Fe-91 at.%Cr alloy.

Figure 10 shows the observed and calculated rate constants for the Fe-Cr alloys sulfidized at 1073 K under 10^{-2}(A) and 10^{-5}Pa(B) sulfur pressures. At 10^{-2}Pa sulfur pressure the calculated rate constants increased gradually with chromium contents, because of the doping effect of Cr^{+3} into the $(Fe,Cr)_{1-x}S$,

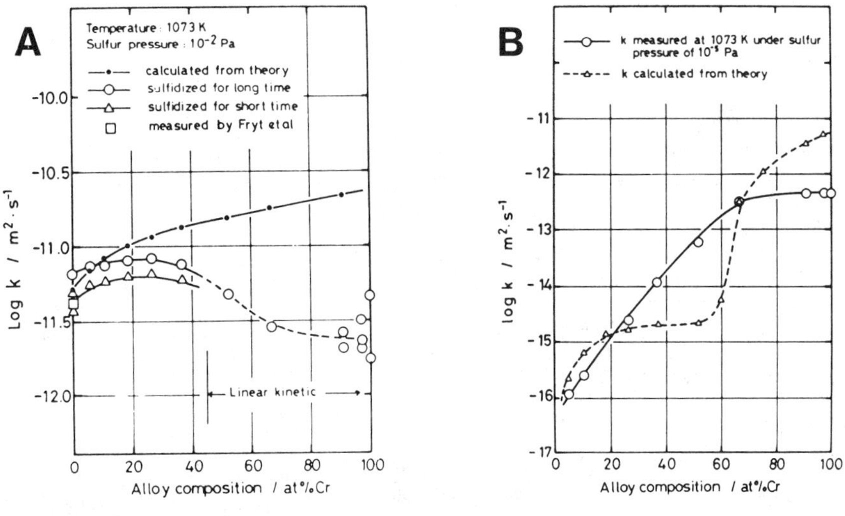

FIG.10 Comparison between the calculated rate constants and observed ones for the Fe-Cr alloys at 1073K under sulfur pressures of 10^{-2}Pa(A) and 10^{-5}Pa(B).

and they are in a good agreement with the observed ones, except for the alloys containing more than about 40%Cr. This is due to the linear sulfidation kinetics for the Fe-high Cr alloys and pure Cr metal.

When the Fe-Cr alloys were sulfidized at 10^{-5}Pa sulfur pressure, which is lower than the dissociation pressure of ferreous sulfide, the calculated rate constants were almost constant between 26.6 and 52 at.%Cr and then increased rapidly for the higher Cr alloys. In the stable composition range two factors compete to control the kinetics: Increasing chromium contents decreases the chemical diffusivity rapidly, as given in Fig.12, and chromium enrichment of the sulfide is reduced.

For the Fe-high Cr alloys the composition limitation is already excluded as can be seen from the phase diagram in Fig.4, and the kinetics may be controlled by other processes than alloy diffusion, such as cation diffusion through sulfide scale, gaseous diffusion, and surface reactions. Experimental investigations on chromium sulfidation(12) demonstrated the linear sulfidation kinetics over the wide range of sulfur pressures, and the similar kinetics was observed for sulfidation of Fe-high Cr alloys at low sulfur pressures. Sulfidation mechanism of this linear kinetics, althoguh it seems to be controlled by a surface reaction, will be discussed elsewhere.

It could be concluded that the calulated concentration profiles are in good agreement with the observed ones, while the large difference was observed between the rate constants in the medium Cr-Fe alloys. This may comes from the increased rate constants by the preferential attack along the alloy grain boundary.

Figure 11 shows the rate constants calculated and observed for sulfidation of the Fe-26.6at.%Cr alloy at 973, 1073, and 1173 K at intermediate and low sulfur pressures. The calcualted rate constants are in good agreement with those observed ones, when the chemical diffusivity of the alpha-phase in the Fe-Cr alloy was used for calculation. However, at 1173 K and sulfur pressures lower than 10^{-3}Pa the chromium depleted zone should be formed near the alloy surface due to preferential attack of chromium, causing the gamma-phase with slower chemical diffusivities than those of the alpha-phase. Computation with chemical diffusivity of the gamma-phase gave a slower sulfidation rate, as indicated by a symbol (▲) in Fig.11. It therefore seems that an agreement between these sulfidation rates under low sulfur pressures at 1173 K was obtained by chance, because of formatioon of copious internal sulfidation.

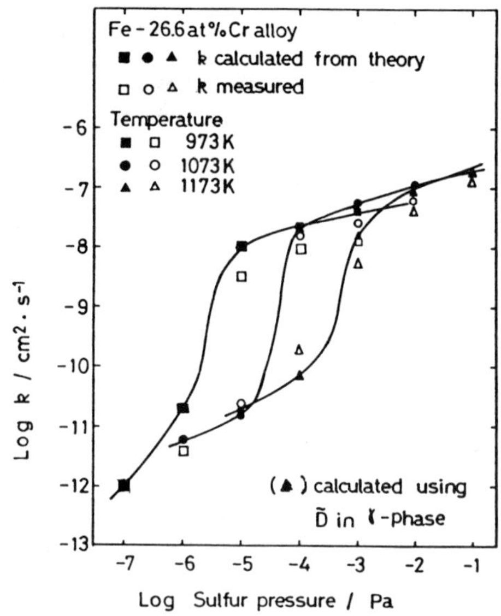

FIG.11. Parabolic rate constants calcualted and observed for the Fe-26.6Cr alloy at 973, 1073, and 1173 K in various sulfur pressures.

This ugument may allow us to conclude that for sulfidation accompanied with copious internal sulfidation chromium diffusion from the bulk alloy toward internal sulfidation front becomes a rate determining step, because mass transport through the internal sulfidiation zone was assumed to be fast, although further investigation could be required for quantitative discussion.

Internal and Grain Boundary Sulfidation

Basing on the experimental results two necessary conditions have been proposed for the commencement of the grain boundary sulfidation(4): selective sulfidation and rapid diffusion of chromium in the sulfide scale. When the scale growth is controlled by the alloy diffusion, as described by Wagner(20), the scale-metal interface can become unstable and then the grain boundaries are the inward protruding sites.

A rapid sulfur migration is required for the propagation of preferential sulfidation along grain boundaries, it can be explained as follows: Because chromium diffuses along grain boundary faster than from the bulk alloy, the counter-flowing vacancies will inject and nucleate voids at the grain boundary, and sulfur can migrate rapidly through these voids, probably as H_2S gas molecules in this case. The sulfide morphology of the grain boundary resembles the so-called inner scale, which often formes the inner-most layer of the surface scale on the alloys and even on pure chromium (12).

Figure 12 shows chemical diffusivities of the Fe-Cr alloy plotted against the alloy chromium content at both temperatures of 1073 and 1173 K. It also gives the alloy composition ranges for each of the three modes of internal sulfidation, and the alloy concentration gradients at the alloy/scale interface, dN'_{Cr}/dr. It was found that with increasing alloy chromium content the chemical diffusivity decreased, whereas the values of dN'_{Cr}/dr increased rapidly.

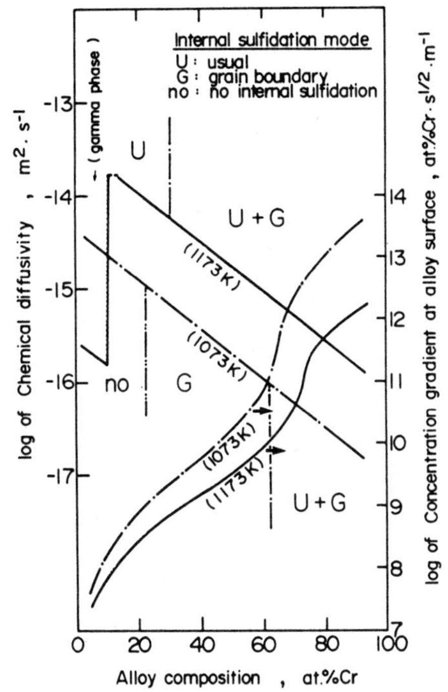

FIG.12. Chemical diffusivity of Fe-Cr alloys at 1073 and 1173 K and calculated concentration gradient, dN_{Cr}/dr.

Usual internal sulfidation has a considerable solubility of sulfur in the alloy and an inward diffusion of sulfur. According to the teranry oxidation model (21), the flux equation describing sulfur diffusion in the alloy may be written assuming dilute solution approximation for the off-diagonal diffusion coefficient

$$J_S = - D_S^S [dN_S/dr + F_{Cr}^S \cdot N_S \cdot dN_{Cr}/dr] \qquad [12]$$

where N_S and D_S^S are concentration and diffusivity of sulfur in the alloy, F_{Cr}^S is a correlation factor between sulfur and chromium.

The second term in eq.[12] makes it possible to determine the diffusion flux of sulfur in addition to the sulfur concentration gradient, by assuming negative values of F^S_{Cr} and a steep chromium concentration gradient dN_{Cr}/dr. Computer analysis indicated that increasing alloy chromium contents the values of the dN_{Cr}/dx increased by several orders of magnitude as given in Fig.12. Hence, when eq.[12] was applied to Fe-high Cr alloys at 1073 K, sulfur diffusion flux is so large that sulfide particles can precipitate internally, which is unlikely for the low Cr-Fe alloys.

At a temperature of 1173 K, however, usual internal sulfidation was formed for low Cr-Fe alloys and with increasing chromium composition grain boundary sulfidation was appeared in front of the usual internal sulfidation zone. These are different from exclussive formation of grain boundary sulfidation at 1073 K.

Figure 13 shows the cross-sectional microstructures of the usual internal, grain boundary sulfidations, and their mixtures.

The Fe-Cr diagram clearly shows that chromium depletion leads to the formation of a gamma phase with a f.c.c. structure near the alloy surface, irrespective of bulk alloy compositions. Because the gamma phase has a very low diffusivity, as indicated in Fig.12, the sulfide forms in the the usual internal sulfidation zone in the way described above. The EPMA analysis demonstrated that the chromium concentration in the alloy matrix was about 10 at% in the internal sulfidation zone, ahead of which it increased steeply to the bulk composition.

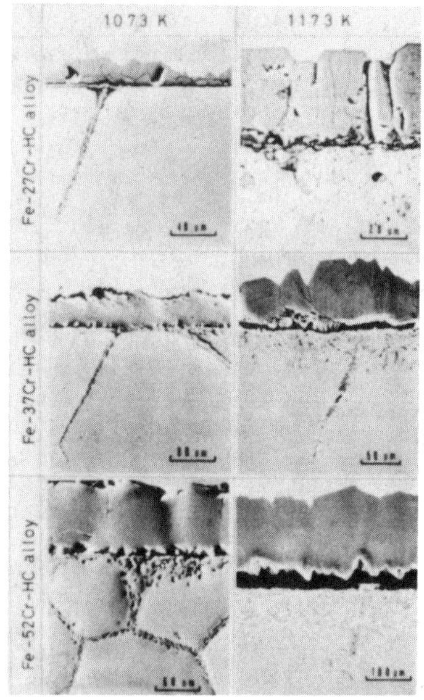

FIG.13. Cross-sectional microstructures of Fe-Cr Alloys sulfidizied at 1073 K; 10^{-5}Pa and 1173 K; 10^{-4}Pa.

References

(1) K.N.Strafford, Metall. Rev., **138**, 409(1969).

(2) D.J.Young, Rev. High Temp. Mater., **4**, 299(1980).

(3) S.Mrowec and K.Przybylski, Oxid. Met., **23**, 107(1985).

(4) T.Narita and K.Nishida, High Temperature Corrosion, Proc. 3rd JIM Int. Symp., in Trans. Jpn. Inst. Met., Suppl., **24**, 457(1983).

(5) T.Narita and T.Ishikawa, Mat. Sci. Engg., **87**,51(1987).

(6) T.Narita and T.Ishikawa, Proc. Symp. High Temp. mat. Chem., Electrochem. Soc., Vol.**88-5**, p.71(1988).

(7) T.Narita and T.Ishikawa, Mat. Sci. Engg., **A120**,31(1989).

(8) T.Narita and W.W.Smeltzer, Oxid. Met., **21**, 39(1984).

(9) T.Narita and W.W.Smeltzer, Oxid. Met., **21**, 57(1984).

(10) T.Narita, T.Ishikawa, and K.Nishida, Oxid. Met., **27**, 239(1987).

(11) K.Nishida, T.Narita, and A.Notoh, Rep. ASAHI GLASS FOUNDATION, **29**, 123(1976).

(12) T.Narita, W.W.Smeltzer, and K.Nishida, Oxid. Met., **17**,299(1982).

(13) C.Wagner, Corr. Sci., **9**,91(1969).
(14) D.J.Young, T.Narita, and W.W.Smeltzer, J. Electrochem. Soc., **127**, 679 (1980).
(15) E.M.Flyt, W.W.Smeltzer, and J.S.Kirkaldy, J. Electrochem.Soc., **126**, 673(1979).
(16) E.M.Flyt, V.S.Bhide, W.W.Smeltzer, and J.S.Kirkaldy, J.Electrochem. Soc., **126**, 683(1979).
(17) H.Rau, J.Less-Common Met., **55**,205(1977).
(18) T.Narita, unpublished data.
(19) B.D.Bastow, G.C.Wood, and D.P.Whittle, Corro. Sci., **25**, 253(1985).
(20) C.Wagner, J. Electrochem. Soc., **103**,571(1956).
(21) W.W.Smeltzer and D.P.Whittle, J. Electrochem. Soc., **125**,1116(1978).

Surface effects in sulphidation reactions

D.J. Young and J.P. Orchard
School of Materials Science and Engineering, University of New South Wales, Kensington, N.S.W. 2033, Australia

Abstract

Pure iron and an Fe-41 weight percent (w/o) Ni alloy were reacted at temperatures in the range 793 to 1073 K with hydrogen-hydrogen sulfide-nitrogen atmospheres corresponding to equilibrium $p(S_2)$ levels from 6.5:E-5 to 0.65 Pa. The kinetics of iron sulfidation were intermediate in form to linear and parabolic rate laws. The instantaneous parabolic rate constant was found to increase with extent of reaction until a constant value was reached. For fixed equilibrium sulfur pressures, the instantaneous rate increases with hydrogen sulfide partial pressure; for fixed hydrogen sulfide partial pressure, the instantaneous rate decreases as the equilibrium sulfur pressure is increased. It is demonstrated that hydrogen sulfide is the reactant species. A Langmuir-Hinshelwood kinetic model based on the slow dissociation of adsorbed hydrogen sulfide accounts satisfactorily for the unusual gas-phase compositional effects, and also for the rate at which the reacting system approaches the steady state. Similar effects were found for the Fe-Ni alloy, where nickel sulfide whisker formation results from localized catalysis of the hydrogen sulfide dissociation reaction.

Introduction

The use of austenitic stainless steels in sulfidising atmospheres at intermediate and high temperatures has frequently been described [1], and a diversity of results reported. In an attempt to resolve some of the disagreements found in the literature, Rao et al. [2,3] examined the thermodynamics of the Fe-Cr-Ni-S system and discussed the sulfidation of Fe-Cr-Ni alloys in this content. Sulfide scale morphologies are duplex, consisting of an Fe-Ni-S layer above an Fe-Cr-S layer. Rao et al. showed that the phase constitutions of these layers were consistent with local equilibrium within the relevant ternary systems. In this way they showed that the higher sulfidation rates associated with increased nickel levels at high temperatures ($T > 973$ K) were due to liquid sulfide formation. At lower temperatures, where no liquid forms, information was much less complete.

The behaviour of these sub-systems is most conveniently studied by examining the sulfidation behaviour of the binary Fe-Cr and Fe-Ni alloys. Considerable information is already available for the Fe-Cr alloys, including data for low sulfur partial pressures which are the conditions of practical interest [4 - 6]. Most work on Fe-Ni alloys [7 -9] has been carried out using a sulfur pressure of 100kPa, which is not relevant to industrial conditions. An examination [10] of Fe-Ni alloy sulfidation behaviour in hydrogen-hydrogen sulfide atmospheres at 793K showed that the phase constituion of the reaction products was in accord with thermodynamic prediction, but reaction morphologies and kinetics were extremely irregular. Those results have now been much more closely defined [11,12], but their interpretation requires a re-examination of the sulfidation of pure iron itself.

The sulfidation of iron in hydrogen-hydrogen sulfide has been studied by many workers, and there is general agreement that steady-state kinetics are parabolic:

$$(\Delta W/A = (k_p t)^{1/2} \qquad (1)$$

where ΔW/A is the weight gain per unit area formed in time, t, and k_p is the rate constant. Parabolic kinetics indicate rate control by solid-state diffusion, and the rate constant can be predicted from Wagner's theory [13] using the expression

$$k_p = \frac{C\,M^2}{4V} \int_{p'}^{p''} \frac{Z_{Fe}}{|Z_S|} D_{Fe}\, d\,\ell n\, p_{S_2} \qquad (2)$$

Here C is the average concentration, in equivalents, of sulfide in the scale, M is the atomic weight of sulfur, V the volume per equivalent of sulfide, Z the effective valence of the indicated species, D the self-diffusion coefficient, p the partial pressure of S_2, and p' and p" are the surface pressures at the metal-scale and scale-gas interfaces.

A major disagreement exists in the literature as to the nature of the kinetics in the early stages of reaction in hydrogen sulfide-hydrogen atmospheres. In some cases [14,15] parabolic kinetics were observed almost from the commencement of an experiment. Others [16-19] report significant, and occasionally very lengthy periods of reaction prior to the establishment of steady-state parabolic kinetics.

The purpose of the present paper is to review recent work [20] on the approach to steady-state sulfidation kinetics of iron in hydrogen sulfide-hydrogen mixtures, and show how this behaviour is reflected also in the sulfidation of an Fe-Ni alloy.

Experimental

The starting materials were iron of stated purity 99.999% (Materials Research Corporation) and nickel of stated purity 99.997% (Alfa Products). The two were alloyed by nonconsumable electrode, argon arc melting to produce a material of 41.1 weight percent nickel, as measured by atomic absorption spectroscopy. After annealing, the alloy was cut to produce rectangular coupons of surface area approximately 1 cm^2. Iron sample were cut as disks of 10 mm diameter from rod stock. Both materials were prepared for reaction by metallographically polishing to a 1μm diamond finish and ultrasonically degreasing immediately before reaction.

Kinetics were measured thermogravimetrically on a McBain balance having a sensitivity in the range 20 to 50 μg cm^{-2} of sample surface area. Reactions were carried out at temperatures ranging from 793 to 1073 K in flowing mixtures of high purity hydrogen, commercial purity hydrogen sulfide, and high purity nitrogen. Gases were pre-dried and their flows controlled using needle valves, and monitored using rotameters. Total gas flow rates were in the range 0.8 - 2.4 litres/min. (at STP). Use was made of thermodynamic data for the reaction

$$H_2S(g) = H_2(g) + \tfrac{1}{2}S_2(g) \qquad (3)$$

to calculate the equilibrium sulfur partial pressure. The use of diluent nitrogen permitted the independent variation of sulfur and hydrogen sulfide pressures.

Sulfidation of Iron

Scales were invariably made up of large columnar grains of hexagonal $Fe_{1-x}S$, and showed no signs of porosity. The effects of gas composition on reaction rates at 793 K are shown in Figs 1 and 2. Parabolic plots with ΔW/A as ordinate are concave upwards. The rate is seen to increase with increasing hydrogen sulfide partial pressure when the equilibrium sulfur partial pressure is fixed. The same effect was also observed at the much higher sulfur partial pressure of 0.65Pa. Even more interestingly, the rate is seen to decrease as the equilibrium sulfur partial pressure increases.

To demonstrate that gas-phase diffusion was not involved, the gas flow was varied from 1.1 to 3.3 (normal) litres/min. at fixed gas composition. No change in kinetics resulted. In another experiment, the nitrogen diluent was replaced with helium. Again, no effect on reaction rate could be detected.

Similar gas composition effects were found at temperatures of 848, 888 and 938 K, where again changing gas flow rates had no effect. Results obtained at 1073 K are shown in Fig. 3. At an equilibrium sulfur pressure of 8.5E-3Pa, the kinetics are parabolic after about 1h of reaction, and independent of hydrogen sulfide pressure. However, at an equilibrium sulfur pressur of 0.65Pa, the rate depends strongly on the hydrogen sulfide pressure. At a hydrogen sulfide pressure of 1.1E4Pa, the kinetics quickly reach a steady-state parabolic condition, but when the hydrogen sulfide pressure is 2.4E3Pa, the slope of the curve is still increasing after 8 hours.

In analysing these results it is useful to start with the observation that the scale was, compact and must have thickened via solid-state diffusion. In fact, the instantaneous rate of scale thickening is directly related to the diffusion flux of iron cations, J. Integration of the relationship

$$dX/dt = V J \tag{4}$$

leads to the parabolic rate equation (1) with k_p given by Eq (2), and k_p is, of course constant for fixed temperature and p". Since the kinetics were not in fact parabolic, it is concluded that p" varied with time. (The possibility of changing crystallographic orientation of the scale cannot account for the observed changes in rate.)

An instantaneous parabolic rate constant can be defined

$$k_p^* = 2\,\Delta W/A \,\frac{d\,(\Delta W/A)}{dt} \tag{5}$$

and is seen from Figures 1 - 3 to increase continuously with time until, in some cases, a steady value is reached. Using these values in Eq (2), together with the data [21,22] necessary to evaluate the integral, values for p" can be calculated. It was found in this way that

(a) p" increases with ΔW/A until a steady-state is reached, where p" equals the gas phase equilibrium value.

(b) The rate of change in p" increases with T, increases with hydrogen sulfide pressure and decreases with the equilibrium gas phase sulfur activity.

In considering these results it is obviously necessary to identify the source of reactant sulfur.

The gas phase content of S_2 molecules can easily be shown to be far too low to account for the observed reaction rates. Moreovoer, the rate at which these species are replenished via the dissociation reaction

$$2H_2S(g) = 2H_2(g) + S_2(g) \tag{6}$$

is slow [23]. Using the known rate of this process, the rates of sulfur regeneration were calculated, and were found to be much less than required by the corrosion process. It is therefore concluded that the reactant species is not sulfur, but is instead hydrogen sulfide. As gas-phase diffusion has been shown not to contribute to rate control, a heterogeneous process must lead to the production of the sulfur which is incorporated into the scale. This surface process cannot be rate controlling, because the linear sulfidation kinetics which would thereby be predicted are not, in fact, observed.

The catalysis of hydrogen sulfide decomposition on transition metal sulfides has been studied in depth [24,25] and it is known that FeS of a stoichiometry in equilibrium with the gas phase does not catalyze the reaction. We proceed therefore to construct a Langmuir-Hinshelwood [26] model based on associative adsorption followed by a slow surface dissociation:

$$H_2S(g) + O = H_2S|O \tag{7}$$

$$H_2S|O \underset{k'}{\overset{k}{\rightleftharpoons}} S|O + H_2(g) \tag{8}$$

$$S|O = S_S^x + V_{Fe}^x + O \tag{9}$$

Here the symbols O, $H_2S|O$ and $S|O$ represent, respectively, an unoccupied surface site, an adsorbed H_2S and an adsorbed S; the Kroger-Vink notation [27] is employed for lattice species. The vacancy concentration is supposed, as usual, to be set by equilibrium:

$$\left[V_{Fe}^x\right] = K_9 \; [S|O]/[O] \tag{10}$$

but the process (8) is slow. To a first approximation, then

$$[H_2S|O = K_7 \, p_{H_2S}[O] \tag{11}$$

$$\frac{d[S\ O]}{dt} = k\,[H_2S|O] - k'\,[S|O]\,p_{H_2} \tag{12}$$

Making use of the conservation of surface sites

$$[O] + [S|O] + (H_2S|O] = \text{constant} \tag{13}$$

Eqns (11), (12) and (13) can be solved to yield a differential equation for $d[S|O]/dt$. Integration followed by some algebra yields for the case where $[S|O]$ is small

$$\ell n \left[\frac{[S|O]_{eq} - [S|O]}{[S|O]_{eq} - [S|O]_o} \right] = k' K_3 \frac{p_{H_2S}}{(p_{S_2}{}^{\{eq\}})^{1/2}} \, t \tag{14}$$

where the subscripts eq and o denote, respectively, the equilibrium and initial values.

The above expression predicts that the rate of approach to surface equilibrium is dependent on hydrogen sulfide partial pressure but inversely dependent on the equilibrium sulfur pressure. These were, in fact, the trends observed. The relationship is tested by making the substitution

$$[S|0] = \alpha(p'')^{\frac{1}{2}}$$

and assuming α independent of surface coverage. Figure 4 illustrates the success of this description. The gradients of these lines have slopes proportional to hydrogen sulfide partial pressure, as seen in Figure 5, again in agreement with the model.

In summary it is concluded that the reaction of iron with hydrogen sulfide-hydrogen atmospheres at these intermediate temperatures is strongly dependent on gas phase composition. The reactant species is hydrogen sulfide, and the slow dissociation of this species on the surface leads to a continuously varying surface sulfur activity. This, in turn, provides a changing boundary value to the diffusion field responsible for mass transport, and hence unusual kinetics until a steady state is eventually reached. This slow increase in surface sulfur activity means that short term laboratory tests are of dubious value. It also has implications for the sulfidation behaviour of ferrous alloys.

Sulfidation of Fe-41Ni Alloy

Reaction kinetics at 793K were irregular and irreproducible [10]. This behaviour was found to be associated with local separation of the scale from the alloy, allowing direct gas access to unprotected metal. This phenomenon did not occur at higher temperatures, and reproducible results were obtained.

Typical sulfidation kinetics are shown in Fig.6 where it is seen that strictly parabolic kinetics obtain at high hydrogen sulfide pressures, but at low pressures an initial period of accelerating rate precedes the parabolic kinetic regime. Unlike the case of pure iron, there is a change in scale morphology. During the initial period of slow reaction, the scale consists entirely of a compact layer of essentially pure FeS. This occurs despite the fact that the equilibrium gas phase sulfur activity far exceeds the value necessary to form nickel sulfide. At longer times, when parabolic kinetics are observed, more complicated reaction products form. An inner, compact layer of (Fe-Ni)S is always formed. At low hydrogen sulfide pressures an outer zone of very fine whiskers is formed, whereas at high hydrogen sulfide pressures the outer zone is compact but of irregular thickness. These two different morphologies could be obtained at the same nominal equilibrium sulfur partial pressure, by adjusting the hydrogen sulfide pressure.

Despite the unusual morphologies and growth kinetics of these reaction products, their phase constitutions are close to the local equilibrium which characterises steady-state, diffusion controlled reactions. Microprobe analysis data obtained on scales producted at 888K are mapped onto an isothermal phase diagram in Fig 7. The composition paths representing traverses across reaction product thicknesses are seen to be consistent with local equilibrium. It should be noted, however, that paths (a) and (b) correspond to the same nominal equilibrium sulfur partial pressure but different hydrogen sulfide pressures. Evidently the effective surface sulfur activities are not the same. Path (a) corresponds to the higher hydrogen sulfide pressure and represents a compact sulfide scale. Path (b) maps compositions of product formed at a lower hydrogen sulfide pressure. In this case, an inner (Fe,Ni)S compact layer is overlaid by two zones of nickel rich sulfide whiskers [12].

Phase assemblages in the Fe-Ni-S system are complicated at lower temperatures by the existence of the ternary phase, pentlandite. Nonetheless, the reaction products have phase constitutions consistent with local equilibrium as shown in Fig 8. In the most complicated case, an inner compact monosulfide layer was overgrown by needles of pentlandite which were covered in turn by whiskers of Ni_7S_6, covered finally by whiskers of NiS.

In order to gain a better understanding of the reaction mechanism, the thickening kinetics of the inner compact layer were measured metallographically. They were found to be parabolic. Since the steady state kinetics of overall growth were also parabolic, it follows that the growth of the whisker zones also proceeded via parabolic kinetics. This is illustrated in Fig 9, where the kinetics of FeS growth on pure iron are also shown for comparison purposes.

Most of the unusual features of the Fe-41Ni sulfidation reaction can be explained in the light of the earlier finding that hydrogen sulfide dissociates very slowly at these temperatures. Thus the observation that only FeS is formed in the early stages of reaction is due to the initially low surface sulfur activity. As the activity increases with time, a critical value at which nickel rich sulfides nucleate is reached. At this point the reaction accelerates and steady-state parabolic kinetics are achieved, indicating that the nickel-rich sulfides are better catalysts for hydrogen sulfide dissociation than is FeS.

The observation that whiskers are apparently at local equilibrium is at first sight remarkable. A calculation [11] for a typical example indicates that the volume fraction of solid in the whisker zone is about 0.05. Obviously the whisker zone is mainly empty space. Even if the whiskers are as fine as 15nm radius a simple geometrical calculation shows that the average whisker spacing is ~ 110 nm. Although the whiskers in the present case are much more numerous and densely arrayed than is usual for corrosion product whiskers [28], the whisker zone is nonetheless quite permeable to gas.

Despite the openness of the whisker zone, the effective sulfur activity near the base of the whiskers is much less than at the whisker tips. This is clearly evident at 793 K, where the whisker tips are nickel sulfide, whereas the whisker bases are pentlandite (see Fig 8). No such sulfur activity gradient could possibly exist if the gas phase were functioning as a sulfur transport medium. Again, it must be concluded that the hydrogen sulfide dissociation reaction does not reach equilibrium in the gas phase.

The phase constitutions of the whisker zones are explained if the whisker tips, but not their sides, are catalytically active to hydrogen sulfide dissociation. In this way, the effective sulfur activity is brought close to the equilibrium value at the whisker tip, and the local phase constitution is thereby determined. However, the hydrogen sulfide dissociation reaction plays little or no part in establishing the sulfur activity along the whisker length. Instead, the sulfur activity is controlled by the outward flux of nickel along the whisker, and consequently decreases toward the interior of the scale. For this to happen, the surface structures of the whisker tip and sides must be very different.

Catalysis of hydrogen sulfide dissociation at whisker tips also explains the observed steady-state parabolic kinetics. A steady-state condition requires that the effective sulfur activity, which initially increased slowly on the FeS surface, becomes more or less constant when the nickel sulfides are formed. The nickel sulfides formed are those expected for equilibrium with the gas, indicating that sulfur activities are much higher than on an FeS surface and that the nickel sulfides must catalyse the hydrogen sulfide dissociation much more effectively than does FeS. Nonethe-less, this catalytic activity is often localized, as shown by the fact that uniform layers of nickel sulfide usually do not form. Furthermore, the catalytic activity of Ni3S2 is specific to the whisker morphology; when this phase was produced as a compact layer on pure nickel in dilute hydrogen sulfide, slow linear scaling kinetics were found [29]. Obviously, the structure of the whiskers is an important factor.

Since the whiskers do not thicken appreciably with time at low hydrogen sulfide pressures, it follows that sulfur uptake occurs at the whisker tip. Thus, the flux of nickel in the direction of the whisker axis is very much greater than it is laterally. This stable whisker morphology is itself due, at least in part, to the localized catalysis phenomenon, which provides a strong sulfur activity gradient along the whisker axis, but not in the transverse directions. The other factor contributing to the highly directional character of the flux would be the relative diffusion coefficients in the axial and transverse directions. Screw dislocations running along the whisker axis would provide a highly favourable diffusion path. Furthermore, the dislocation emergence site at the whisker tip could provide the catalytic site for hydrogen sulfide dissociation.

There remains the need to explain why the inner compact layer of monosulfide on the alloy grows so much faster than does the FeS layer on pure iron. The relatively small difference in self-diffusion coefficients does not indicate such a large effect. This observation is correct only at low sulfur activities, because in 100kPa of sulfur [9], mixed (Fe,Ni)S scales grow at about the same rate as does FeS on pure iron.

It is seen from Figs 7 and 8 that the composition paths through the inner monosulfide layer are unusual in that they both enter and leave the sulfide phase field from the low sulfur activity side. This suggests a strong deviation from ideality within this solid solution. Microprobe analysis of the layer yielded results such as are shown in Fig 10. Obviously the deviation from stoichiometry varies in a non-monotonic fashion. Indeed, the deviation from stoichiometry exceeds that found in pure FeS under the same conditions.

The growth of the solid solution layer (Fe,Ni)S may be described using Wagner's theory [30] of alloy oxidation. If it is assumed that the psuedobinary solution FeS-NiS is ideal, that deviations from stoichiometry can be ignored, and further that the ratio

$$p = D_{Fe}/D_{Ni}$$

is constant, independent of composition and sulfur activity, then [31] Wagner's expression simplifies to

$$(1-p)\,\frac{d\xi}{dy} + [1 - \xi + p\xi]\,\frac{d\ln a_s}{dy} = k_p\,\frac{V^2}{32}\,\frac{p}{D_{Fe}} \tag{15}$$

where ξ is the mole fraction of FeS in the scale and y is the position co-ordinate normalized to the scale thickness. If it is further assumed that D_{Fe} in the (Fe,Ni)S layers can be calculated from the expression due to Condit et al. [22] for FeS

$$D_{Fe} = D_o\,\delta\,\exp[-(81 + 84\delta)RT] \tag{16}$$

where δ is the deviation from stoichiometry, $D_o = 1.7\times10^{-2}\,cm^2 s^{-1}$ and the exponent units in kJ, then Eq [15] may be applied to the microprobe data for ξ and δ as functions of y. This procedure permitted the evaluation of the gradient d ln a/dy. The value of p was then varied to match the sulfur activity profiles to the boundary values.

The results of this calculation are shown in Fig.11 where it is seen that d lna/dy is constant across the width of the scale, despite the unusual behaviour of δ. From the optimization procedure it was found that all the data were best described by the value p = 0.4. This is in reasonable agree-ment with the value of p = 0.6 calculated by Narita and Nishida [9] for a sulfur pressure of 100kPa when a monophase scale was formed. The success of this description derives from the fact that nonideal sulfur activities (and values of δ) are permitted. The treatment is somewhat inaccurate, however, as Eq (15) relies on the approximation $\delta = 0$.

References

1. D.J. Young, Rev.High Temp. Mater., 14, 299 (1980).
2. D.B. Rao and H.G. Nelson, Oxid. Met., 12, 111 (1979).
3. K.T. Jacob, D.B. Rao and H.G. Nelson, Oxid. Met., 13, 25 (1979).
4. T. Narita, W.W. Smeltzer and K. Nishida, Oxid. Met., 17, 299 (1982).
5. T. Narita and W.W. Smeltzer, Oxid. Met., 21, 39 (1984).
6. T. Narita and W.W. Smeltzer, Oxid. Met., 21, 57 (1984).
7. Z. Skrzypeck and T. Weber, Arch. Hutn., 21, 115 (1976).
8. T. Weber, T. Biegun, Z. Skrzpek and J. Podhoracki, Rocz.Chem., 44, 915 (1970).
9. T. Narita and K. Nishida, Denki Kagaku, 43, 443 (1975).
10. D.J. Young, C.L. Tasker and J. P. Orchard, Proc. JIMIS-3, Trans. Jpn. Inst. Met. Supp., 24, 491 (1983).
11. J. P. Orchard and D.J. Young, Oxid. Met., 31, 105 (1989).
12. J. P. Orchard and D.J. Young, J. Electrochem. Soc., 136, 545 (1989).
13. C. Wagner, in "Atom Movements", p.153, ASM, Cleveland (1951).
14. E.M. Fryt, V.S. Bhide, W.W. Smeltzer and J.S. Kirkaldy, J. Electrochem. Soc., 126, 683 (1979).
15. D.J. Young and W.W. Smeltzer, J. Electrochem. Soc., 123, 229 (1976).
16. E.T. Turkdogan, Trans. AIME, 242, 1665 (1968).
17. P.D. Zelanko and G. Simkovich, Oxid. Met., 8, 343 (1974).
18. G. Simkovich, Werkst. Korros. 21, 973 (1970).
19. D.F. Wilson and O.F. Devereaux, in "Corrosion in Fossil Fuel Systems", I.G. Wright, Editor, p.247, The Electrochemical Society, Pennington, NJ (1983).
20. J.P. Orchard and D.J. Young, J. Electrochem. Soc., 133, 1734 (1986).
21. P. Toulmin and P.B. Barton, Geochim. Cosmochim. Acta, 28, 641 (1064).
22. R.H. Condit, R.R. Hobbins and C.E. Birchenall, Oxid.Met.,8, 409,(1974).
23. B. de B. Darwent and R. Roberts, Proc. R. Soc., London, Ser.A, 216, 344 (1953).
24. H. Kiuchi, T. Iwasaki, I. Nakamura and T. Tanaka, "H_2 Production and Marketing", p.349, A.C.S. Symposium Series, 116 (1979).
25. T.Chivers, J.B.Hyne and C.Lau, Int.J.Hydrogen Energ., 5, (6) 499,(1980).
26. C.N. Hinshelwood in "The Kinetics of Chemical Change", p.178, Clarendon Press, Oxford (1940).
27. F.A. Kroger and H.J. Vink, Solid State Phys., 3, 307 (1956).
28. R.A. Rapp, Met. Trans.A, 15A, 765 (1984).
29. A. Stoklosa and J. Stringer, Oxid. Met., 11, 263 (1977).
30. C. Wagner, Corros. Sci., 9, 91 (1969).
31. P. Mayer and W.W. Smeltzer, J.Electrochem. Soc., 121, 538 (1974).

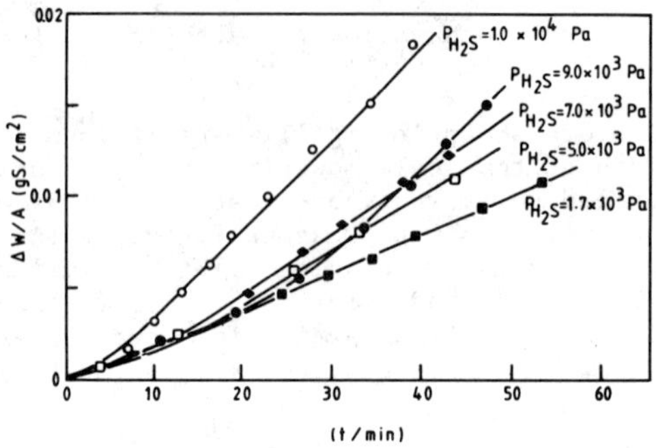

Fig.1. Effect of $P(H_2S)$ on sulfidation of Fe at 793K, $p(S_2)$ = 3.3E-4 Pa.

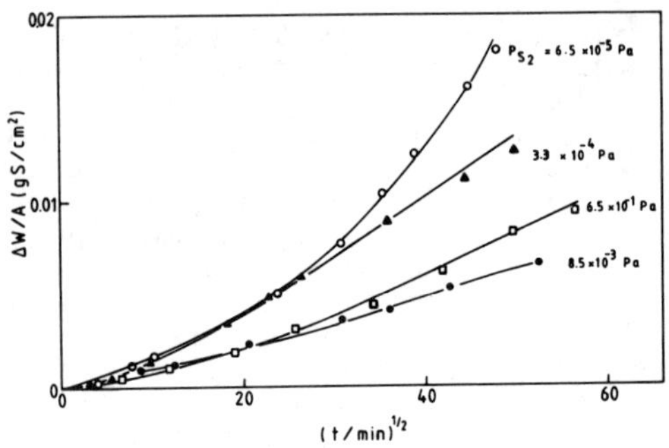

Fig.2. Effect of $P(S_2)$ on sulfidation of Fe at 793K, $P(H_2S)$ = 5E3Pa.

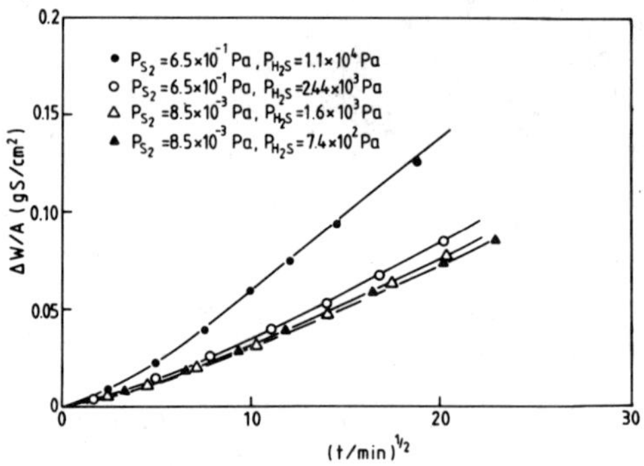

Fig.3. Sulfidation kinetics of Fe at 1073K.

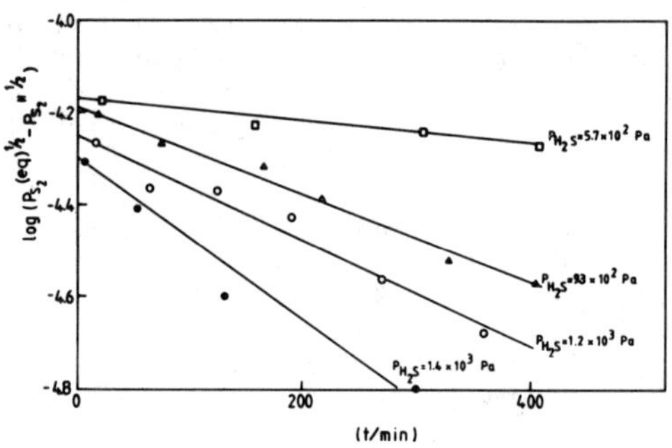

Fig.4. Fit of Fe sulfidation data to Eq.(14), T = 938K, $P(S_2)$ = 3.3E-4Pa

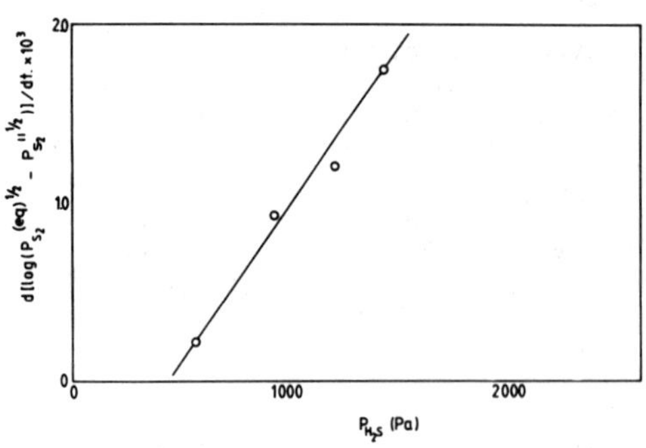

Fig.5. Slopes of lines in Fig.4 vs $P(H_2S)$

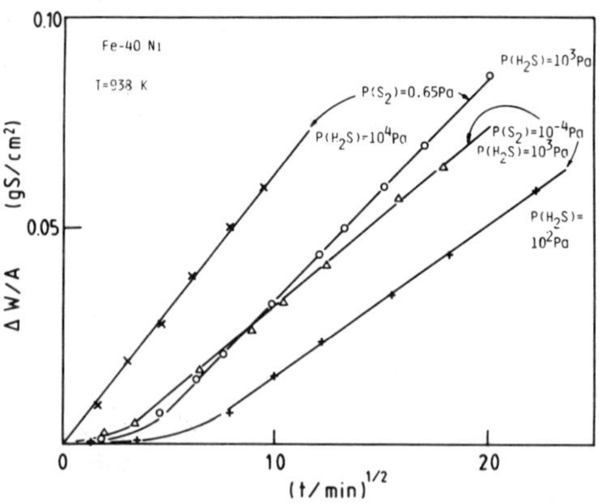

Fig.6. Sulfidation kinetics for Fe-41Ni at 938 K

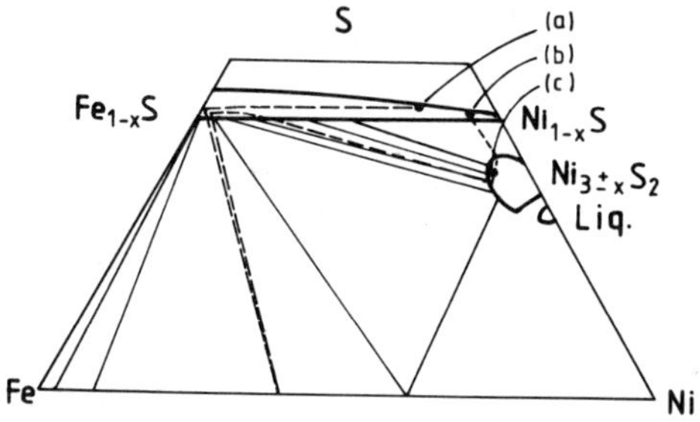

Fig.7. Isothermal section of phase diagram at 973K showing composition paths for scales grown at 888K (a) $P(H_2S)$ = 6.2E4 Pa, $P(S_2)$ = 0.1Pa; (b) $P(H_2S)$ = 3.2E3 PA, $P(S_2)$ = 0.1 Pa; (c) $P(H_2S)$ = 3.2E3Pa, $P(S_2)$ = 0.02Pa.

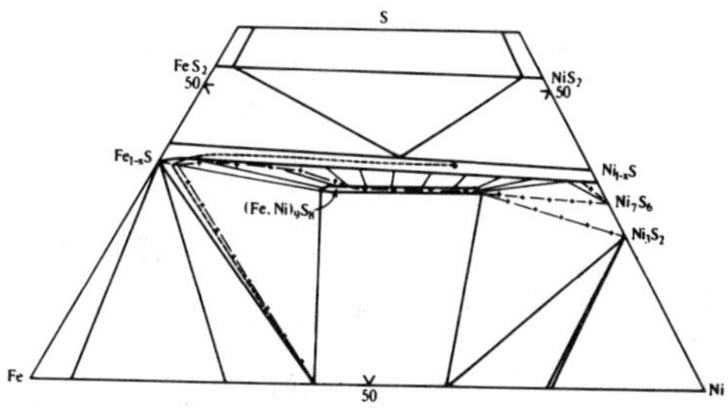

Fig.8. Isothermal section of phase diagram at 793K showing composition paths: dashed lines at $P(H_2S)$ = 5E3Pa; upper line $P(S_2)$ = 0.12Pa, lower line $P(S_2)$ = 1.6E-5Pa. Dotted line $P(H_2S)$ = 2.2E4Pa, $P(S_2)$ = 1.8E-5 Pa.

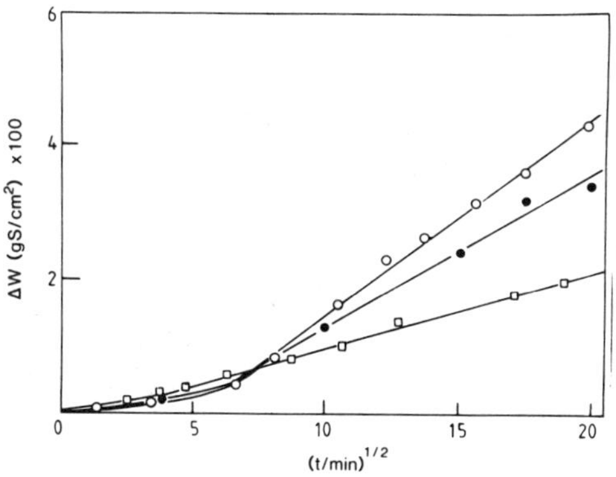

Fig.9. Sulfidation kinetics at T=888K, $P(H_2S)$=3.2E3Pa, $P(S_2)$=3.3E-4Pa: (o) Fe-41 Ni total weight gain, (•) Fe-41Ni inner layer weight gain, (□) Fe total weight gain.

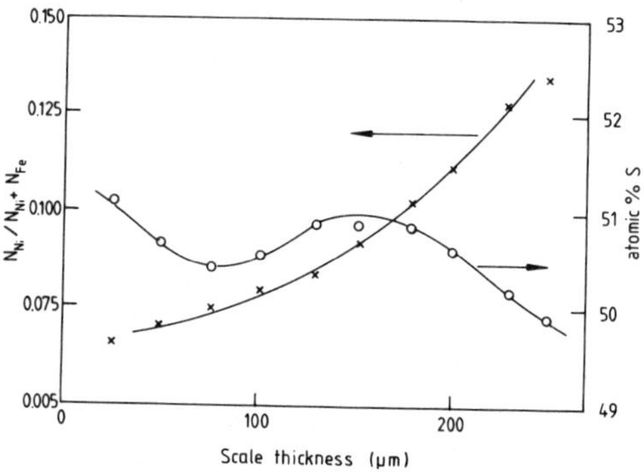

Fig.10. EPMA results for inner layer on Fe-41Ni at 938K, $P(H_2S)$ = 1.4E3 Pa, $P(S_2)$ = 3.3E-4 Pa

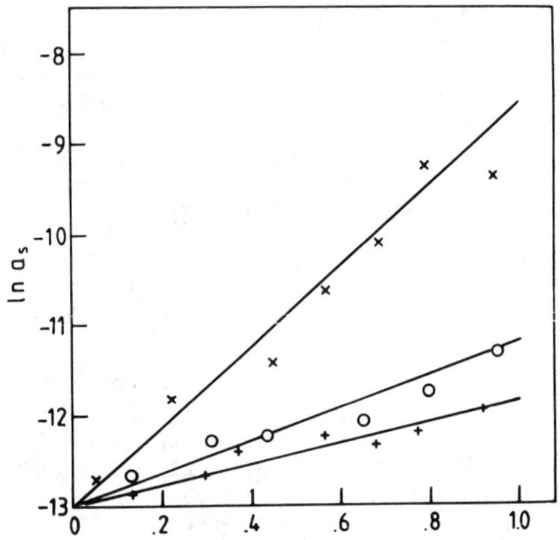

Fig.11. Variation of sulfur activity with normalized position in inner layer from Eqns (15),(16): (x)$P(S_2)$= 0.65Pa, $P(H_2S)$=3.9E4Pa; (o)$P(S_2)$ = 0.65Pa, $P(H_2S)$=1.4E3Pa; (+) $P(S_2)$ = 3.3E-4Pa, $P(H_2S)$=1.4E3Pa.

High-temperature oxidation of Fe-Cr-Al alloys: influence of Al content

M. Lambertin and G. Beranger

Département de génie mécanique, Université de technologie de Compiègne, F-60206 Compiègne, Cedex, France

ABSTRACT

Four different alloys containing chromium and aluminum were oxidized at high temperature (800-1100°C). Three of them were cast alloys containing 20wo%Cr and x%Al (x = 2 , 4 and 6). The last material was prepared from a Fe-17Cr alloy which was pack - aluminized. The oxidation kinetics was studied by thermogravimmetry. Analyses were performed by using X ray diffraction, SEM + EDS and glow discharge optical spectroscopy.

Classical results were exhibited by the alloys containing respectively 4 and 6%Al. The parabolic law was obeyed during entire oxidation runs. A circonvoluted α-Al_2O_3 scale is formed on the alloys. When the Al content is as low as 2%, the results are more complicated. Two different parts are observed on the kinetic curves. During the initial part (high oxidation rate) an external scale containing Cr and Fe oxides grows on the alloy. Alumina forms by precipitation at the metal- scale interface. When a continuous Al_2O_3 barrier is obtained, the kinetics is modified; the oxidation rate is decreased and the parabolic law is obeyed again.

Concerning the pack aluminized alloy, the analyses show that the surface Al content is very high (>50%). A unique parabolic law is never observed during the oxidation, whatever is the temperature. But linear short parts can be plotted one after another on the $(\Delta m/S)^2$ vs t curves. The parabolic rate constants which can be then calculated are decreasing when the oxidation time is increasing. A limit value is attained after about 50 hours (depending on the temperature). The analyses show that only pure α-alumina is formed during the tests. These results can be explained if one consider that the scale, which contains initially numerous deffects like cracks, becomes more and more compact as the oxidation proceeds.

INTRODUCTION

Alloys containing aluminum and forming alumina during their high temperature oxidation are more and more widely used and have a large number of potential applications (1,2). Thus they were very often studied. They oxidize more slowly than the classical stainless steels which are Cr_2O_3 former and can be used at higher temperature because volatile species like CrO_3 are not obtained.

However the alloys are embrittled by Al additions (3) and it is necessary to limit them. But a minimum aluminum level is required to make possible the alumina formation. A satisfactory solution may be a surface enrichment (using thermochemical or other appropriate treatments) of alloys which don't contain aluminum in the bulk

By another way, the oxidation rate may be modified (and decreased) by the addition of active elements which are also well known to enhance the scale adherence(4,5).

The present work concerns the study of the high temperature oxidation of different Fe-Cr-Al alloy steels which contain various amounts of aluminum: some of these alloys were obtained by melting the elements (melted alloys), the last one being a Fe-Cr alloy which was superficially enriched by thermochemical treatment (cemented or aluminized alloy).

EXPERIMENTAL

1) Oxidation tests and analyses

The oxidation kinetics was studied by continuous weight gain recording using a micro-thermobalance. The specimens were analyzed by X ray diffraction, glow discharge optical spectrometry (GDOS) and X ray energy dispersion analysis (EDS) coupled with a scanning microscope (SEM).

The kinetic runs were performed in pure oxygen under atmospheric pressure (10^5 Pa). The temperature was ranging between 1073 and 1373K. The experimental procedure was as follows: after polishing, cleaning and drying, the sample was introduced in the reaction tube. The vessel was then evacuated and the oxygen pressure was established. The furnace, which was hot, was raised up and placed around the specimen and the test started. At the end of the experiment, the furnace was removed and the tube evacuated again.

2) The alloys

2-1 Melted alloys

Three different Fe-20Cr-xAl steels were studied. Their actual composition is given in Table I. Account taken on it, they have a ferritic structure whatever is the temperature. Samples which were used to perform the tests were discs (Diameter: 20mm - Thickness: 0.7mm). They were polished with grit paper (up to 800 grade) and then with diamond paste (up to 1mm), rinced with water and alcohol and dried before to be placed in the reaction tube and oxidized.

	Fe	C	Mn	Si	S	P	Cr	Al
Alloy I	bal.	0.002	0.54	0.080	0.011	-	20.61	2.0
AlloyII	bal.	0.0035	0.525	0.100	0.013	0.008	20.60	3.82
AlloyIII	bal	0.002	0.530	0.140	0.011	0.009	20.33	20.60

Table I: Composition of melted alloys (weight %)

2-2 Aluminized alloy

The base alloy is a Fe-17Cr steel whose actual composition is given in Table II.Samples were 10x10x2.5mm platelets. They were polished with grit paper (up to 800 grade) then sand-blasted immediately before the aluminization.The main features of the pack-cementation treatment are summarized in Table III.

Element	Fe	C	Cr	Mn	Si	Ti	Ni	Al	N_2
Weight %	bal.	0.075	17.45	0.42	0.41	0.76	<0.3	0.82	0.019

Table II: Composition of the base alloy

Cement composition	Temperature (K)	Duration (h)	Thickness (μm)
$Al+Si+NH_4Cl$	773-1073	2.5	25

Table III: Main features of the pack-cementation treatment

The EDS analysis of the aluminized layer is shown on Fig.1. One immediately remarks that a high Al level was obtained. By X ray diffraction, the (Fe,Cr)Al_3 compound is identified. Its structure is comparable to that of the $FeAl_3$ intermetallic phase (6).

The oxidation experimental procedure is similar to that which was previously described.

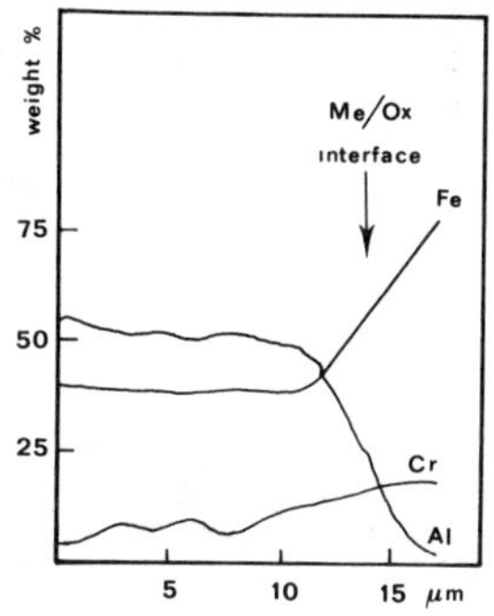

Fig.1: EDS analysis of the aluminized scale

OXIDATION RESULTS

1) Fe-Cr-Al alloys

The oxidation tests were performed between 1073 and 1373K. As shown on Fig.2a to 2c, when plotting the weight gain by unit area versus the square root of the time (7), it is thus evident that the oxidation kinetics follows the classical parabolic law during the major part of the runs, except for the alloy containing only 2%Al. In that last case, a very fast (but short) initial stage is followed by a slower parabolic second stage; the main part of the weight gain was acquired during the first stage. The parabolic constants can be calculated from these curves. They are reported in Table IV. It should be observed that the values are very close for alloys II and III but different for alloy I. The knowledge of the rate constants allows to calculate the experimental activation energies. It should be remarked (Fig.3) that they are comparable for the three alloys which means that the rate determining process is probably the same in all cases.

The various analyses show that only pure α-alumina forms on the surface of alloys II and III. On the third steel, three different subscales are observed: the external layer is Fe_2O_3, the intermediate one is Cr_2O_3, the innermost layer being Al_2O_3. This result is in good accordance with the thermodynamical stability of the compounds (8). We shall see later that the composition differences of scales allow to justify the various kinetic behaviors which were observed.

	kp ($mg^2.cm^{-4}.s^{-1}$)				E (kJ)
Temperature(K)	1073	1173	1273	1373	
Alloy I	$1.69\ 10^{-6}$	$1.16\ 10^{-5}$	$8.10\ 10^{-5}$	$3.84\ 10^{-4}$	228.5
Alloy II	$2.50\ 10^{-5}$	$1.30\ 10^{-4}$	$8.55\ 10^{-4}$	$5.87\ 10^{-3}$	213.2
Alloy III	$5.50\ 10^{-5}$	$9.00\ 10^{-5}$	$8.35\ 10^{-4}$	$5.76\ 10^{-3}$	

Table IV: Oxidation of melted alloys: parabolic rate constants and activation energies

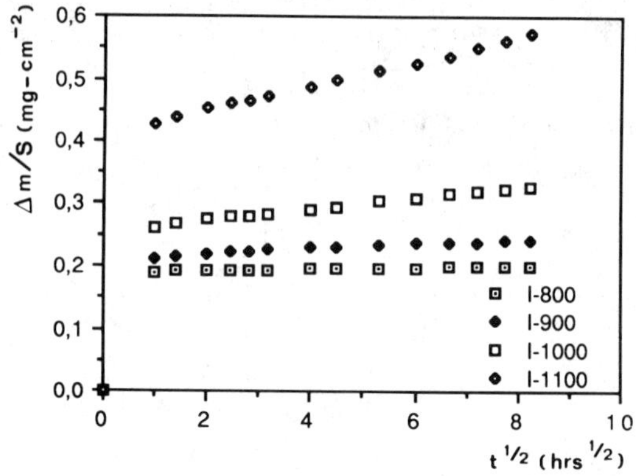

a - Oxidation of alloy I

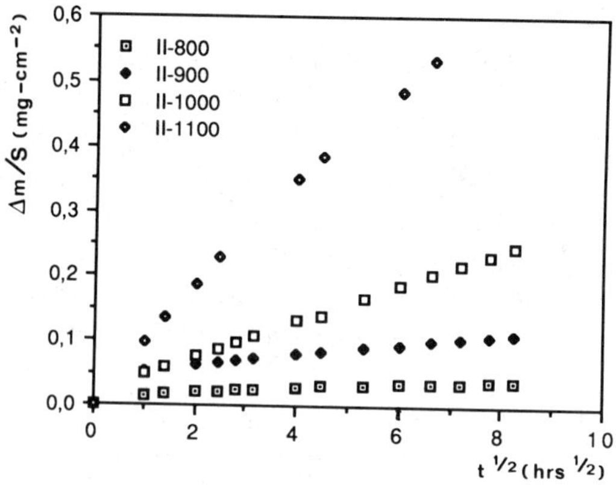

b - Oxidation of alloy II

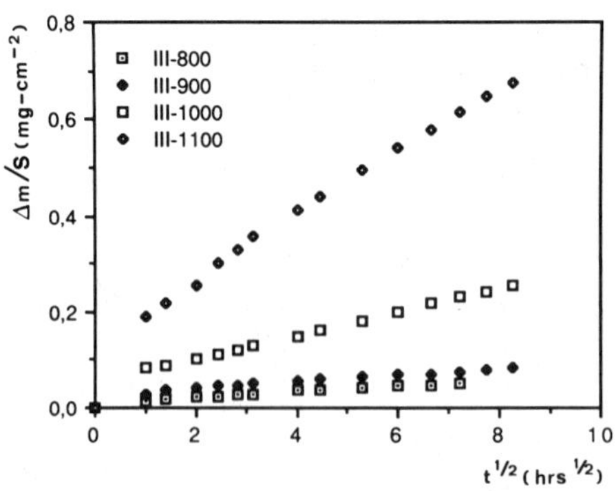

c - Oxidation of alloy III

Fig.2: Oxidation kinetic curves of FeCrAl melted alloys

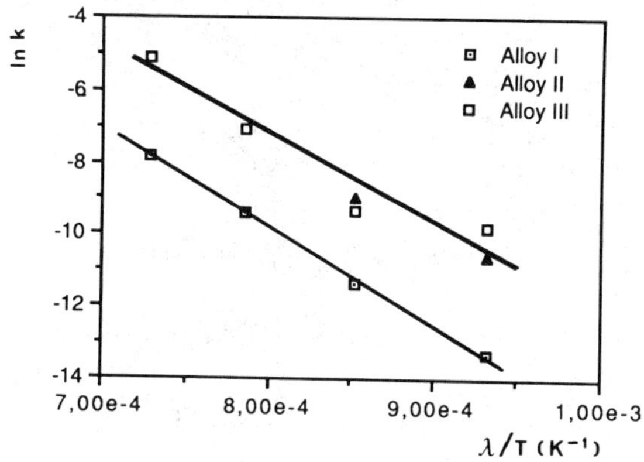

Fig.3: Oxidation of melted alloys: activation energies

2) Aluminized alloy

On this alloy, oxidation tests were performed only in the temperature range 1073 - 1273K. As shown on Fig.4, it may be estimated that at 1073K the classical parabolic law is obeyed except during a short stage at the beginning of the run. At higher temperature the results are different: the parabolic law is never obeyed, the slope of the curves mass gain per unit area versus square root of the time being continuously decreasing. It seems that in this range the temperature does not influence the oxidation rate: the curves at 1173 and 1273K are superimposed. It must be observed that an important part of the total weight gain occurs during the very short and fast initial stage, whatever is the temperature.

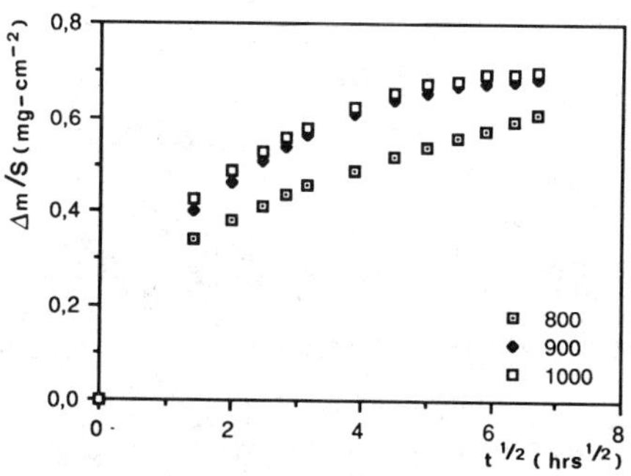

Fig.4: Oxidation of the cemented alloy: kinetic curves

The analyses show that only pure α-alumina forms during the oxidation. Microscopical observations with SEM show that whiskers are grown on the oxide surface (Fig.5) which denotes an important stress level in the scale.

DISCUSSION

If one consider the oxidation of melted alloys, it was observed that the parabolic law was the most general experimental law. Moreover the values of the activation energy which were calculated were very close. Thus it could be concluded that the rate controlling process was the same for the three alloys. But in this case it has to be established why the parabolic law is not followed during

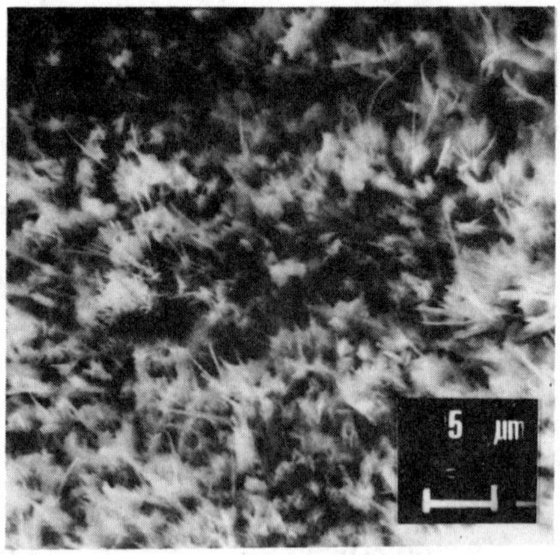

Fig.5: Alumina scale formed on the aluminized alloy at 1173K

the first stage of the oxidation of alloy I (containing 2%Al). Comparable results were already observed (9) on similar alloys. It was observed that after long term runs three different oxides were grown on this steel. Complementary microscopical observations were carried out after shorter oxidation times. They revealed that after 10 minutes continuous iron and chromium oxide layers were formed, alumina being only present as discontinuous internal precipitates. After longer reaction times, the situation was as shown on Fig.6: in certain zones, a continuous alumina scale is established, in other ones it is discontinuous.

Fig.6: Oxide scale formed on alloy I after a short oxidation time (<1 hr)

The oxidation mechanism can be imagined as follows: at the beginning of the reaction, iron and chromium form immediately a continuous oxide barrier on the steel; alumina precipitates at the internal interface. The rate is controlled by the outward cationic diffusion and the oxidation is very fast. As it proceeds, Al_2O_3 precipitates are more and more important and are joining in some zones (Fig.6). The access of Fe and Cr at the metal-scale interface becomes more and more difficult. The chemical gradient in the external oxides (and subsequently the rate constant) decreases and the parabolic law is not followed. Finally, after a more or less long time (depending on the temperature) a continuous alumina layer is established and the oxidation rate becomes parabolic.

A second problem has thus to be solved: if, as assumed, the rate controlling step is the same for the three alloys during the parabolic part, the rate constants should be theoritically identical. It is well verified for alloys II and III but not for alloy I. In this last case, the parabolic constant is much lower. It can be easily explained if one consider that the external interface of the alumina scale is not exposed directly to the oxidizing atmosphere; the oxygen pressure is fixed by the dissociation pressure of Cr_2O_3. Whatever is the diffusing species (anion or cation), its diffusion constant depends on the oxygen gradient in the scale. That gradient being considerably lower in the case of alloy I, it justifies the lower parabolic rate constant.

In these conditions, it should be theoritically possible to quantify the modification of the kinetic constants (parabolic rate constant and activation energy) which is due to the presence of the iron and chromium oxide scales. The oxidation rate being diffusion controlled and dependent on the oxygen pressure, two different processes have to be envisaged:

- Inward oxygen diffusion by interstitial ions

$$1/2\ O_2 = O_i'' + 2h^\bullet$$

The growth rate can be written $v \sim P^{1/6}(O_2)$.

- Outward diffusion by metals vacancies

$$1/2\ O_2 = O_o + 2/3V_{Al}''' + 2h^\bullet$$

Thus the growth rate is $v \sim P^{3/16}(O_2)$.

The most general form of the rate is:

$$v = v_o P^n(O_2)\exp(-E/RT)$$

where n=1/6 or 3/16.

When the alloys II or III are oxidized, the oxygen pressure is $P(O_2)$=1atm (10^5Pa). In the case of the experiments which are carried out with alloy I, during the parabolic part, the oxygen pressure is fixed by the chromia dissociation:

$$2/3Cr_2O_3 = 4/3Cr + O_2 \qquad \Delta G° = 772436 - 148{,}39T \quad (J) \quad (8)$$

When the equilibrium is established, the oxygen pressure can be calculated by:

$$P(O_2) = \exp(-\Delta G°/RT) \quad (atm)$$

Thus the reaction rate becomes:

$$v = v_o\exp(n\Delta S°/R)\exp(-(E+n\Delta H°)/RT)$$

The experimental activation energy is theoritically:

$$E_{exp} = E + n\Delta H°$$

E being the experimental value which has been determined with alloys II and III and $\Delta H°$ the enthalpy of the dissociation reaction. Thus, depending on the values of n, one may calculate the values of E_{exp}.

n=1/6 $\qquad E_{exp}$=341939 J

n=3/16 $\qquad E_{exp}$=358032 J

It can be immediately seen that there is no agreement between these values and the experimental ones. It is thus possible to determine an experimental value for n which is 0.027. It indicates that the influence of the oxygen pressure on the oxidation rate is very low, much lower than the influence which could be theoritically envisaged.

To complete these results, one may try to compare the experimental and theoritical differences in the rate constants at a given temperature. From the previous equations, it can be written:

$$k_{III}/k_{I,II} = P^n(O_2)$$

with $P(O_2) = \exp(-\Delta G°/RT)$

When using the experimental results, it can be calculated that n=0.027, its value being perfectly comparable with the previous one. Two hypotheses should be envisaged. The first one is to admit that the oxygen pressure is as low as calculated and in this case it is necessary to envisage an other growth mechanism to understand the pressure influence. The second hypothesis consists to admit that, in fact, the actual value of the oxygen pressure is higher than that which has been calculated. It means that gaseous oxygen can penetrate by inward molecular diffusion throughin the Fe_2O_3 and Cr_2O_3 scales. That second hypothesis is more probable but unfortunately the results don't allow to precise what is the diffusion mechanism (anionic or cationic).

Concerning now the pack-aluminized alloy, the problem is different. At 1073K it can be considered that the parabolic law is followed except during the beginning of the run (1 hour). As previously mentionned, it must be noted that during this initial stage a non-negligible weight gain is observed. At higher temperature, the classical parabolic law is never obeyed (Fig.4). Moreover it looks like if above 1173K the reaction rate was independent on the temperature (the curves are superimposed).

As observed in the case of melted alloys, it could be imagined that the abnormalous mass gain during the initial stage was due to the formation of other oxides, less protective than alumina (9).

But the analyses after short reaction times prove that only pure α-Al_2O_3 grows. Nevertheless SEM observations of the scale morphology show that its protectivness is poor at the beginning, numerous cracks being remarked. However, when the oxidation proceeds, the alumina compacity increases more and more, leading to an increased protective behavior. Thus it can be understood that the diffusion coefficient (and consequently the rate constant) decreases progressively.

In a previous paper, we have proved that the parabolic rate constant decreases continuously during more than the first fifty hours and reaches finally a limit value; it was estimated at 1173K ($6.10^{-14} mg^2 cm^{-4} h^{-1}$) and is very close to the rate constants determined with both melted alloys II and III (Table IV) at the same temperature.

CONCLUSIONS

In the present work, which concerns the high temperature corrosion (oxidation) of different Fe-Cr-Al alloys, it was shown that in most cases the oxidation rate is controlled by a diffusion process through the alumina scale which grows. After long term runs, the classical parabolic law is obeyed except with the aluminized steel. In that last case a continuous decrease of the rate constant is evidenced, a limit value of the same order of magnitude than the corresponding value determined with melted alloys being finally reached.

The main oxide which forms is α-Al_2O_3. When the Al content of the alloy is too low (2%) a continuous and protective alumina scale is not immediately obtained. Iron and chromium oxides grow initially and alumina forms only as internal precipitates. After a few hours a compact internal Al_2O_3 layer is finally observed whose growth rate is lower (at the same temperature) than on the melted alloys with higher Al content. This results is due to the presence of the iron and chromium oxides. It should be theoritically possible to determine precisely the rate determining process but unfortunately the experimental results don't allow to do it.

REFERENCES

1. S.L. CASE and K.R. VAN HORN, Aluminium in iron and steels, p.265, John Wiley and sons, New York (1953).
2. J. CASTEL and P. BOURGAIN, Revue de Métallurgie-CIT, Août-Septembre, 667 (1986).
3. M. LAMBERTIN, J.P. LARPIN, J.C. COLSON, E.BOLLINGER and L. VINCENT, DGRST Report 78-7-1047 (1980).
4. D.P. WHITTLE and J. STRINGER, Phil. Trans. Roy. Soc. London, A295, 309 (1980).
5. The role of active elements in the oxidation behaviour of high temperature metals and alloys,Elsevier Applied Science, London (1989).
6. J.N. PRATT and G.V. RAYNOR, J. Inst. Met., 80, 449 (1951).
7. B. PIERAGGI, Oxid. Met., 27, 3/4, 177 (1987).
8. O. KUBASCHEWSKI and E.LL. EVANS, Metallurgical chemistry, Pergamon Press, London (1955).
9. I.A. KVERNES, Oxid. Met., 6, 4 (1973).

Segregation of cerium in chromia scales

D.A Downham, R.J. Hussey, D.F. Mitchell and M.J. Graham
Division of Chemistry, National Research Council, Ottawa, Ontario, Canada, K1A 0R9

Abstract

The oxidation rate of Fe-Cr alloys at 900-1100°C is significantly decreased by the presence of a thin (~40Å) sputter coating of CeO_2. ^{18}O/SIMS experiments show that Ce found within the scale changes the mechanism of Cr_2O_3 growth from predominantly cation to predominantly anion diffusion, both of which are considered to occur along oxide grain boundaries. For oxides < 1000 Å thick, EDX, EELS, TEM and electron diffraction data indicate that Ce segregates to form particles of $CeCrO_3$ which likely retard grain boundary diffusion of Cr cations. Analogous results are observed for Y_2O_3-coated and Y-implanted alloys.

Introduction

The addition of small amounts of so-called 'reactive elements' such as Ce, Y, Hf, Th (or their oxide dispersions) to Fe-Cr alloys greatly increases the alloy's oxidation resistance under both isothermal and cyclic conditions (1-5). Beneficial effects also result from ion implantation of reactive elements (6-9), or from surface-applied coatings (10-17). We have applied very thin (40 Å) reactive element coatings by sputtering CeO_2 directly onto Fe-Cr alloys, and have used SIMS to determine the location of Ce within the oxide scale and to study its effect on transport processes. We have also used TEM/EDX, electron diffraction and EELS microscopy to examine Ce segregation in the oxide and to identify the presence of particles of $CeCrO_3$ which likely influence oxide grain boundary transport processes. This is an analogous situation to that reported in the literature for Y-implanted Fe-Cr alloys where evidence was presented for the existence of $YCrO_3$ particles incorporated into chromia scales (18-20).

Experimental

High-purity Fe-26 wt.% Cr alloy samples (~ 3 cm^2) containing the following impurities in ppm wt: N 12, O 44, Si 4, Co 11, Ni 3, Cu 58 (all other impurities were < 1 ppm) were polished to 1 μm diamond, annealed at 900°C and repolished to a 1 μm diamond finish. Samples were coated with 40 Å CeO_2 by reactively sputtering a Ce target with oxygen ions and then oxidized at 900°C in $5x10^{-3}$ torr oxygen in a manometric apparatus (21). Additional details have been reported elsewhere (22).

After oxidation the specimens were analyzed by SIMS profiling to evaluate the scale composition using a Physical Electronics Inc. SIMS II system in conjunction with a differential ion gun operated with a 4 keV xenon beam (0.5 mm dia.) at 56° off the specimen normal. The beam was rastered over an area of 2.5 x 3 mm and was electronically gated to analyze the central 10% of the sputter crater area. The principal species analyzed by SIMS were $^{53}Cr^{16}O^+$, $^{56}Fe^{16}O^+$, $^{53}Cr^{18}O^+$, $^{56}Fe^{18}O^+$, $^{140}Ce^{16}O^+$ and $^{140}Ce^{18}O^+$. The profile for $^{52}Cr_2{}^{56}Fe^+$ is a convenient indicator of the alloy/oxide interface. The results are presented as atomic concentration versus sputter time. For some specimens, after SIMS profiling, one side of the specimen was abraded to remove the oxide scale. Specimens were then placed in a solution of 100 g of iodine in 1 l of methanol for 1-2 days until all the metal had dissolved. SIMS profiling was then carried out on the resultant scale. Examination of the oxide scales was performed in the following electron microscopes: Zeiss EM 902 at 80 kV, Philips EM 400, EM 430 at 120 and 300 kV, respectively, and JEOL 2000 FX at 200 kV. A Kontron image analysis system with a 25 MHz pipeline array processor was used for data analysis from the EM 902 EELS imaging microscope. EDX was carried out on Link Analytical Systems AN 10,000 fitted to both the EM 430 and 2000 FX TEM's.

Results and Discussion

SIMS Data

Sequential oxidation of uncoated Fe-26Cr at 900°C in $^{16}O_2$ and $^{18}O_2$ shows that the second oxide forms a discrete layer on top of the first, illustrating that cation diffusion has been the predominant oxidation mechanism (23). The scale is principally Cr_2O_3, containing a very small amount of Fe. The oxidation rate of Fe-Cr alloys is found to be reduced with increasing thickness of ceria coatings up to 40 Å; at a coating thickness > 40 Å there is no further reduction in the extent of oxidation (22,24). Fig. 1(a) shows a SIMS composition-depth profile for an oxide scale formed at 900°C on Fe-26Cr with 40 Å CeO_2, when oxidized first in $^{16}O_2$ and then in $^{18}O_2$. The CeO_2 coating can be considered to be an inert marker whose position indicates the transport mechanism for scale growth. The position of the Ce maximum in the middle of the scale would suggest that inward oxygen transport is now becoming predominant. This is confirmed by the $^{16}O/^{18}O$ data in Fig. 1(a) which shows there are ^{18}O maxima ($^{53}Cr^{18}O^+$) at the gas/oxide interface, within the scale at the position of the Ce maxima, and at the alloy/oxide interface showing that oxygen diffusion has occurred during scale growth. It is also observed in Fig. 1(a) that the maxima of the $^{140}Ce^{16}O^+$ and $^{140}Ce^{18}O^+$ signals occur at the same position. If the SIMS data are represented in the form of the fraction of ^{18}O associated with the Ce- and Cr-bearing species through the scale, it is shown in Fig. 1(b) that a higher fraction is associated with the Ce-bearing species. If the Ce in the scale is located at oxide grain boundaries, the association of ^{18}O with Ce strongly suggests that transport of ^{18}O through the scale occurs along grain boundaries. These SIMS results and those at 1100°C (22,24), on Fe-20Cr (22,25) and on Fe-16Cr alloys (26) indicate the Ce found within the scale changes the growth mechanism of Cr_2O_3 from predominantly cation to predominantly anion diffusion.

SIMS profiles for a thinner oxide formed during short-term oxidation of coated Fe-26Cr at 900°C are shown in Fig. 2. Fig. 2(a) shows the profile for the oxide produced after ~1 min. oxidation. It is apparent, even for this short-term oxidation, that Ce is already distributed within the oxide scale. Fig. 2(b) shows the profile obtained for the specimen used for Fig. 2(a) after dissolving the metal in the iodine/methanol solution. The profile indicates that the distribution of elements in the oxide is basically unaltered. Oxides, like the one whose SIMS profile is shown in Fig. 2, are sufficiently thin to be examined by TEM to determine the location and phase formed by Ce segregation in the oxide.

TEM/EDX, EELS and Electron Diffraction Data

The 440 Å-thick oxide formed during heating to 900°C and the 760 Å-thick oxide formed after 1 min. oxidation exhibit a similar polycrystalline Cr_2O_3 grain structure as shown in Figs. 3 (a) and (b). Fig. 3(a) shows that the initial oxide is incomplete with islands containing individual grains of Cr_2O_3 and a Ce-rich oxide phase. The grain size of the Ce-rich phase varied from ~150-600 Å, whereas Cr_2O_3 varied from ~150-1200 Å. Figs. 4 and 5 show both bright field and EELS images (27) for Ce from similar areas of oxides as shown in Fig. 3. The distribution of Ce-containing oxide is shown in Fig. 4(b) and Fig. 5(b) by the light coloured regions. In Fig. 5(b) it is clear that the Ce-rich particles are much smaller and appear to be more finely dispersed in the scale than in Fig. 4(b). Areas similar to those shown in Figs. 3-5 have also been investigated by conventional analytical TEM techniques. The Ce-rich particles shown in Figs. 4(b) and 5(b) were analyzed by EDX. A typical spectrum of the particles shown in Fig. 4(b) is given in Fig. 6. From comparison with a standard, it is clear that the Ce:Cr ratio is approximately 1:1. The finely dispersed particles shown in Fig. 5(b) were more difficult to isolate since the analysed volume is large compared to the size of the particles. The Ce:Cr ratio of unity illustrated in Fig. 6 would suggest that the particles in Fig. 4(b) are $CeCrO_3$ and this has been confirmed by the following convergent beam electron diffraction (CBED) data.

Diffraction parameters for the phases $CeCrO_3$, Cr_2O_3 and CeO_2 are shown in Table1.$CeCrO_3$ belongs to the Pbnm orthorhombic space group (28), and has lattice parameters a=b=5.475 Å, c=7.74 Å (29). A CBED pattern from the oxide shown in Fig. 4 is given in Fig. 7, together with the indexing which is consistent with the phase $CeCrO_3$. Diffraction patterns taken from other grains in the $<3\bar{3}2>$ and $<2\bar{1}0>$ zone axes also showed the presence of $CeCrO_3$. In order to confirm the indexing of each pattern, Higher Order Laue Zone (HOLZ) information (30) was obtained and a calculation made of the reciprocal lattice layer spacing. From this a calculation of the unit cell volume was made. Table 2 shows the unit cell volumes calculated from the diffraction pattern shown in Fig. 7 as well as from other patterns with the respective HOLZ and lattice vectors used from the CBED patterns. The calculated cell volumes are in good agreement with that for $CeCrO_3$ given in Table 1. To verify the accuracy of the technique, a comparison calculation was made on a diffraction pattern of CeO_2 in one of the instruments (EM 400). The calculated cell volume of 37 $Å^3$ is in good agreement with that for CeO_2 given in Table 1. $CeCrO_3$ was also proposed to be the Ce-rich phase observed at Cr_2O_3 grain boundaries on oxidation of Ce-implanted Ni-Cr alloys (31). This was based on the Ce^{3+} valence state found by the XANES (X-ray Absorption Near Edge Spectrometry) technique and on EDX and selected area electron diffraction patterns.

TABLE 1

Structure and Diffraction Data for $CeCrO_3$ Cr_2O_3 and CeO_2

Oxide	Crystal Structure	Lattice Parameter (Å)	Cell Volume $(Å)^3$
$CeCrO_3$	Orthorhombic	a=b=5.475, c=7.74	232
Cr_2O_3	Rhombohedral	a=5.35, α=54.83°	96*
CeO_2	Cubic (fluorite)	a=5.41	40*

TABLE 2

Data Used to Calculate Cell Volumes from the CBED Patterns

Zone Axis	Lattice Spacing (Å)	Cell Vectors (Å)		Angle (°)	Volume ($Å^3$)
$\langle 1\bar{1}0 \rangle$ (Fig. 7)	7.74 7.497	7.26	3.62	90	204
$\langle 3\bar{3}2 \rangle$	27.56	4.07	2.21	90	248
$\langle 2\bar{1}0 \rangle$	13.1	7.87	2.38	90	245
CeO_2 $\langle 001 \rangle$	4.8 4.3	2.77	2.77	90	37

* Note the volume of the primitive cell is that of the conventional Bravais cell divided by the integer constant of 2 for A,B,C or I-centered lattices, 4 for F-centered lattices and 3 for rhombohedral cells referred to a hexagonal set of axes.

$CeCrO_3$ can also be confirmed from the moiré fringe spacing produced by the interference between diffracted beams arising from Cr_2O_3 and $CeCrO_3$. Fig. 8, a bright field image of a 1280 Å-thick oxide, shows the moiré fringes and Fig. 9 gives the diffraction information and the indexing of the pattern. (X-ray microanalysis from the moiré fringe region showed a greater than 10 fold increase in Ce concentration compared with adjacent areas). The moiré fringe spacing calculated from the interference between the Cr_2O_3 $(\bar{1}14)$ and $CeCrO_3$ (200) diffracted beams was found to be in reasonable agreement with the values measured in the bright field image. This approach is based on that used by Przybylski et al. (18-20) to identify $YCrO_3$ in Cr_2O_3 scales formed on Y-implanted Co-Cr alloys. We have similarly used moiré fringes to identify $YCrO_3$ in oxides formed on both Y-implanted and Y_2O_3-coated materials.

It is pertinent to further compare our data with that of previous work on Y-implanted materials. A 40 Å coating of CeO_2 is equivalent to 1 x 10^{16} Ce atoms/cm^2, which can be compared with a Y dose level of 2 x 10^{16} Y ions/cm^2 known to cause a change in growth mechanism of Cr_2O_3 on Cr from predominantly cation to predominantly anion diffusion (9,32). As discussed, Y was found by STEM X-ray analysis to be segregated to oxide grain boundaries and particles of $YCrO_3$ identified (18-20). ^{18}O/SIMS experiments on this Y-implanted material showed that, as for CeO_2-coated material, inward oxygen diffusion occurred and that ^{18}O was strongly associated with the Y in the scale (9,32,33). Segregation of Y to oxide grain boundaries has also been observed for CrO_3 formed on Y-implanted stabilized stainless steel (6,8,34,35). The above results demonstrate that segregation of Y to oxide grain boundaries likely influences the relative diffusivities of Cr cations and oxygen anions along oxide grain boundaries resulting in a change in growth mechanism. This, we believe, is an analogous situation to that reported here for CeO_2-coated alloys, where particles of $CeCrO_3$ likely impede the grain boundary diffusion of Cr cations while allowing the diffusion of oxygen anions to continue.

Conclusions

A significant reduction in the rate of oxidation of Fe-26Cr alloys at 900°C has been achieved by sputter coating 40 Å of CeO_2 onto the alloy surface prior to oxidation. ^{18}O/SIMS analysis of oxide scales shows that the maximum Ce concentration occurs within the scale and that oxygen diffusion has taken place. EDX, EELS, TEM and electron diffraction data indicate that Ce segregates to form particles of $CeCrO_3$ within the scale. These particles likely impede Cr grain boundary cation diffusion while allowing oxygen diffusion to continue, resulting in a change in oxide growth mechanism from predominantly cation to predominantly anion diffusion.

Acknowledgement

The authors thank Mr. G.I. Sproule for his assistance with the SIMS analysis, and Dr. Y. LePage for helpful discussions regarding interpretation of the CBED patterns.

References

1. G.C. Wood, Werkst. und Korros., 22, 491 (1971).
2. D.P. Whittle and J. Stringer, Phil. Trans. Roy. Soc. London, A295, 309 (1980).
3. D.P. Moon, Mater. Sci. Technol., 5, 754 (1989).
4. M.J. Bennett and D.P. Moon, AERE Rept. R-13368 (1989).
5. J. Stringer and P.Y. Hou, Proc. Symp. on Corrosion and Particle Erosion at High Temp., Ed. V. Srinivasan and K. Vedula, The Minerals, Metals and Materials Soc., 383 (1989).
6. M.J. Bennett, B.A. Bellamy, C.F. Knights, N. Meadows and N.J. Eyre, Mater. Sci. Eng., 69, 359 (1985).
7. M.J. Bennett, H.E. Bishop, P.R. Chalker and A.T. Tuson, AERE Rept. R-12540 (1987).
8. M.J. Bennett, J.A. Desport, M.R. Houlton, P.A. Labun and J.M. Titchmarch, AERE Rept. 13107 (1988).
9. C.M. Cotell, G.J. Yurek, R.J. Hussey, D.F. Mitchell and M.J. Graham, Proc. Symp. High Temp. Mat. Chem. -IV, Ed. Z.A. Munir, D. Cubicciotti and H. Takagawa, Vol. 88-5, The Electrochemical Soc., Inc., 268 (1987).
10. L.P. Pfeil, U.K. Patent No. 574088 (1945).
11. D.P. Moon and M.J. Bennett, AERE Rept. R-12757 (1987).
12. P.Y. Hou and J. Stringer, Proc. Symp. High Temp. Mat. Chem. III, Ed. Z.A. Munir and D. Cubicciotti, Vol. 86-2, The Electrochemical Soc. Inc., 198 (1986).

13. P.Y. Hou and J. Stringer, Mater. Sci. Eng. 87, 295 (1987).
14. P.Y. Hou and J. Stringer, J. Electrochem. Soc., 134, 1836 (1987).
15. G.M. Ecer, B.B. Singh and G.H. Meier, Oxid. Met., 18, 55 (1982).
16. M.J. Bennett, M.R. Houlton and J.B. Price, IAEA Specialists Meeting, High Temp. Metallic Mat. for Appl. in Gas-cooled Reactors, Vienna, Austria, May 1981.
17. T. Fransen, M.M.A. Perik, H. Fikkert, M.A. de Jongh and P.J. Gellings, 8th European Congr. on Corrosion, Paper 53 (1985).
18. K. Przybylski, A.J. Garratt-Reed and G.J. Yurek, J. Amer. Ceram. Soc., 69, C-264 (1986).
19. K. Przybylski, A.J. Garratt-Reed and G.J. Yurek, J. Electrochem. Soc., 135, 509 (1988).
20. K. Przybylski and G.J. Yurek, Mat. Sci. Forum, 43, 1 (1989).
21. M.J. Graham and M. Cohen, J. Electrochem. Soc., 119, 879 (1972).
22. R.J. Hussey, P. Papaiacovou, J. Shen, D.F. Mitchell and M.J. Graham, Proc. Symp. on Corrosion and Particle Erosion at High Temp., Ed. V. Srinivasan and K. Vedula, The Minerals, Metals and Materials Soc., 567 (1989).
23. M.J. Graham, D.A. Downham, R.J. Hussey, D.F. Mitchell and G.I. Sproule, Proc. 4th Berkeley Conf. on Corrosion-Erosion-Wear of Materials at Elevated Temp., Ed. A.V. Levy and J. Stringer, NACE (in press).
24. R.J. Hussey, P. Papaiacovou, J. Shen, D.F. Mitchell and M.J. Graham, Mater. Sci. Eng. A120, 147 (1989).
25. P. Papaiacovou, R.J. Hussey, D.F. Mitchell and M.J. Graham, Corros. Sci., (in press).
26. R.J. Hussey, P. Papaiacovou, J. Shen, D.F. Mitchell and M.J. Graham, Proc. 11th Int. Corros. Congr., April 1990, Florence, Italy (in press).
27. D.F. Mitchell and M.J. Graham, J. Electrochem. Soc., 133, 2433 (1986).
28. S. Geller, Acta. Cryst. 10, 243 (1957).
29. S. Quezel-Ambrunaz and M. Mareschal, Bull. Soc. Franc. Minér. Crist., LXXXVI, 204 (1963).
30. D.B. Williams, Practical Analytical Electron Microscopy in Mater. Sci, Philips Electron Optical, 127 (1987).
31. N. Patibandla, T.A. Ramanarayan, R. Ayer and F. Cosandey, Proc. Symp. on Corrosion and Particle Erosion at High Temp., Ed. V. Srinivasan and K. Vedula, The Minerals, Metals and Materials Soc., 585 (1989).
32. C.M. Cotell, G.J. Yurek, R.J. Hussey, D.F. Mitchell and M.J. Graham, J. Electrochem. Soc., 134, 1871 (1987).
33. C.M. Cotell, G.J. Yurek, R.J. Hussey, D.F. Mitchell and M.J. Graham, submitted to Oxid. Met.
34. M.J. Bennett, J.A. Desport and P.A. Labun, Proc. 43rd Ann. Meeting of the Electron Microscopy Soc. of America, 270 (1985).
35. M.J. Bennett and A.T. Tuson, AERE Rept. R-13309 (1988).

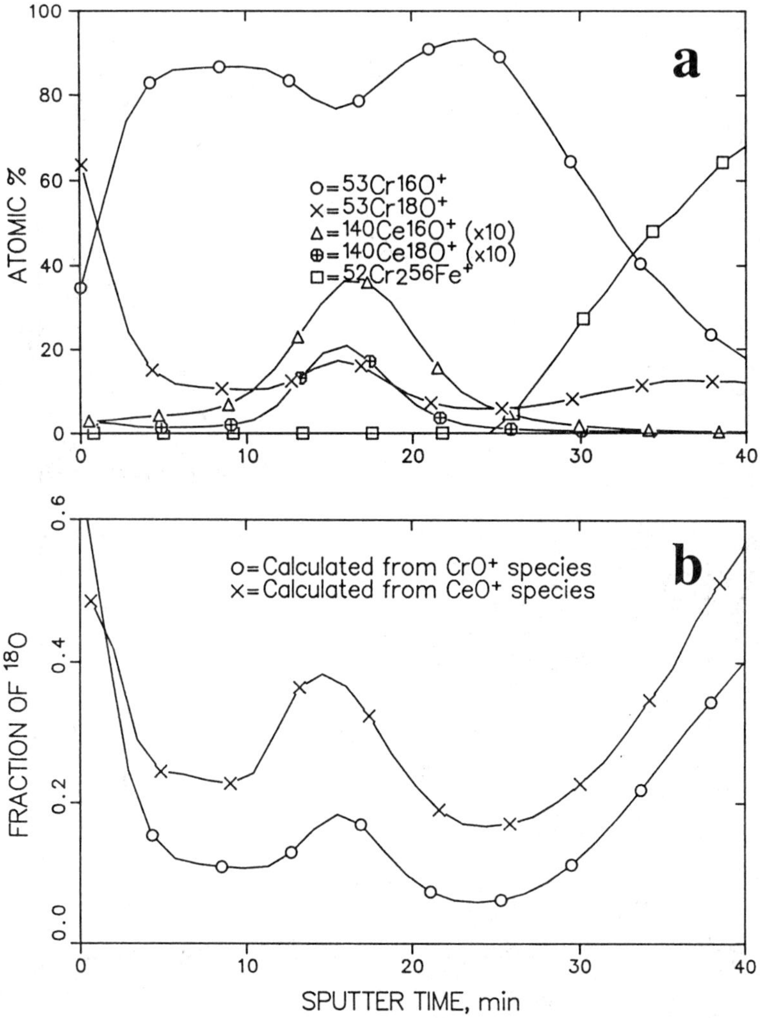

FIG. 1 (a) Positive SIMS profiles for a 2300 Å-thick oxide formed on Fe-26Cr coated with 40 Å CeO_2 at 900°C in 5 x 10^{-3} torr oxygen. 1650 Å of oxide produced in $^{16}O_2$ followed by 650 Å in $^{18}O_2$. (b) Fraction of ^{18}O calculated from (o) CrO^+ species and (x) CeO^+ species. Sputtering was by 4 keV xenon.

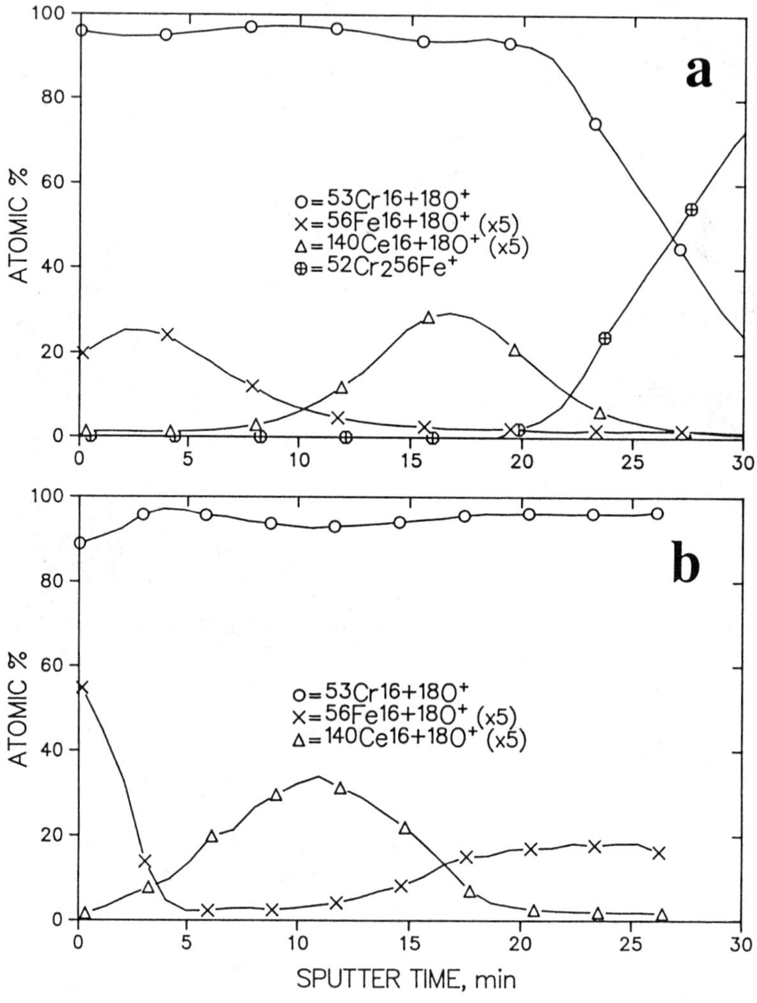

FIG. 2 Positive SIMS profiles for (a) a 760 Å-thick oxide formed on Fe-26Cr coated with 40 Å CeO_2 after ~1 min. oxidation at 900°C and (b) after dissolving the metal from specimen used in (a) in iodine/methanol solution. (Note that profile (b) is the reverse of (a), i.e. the original alloy/oxide interface is at time = 0.) Sputtering was by 4 keV xenon.

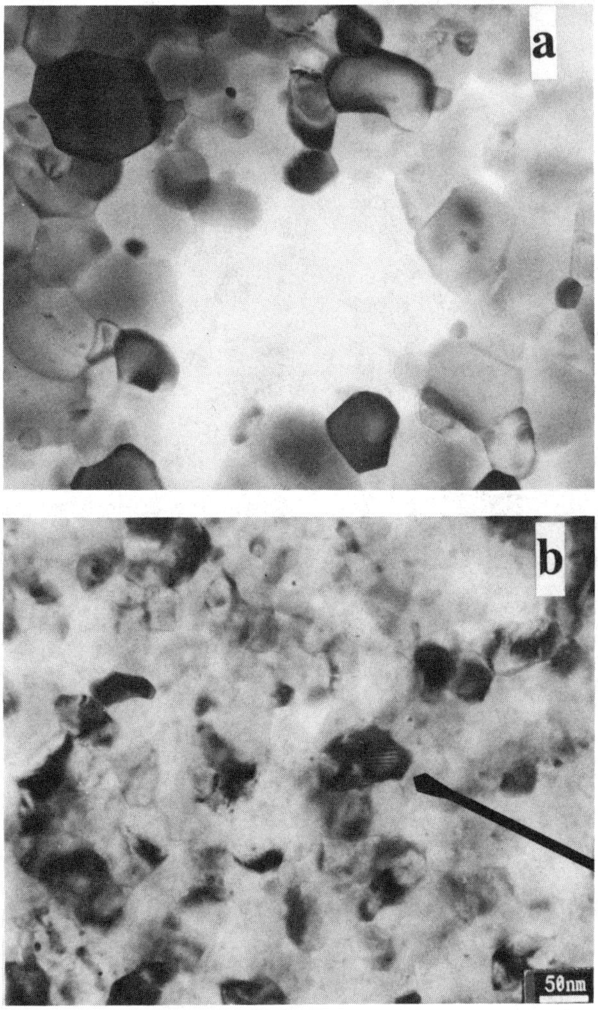

FIG. 3 Overall grain structure of oxide formed on Fe-26Cr coated with 40 Å CeO_2 at 900°C in 5 x 10^{-3} torr oxygen. (a) 440 Å-thick oxide formed on heating to 900°C; (b) 760 Å-thick oxide formed after ~1 min. at 900°C.

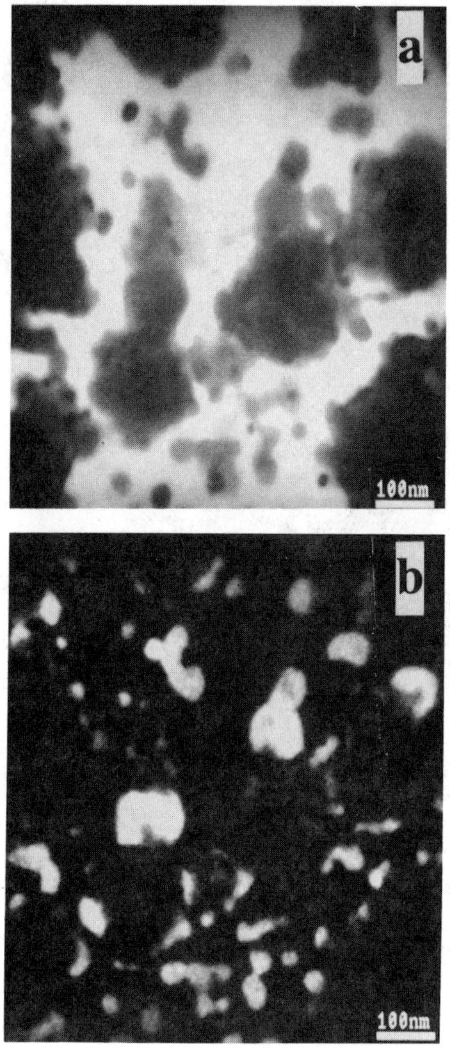

FIG. 4

TEM bright field image (a) and electron energy loss (883 eV) images for Ce (b) from the same area of a 440 Å-thick oxide whose overall grain structure is illustrated in Fig. 3(a)

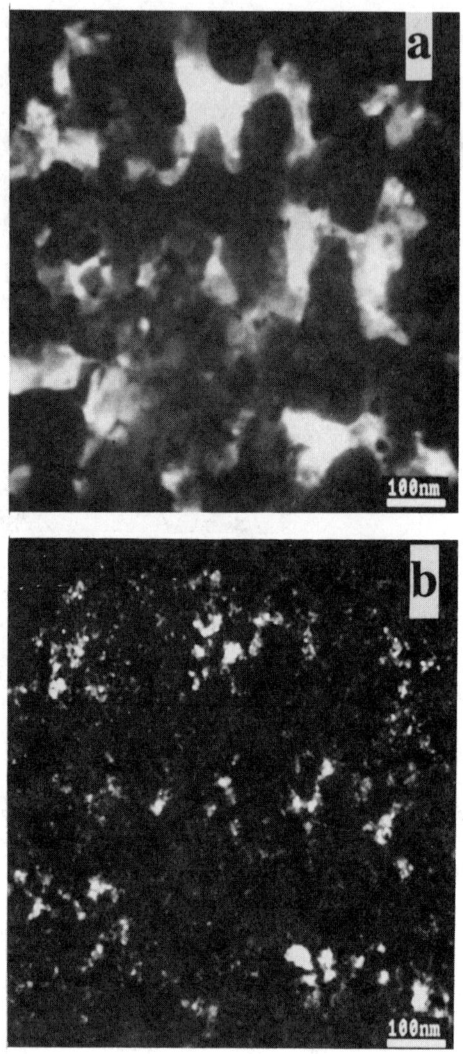

FIG. 5

TEM bright field image (a) and electron energy loss (883 eV) images for Ce (b) from the same area of a 760 Å-thick oxide whose overall grain structure is illustrated in Fig. 3(b).

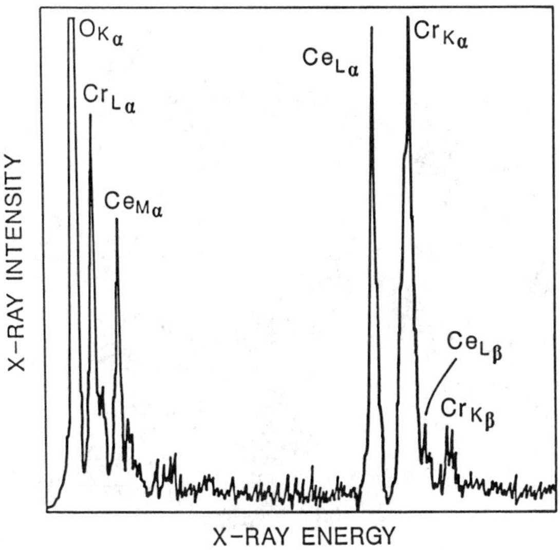

FIG. 6 Typical energy dispersive X-ray spectrum of oxide grains as in Fig. 4 showing O, Cr and Ce signals.

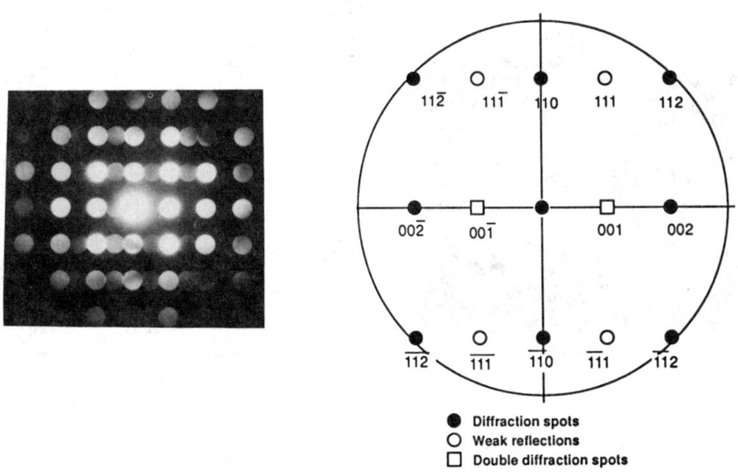

FIG. 7 CBED pattern (and its indexing) from $CeCrO_3$ oxide grains as in Fig. 4. Electron beam direction, $<1\bar{1}0>$.

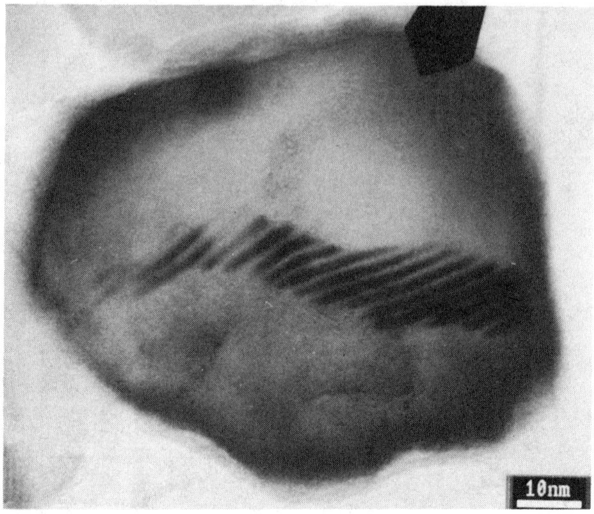

FIG. 8 TEM bright field image showing moiré fringes from a 1280 Å-thick oxide formed on oxidizing Fe-26Cr coated with 120 Å CeO_2 for ~3 min. at 900°C.

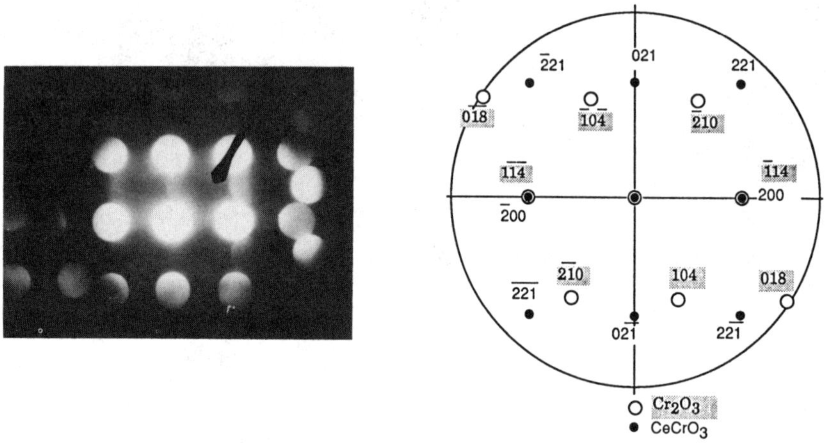

FIG. 9 CBED pattern (and its indexing) from Cr_2O_3 and $CeCrO_3$ oxide grains from a region with fringe-type contrast in Fig. 8. Electron beam direction, $\langle 0\bar{1}2\rangle_{CeCrO_3}$ parallel to $\langle \bar{4}\bar{8}1\rangle_{Cr_2O_3}$.

Breakdown of Cr_2O_3 scales by sulphur

C. Zhou, L.W. Hobbs and G.J. Yurek
H.H. Uhlig Corrosion Center and Department of Materials Science and Engineering, Massachusetts Institute of Technology, Cambridge, Massachusetts 02139, U.S.A.

Abstract

This research was to investigate the growth mechanisms of protective oxide films grown selectively from certain alloys by high temperature oxidation/sulfidation in H_2 - H_2O and in H_2 - H_2O - H_2S atmospheres, the migration of species within predeveloped oxide scales in the presence of sulfur, and breakdown of these oxide scales by a relatively stronger sulfidizing atmosphere. This paper concerns itself with the behavior of Fe-45Cr alloy and the internal oxidation/sulfidation in this system after breakdown of a protective Cr_2O_3 scale.

The reaction kinetics and the microstructure of the reaction products were examined under different conditions by controlling P_{S_2} in the gas mixtures. The measured kinetics are closely related to the observed microstructural state of the scale, particularly for the penetration of sulfur through a protective oxide scale developed by preoxidation in H_2 - H_2O gas mixtures prior to the subsequent oxidation and *sulfidation in H_2 - H_2O - H_2S gas mixtures. The morphological structure of the scale is influenced by the level of P_{S_2} in the gas atmosphere.*

1 Introduction

Because of the poor protection afforded by most metallic sulfides like Cr_xS or Al_xS, (some transition metal sulfides, Nb-, Mo- sulfides for instance, are exceptions), it is preferable to develop dense, adherent oxide scales on metal alloy substrates where ambient conditions allow. When exposed to mixed oxygen and sulfur-bearing gas mixtures, such as $SO_2 - O_2$ or $H_2 - H_2O - H_2S$, pure Cr, M – Cr and M – Cr – Al alloys, (M refers to Fe, Ni or Co), are observed to sulfidize rapidly even though the gas compositions are located in regions of the phase stability diagrams where oxides are thermodynamically stable. It appears that equilibrium thermodynamics are in conflict with experimental evidence, and that kinetic factors are more important in determining stability[1-7]. An example of such a region defined by a kinetic phase boundary is shown in Figure 1.

Thermodynamic diagrams, such as that in Figure 1, are based on the direct interaction between metallic elements and oxygen or/and sulfur in gas mixtures. As long as an oxide film locally or wholly covers the surface of the alloy substrate, continuation of the gas/metal reactions depends not only on the gas/solid conditions, but more importantly on the film/metal interfacial conditions, since both oxygen and sulfur atoms can migrate through oxide films and sulfur can dissolve into oxides. Even dense and protective oxide scales may be penetrated and finally destroyed by sulfur. The transportation of sulfur through the oxide scale may be characterized one of or a combination of the following:

1. *Dissolution and diffusion of atomic sulfur in the oxide lattice;*

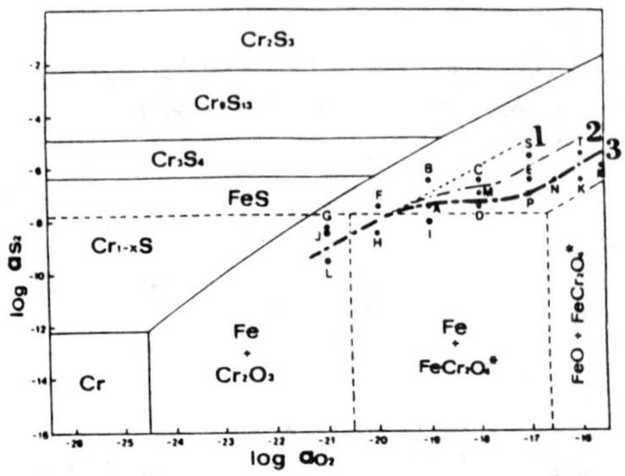

Figure 1: Stability diagram for Fe-Cr-S-O system at 1173K[3]. Thermodynamic boundaries: solid line for Cr and dashed line for Fe; kinetic boundaries: line 1 for pure Cr, line 2 for Fe-45Cr, line 3 for Fe-25Cr.

2. *Short circuit diffusion of atomic sulfur (grain boundary diffusion and dislocation pipe diffusion) in the oxide scale;*

3. *Outward migration of Cr coupled with inward migration of S;*

4. *Sulfur migration in the molecular form of S_2, SO_2 or H_2S along microcracks, grain boundaries or dislocations in the oxide scale.*

This paper reports in part results obtained in the present study for the high temperature oxidation/sulfidation of Cr-bearing alloys and the breakdown of protective oxide scales by sulphur-bearing gases.

2 Experimental Conditions

Fe-45Cr alloy ingots were prepared by Ames Laboratories, Ames, IA. Each as-cast ingot was encapsulated in a quartz tube flushed with high purity argon and with Ta as an oxygen getter, then annealed at 1373K for 50 hours. It was next sliced into disks 12.7mm in diameter and 0.5 - 1.0mm in thickness. A small suspension hole (0.1mm) was drilled in the disks which were polished to 1μm finish and degreased ultrasonically in acetone and methanol prior to oxidation/sulfidation.

The temperature of the furnace was controlled using a Eurotherm Model 919 controller equipped with Pt-Pt+13%Rh thermocouple. A second thermocouple was inserted into the reaction chamber from the bottom to monitor the temperature of the specimen more directly. Since H_2 and H_2S were employed for the present experiments, strict care and precautions were required for the furnace and microbalance. The furnace tube was made removable. After loading the specimen, the tube was reconnected, sealed and the whole system pumped by a mechanical vacuum pump, after which the system was flushed with ultra-high purity argon. The system was then pumped again, and the process repeated three times to further reduce the residual air level. An $H_2 - H_2O$ gas mixture was introduced into the reaction chamber, the reaction gas flow rate reaching steady-state in 10 minutes.

The specimen was next lowered down into the hot zone using an adjustable U-shaped magnet and oxidized for a certain time, after which a controlled $H_2 - H_2S$ gas mixture was let into the reaction zone to start oxidation/sulfidation experiments. Kinetic data were obtained thermogravimetrically by means of a Cahn Model 1000 microbalance. For simultaneous oxidation/sulfidation alone, the $H_2 - H_2S$ gas mixture was let into the reaction tube immediately following introduction of $H_2 - H_2O$.

The microstructure and the constitution of the reaction products were analyzed by various techniques: X-ray diffraction(XRD), both conventional and glancing angle, scanning electron microscopy(SEM) and associated quantitative elemental analysis using X-ray energy dispersive spectroscopy(XEDS), and electron microprobe analysis(EPMA).

Table 1: Oxidation/Sulfidation of Fe - 45 Cr Alloy at 1173K in H_2 - H_2O and in H_2 - H_2S - H_2O Gas Mixtures

#	PreO.	O/S	$log P_{O_2}$	$log P_{S_2}$	R
A	—	20 min	-19	-6.5	CC
B	—	20 min	-19	-7.0	CC
C	1 hr	25 hr	-19	-6.5	BD
D	9 hr	6 hr	-19	-6.5	BD
E	61 hr	—	-19	—	P
F	1 hr	1 hr	-18.9	-9.1	P
G	3 hr	17 hr	-19	-7.5	P
H	3 hr	17 hr	-19	-8	P
I	40 hr	—	-19	—	P
J	10 hr	6 hr	-19	-6.5	T

3 Experimental Results

3.1 Kinetics

The measured kinetic results for the oxidation and/or sulfidation at 1173K of certain groups of Fe - 45Cr specimens are summarized in Figure 2a in linear plots of weight gain/unit area ($\Delta W/A$) against time. The corresponding reaction conditions and relevant comments are presented in Table 1. # stands for specimen number; PreO. for preoxidation time of a specimen in $H_2 - H_2O$ prior to the subsequent oxidation/sulfidation in $H_2 - H_2O - H_2S$; O/S for the time of simultaneous oxidation/sulfidation; R for remarks; CC for catastrophic corrosion; BD for break down of the scale; P for protective and T for transition from protective scale to a state of break down.

For convenience in comparing specimens experiencing preoxidation to those subjected to other conditions, the kinetics for simple oxidation in H_2-H_2O ($P_{O_2} = 10^{-19}$) as well as for initial exposure to H_2-H_2O-H_2S gas mixtures are also presented in this diagram.

Figure 2b is the conventional parabolic plot of the square of weight gain/unit area $(\Delta W/A)^2$ against time. Note that curves A and B follow a linear relation for direct oxidation/sulfidation. Curves C, D and J exhibit breakaway behavior for a preformed Cr_2O_3

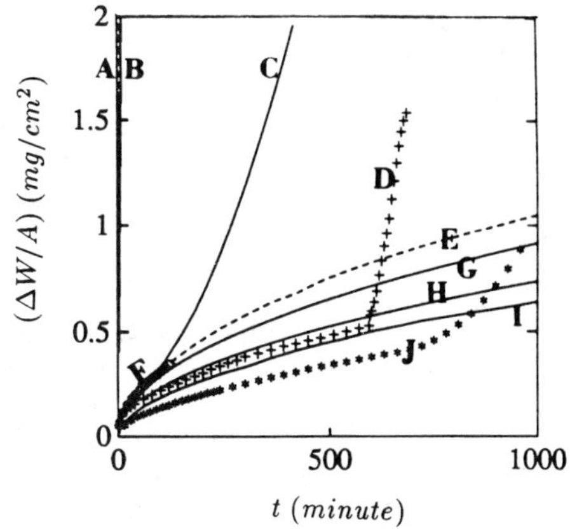

a

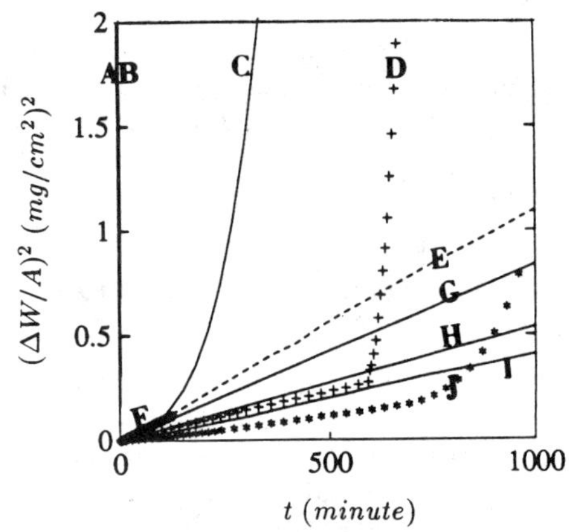

b

Figure 2: (a) Kinetic results plotted as weight gain per unit area against time for $Fe-45Cr$ alloys at 1173K for the various conditions presented in Table 1. (b) Plot of (weight gain)^2 against time for the same data.

scale developed in $H_2 - H_2O$ gas mixtures of $P_{O_2} = 10^{-19} atm$. Curve C has completed a transition from a protective scale to breakaway. Curve J is experiencing such a transition, but the scale is not yet in the breakaway state. Curves C and J both show a continuous transition of the oxide scale from protective to breakaway corrosion, whereas curve D shows an abrupt change which is due to mechanical failure of the oxide scale; the latter phenomenon is not reproducible. Smooth transitions, such as for C and J, are observed to be reproducible for most of the specimens examined under similar conditions. The time before initiation of the transition from protective scale to breakaway varies approximately linearly with the preoxidation time for a constant sulfur level maintained.

Curves E, F, G, H and I exhibit behavior very close to parabolic for long-time exposure. It is noted that the effective parabolic constants for A and B are 10^5 times higher than those for E, F, G, H and I. Variation of parabolic constants was $5 \sim 16 \times 10^{-12} g^2 cm^{-4} sec^{-2}$ for these specimens.

3.2 Structures of Scales

Several groups of specimens were preoxidized in $H_2 - H_2O$ gas mixtures with $Po_2 = 10^{-18.5} \sim 10^{-19} atm$, and then $H_2 - H_2S$ was introduced into the reaction chamber.

During the subsequent exposure to $H_2 - H_2O - H_2S$ atmosphere, an oxygen partial pressure $Po_2 = 10^{-19} atm$ was maintained by controlling the temperature of saturation water.

At $Ps_2 = 10^{-9.1} atm$ and $Ps_2 = 10^{-8} atm$, the scale maintained protection and no sulfur was detected on surface or in cross-section by energy-dispersive X-ray spectrometry (XEDS). A small sulfur content was detected however using Auger electron spectroscopy (AES) on the surface.

At $P_{S_2} = 10^{-7.5}$ atm, detectable sulfur was observed by XEDS at the gas/oxide scale interface, even though the scale continued to maintain protection, and the kinetic results are comparable to those for $Ps_2 = 10^{-9.1} atm$ and $Ps_2 = 10^{-8} atm$.

At $Ps_2 = 10^{-6.5} atm$, substantial changes were observed both kinetically (as seen in curves C, D and J in Figure 2) and macroscopically. The corrosion rate was enhanced by introducing H_2 - H_2S gas mixture into the reaction chamber, and the preformed oxide scale was finally broken down and lost its protectiveness.

Figure 3a shows the surface morphology of a sample preoxidized at $P_{O_2} = 10^{-19} atm$ in H_2 -H_2O gas mixture for one hour and followed by one hour oxidation/sulfidation at $P_{O_2} = 10^{-19} atm$ and $P_{S_2} = 10^{-8.5} atm$ in H_2 - H_2O - H_2S gas mixture. The composition of this atmosphere is located below the so-called kinetic boundary as shown in Figure 1. Whiskers of Cr_2O_3 were observed to grow on the surface. This phenomenon has been observed for the pure oxidation of Cr and other Cr-alloys[1-5]. In dry oxygen, however, no such whiskers were observed. Buttons of Cr_2O_3 spalled off the as-grown surface at random in cooling the specimen to room temperature. Longer exposure to the sulfidizing atmosphere resulted in a denser distribution of spalling sites.

Figure 3b, from a polished transverse section of this specimen, shows a uniform Cr_2O_3 scale which is adherent to the substrate. A much longer time of exposure to this H_2 - H_2O - H_2S gas mixture of 40 hours resulted in the formation of small voids between scale and substrate.

Figure 4 shows a sample which experienced one hour preoxidation at $P_{O_2} = 10^{-19} atm$ in H_2 -H_2O and a subsequent 20 minute exposure to H_2-H_2O-H_2S ($P_{O_2} = 10^{-19} atm$ and $P_{S_2} = 10^{-6.5} atm$). This gas composition is located at point B between the thermodynamic and kinetic boundaries as shown in Figure 1. Figure 4a reveals that CrS_{1+x} particles grow on the top surface of Cr_2O_3 scale preferentially along lines which appear to correspond to the location of grain boundaries in the alloy substrate. This effect of substrate grain boundaries was observed more clearly for shorter time preoxidation treatment. Figure 4c is an enlargement of one of these particles whose XEDS spectrum is given in Figure 4d. The particle size of the

CrS_{1+x} is around $1\mu m$. The polished cross-section(Figure 4b) shows an intact Cr_2O_3 scale, but gaps or voids are seen at the substrate/scale interface. Figure 4e shows an enlarged region from Figure 4(a) surrounded by the CrS_{1+x}-decorated boundaries. Fine CrS_{1+x} particles are densely distributed on the surface at this magnification. These fine particles are attributed to the normal outward grain boundary and lattice diffusion of Cr in the oxide scale. The large particles in Figures 4a and 4c are attributed instead to the short-circuit diffusion of Cr which has ostensibly marked the location of substrate grain boundaries. The scale has spalled in patches during cooling to room temperature, and detectable sulfur is observed at the alloy/oxide interface so revealed (Figure 4f).

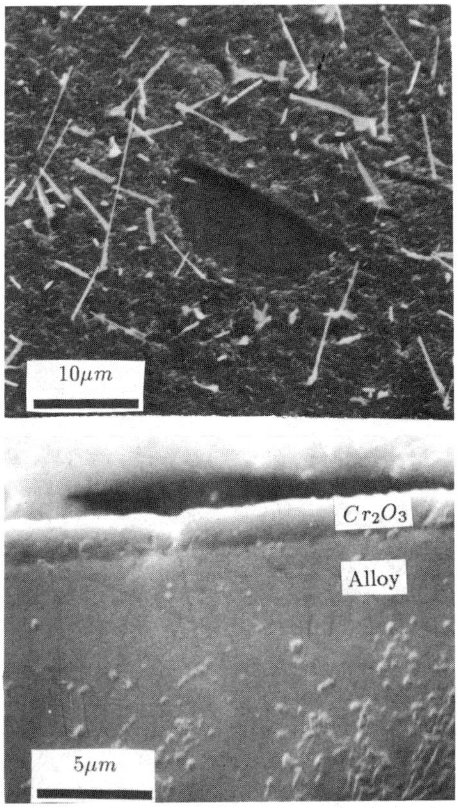

Figure 3: Secondary electron image using SEM. (a) Surface morphology of sample oxidized for 1 hour in H_2 - H_2O ($P_{O_2} = 10^{-19} atm$) and 1 hour in H_2 - H_2S - H_2O ($P_{O_2} = 10^{-19} atm$, $P_{S_2} = 10^{-8.5} atm$) for the subsequent exposure. Whiskers of Cr_2O_3 appear on the surface of the Cr_2O_3 scale. (b) Polished cross-section showing adherent Cr_2O_3 scale.

By maintaining the same conditions for the one hour preoxidation, while extending the subsequent oxidation/sulfidation time to 40 minutes, a relatively higher density of CrS_{1+x} particles was formed on the surface (Figure 5a). The same technique was employed for the preparation of transverse sections for this sample, but many parts of the preformed scale were destroyed in this case in the preparation process, as shown in Figure 5b. This fragility suggests a strongly deleterious effect of the gas mixture on the integrity of the existing Cr_2O_3 scale. Points A, B, C, D and H show the formation of CrS_{1+x}, points E and F are Cr_2O_3 scale. At point E good contact to the substrate is maintained. At point G, considerable S and Fe are detected.

In order to further determine whether the CrS_{1+x} particles on top of the Cr_2O_3 scale are formed by the outward diffusion of Cr or by the interaction between Cr_2O_3 and H_2S, longer preoxidation times were tried for other samples at the same gas compositions as for the specimens illustrated in Figures 4 and 5. Two samples, preoxidized for 8 and 10 hours respectively and subsequently subjected to oxidation/sulfidation at $P_{S_2} = 10^{-6.5} atm$ as above for 2 and 6 hours respectively, are shown in Figures 6a and 6b. That such extended exposure to H_2 - H_2O - H_2S gas mixture does not produce a denser distribution of CrS_{1+x} particles on the top surface of specimens experiencing a longer preoxidation than specimens in Figure 4 and Figure 5 suggests that the controlling process for the formation of CrS_{1+x} particles on top surface of the scale largely depends on outward migration of Cr through the Cr_2O_3 scale.

3.3 Internal Corrosion

The breakdown of the dense, protective Cr_2O_3 scale originally developed resulted in

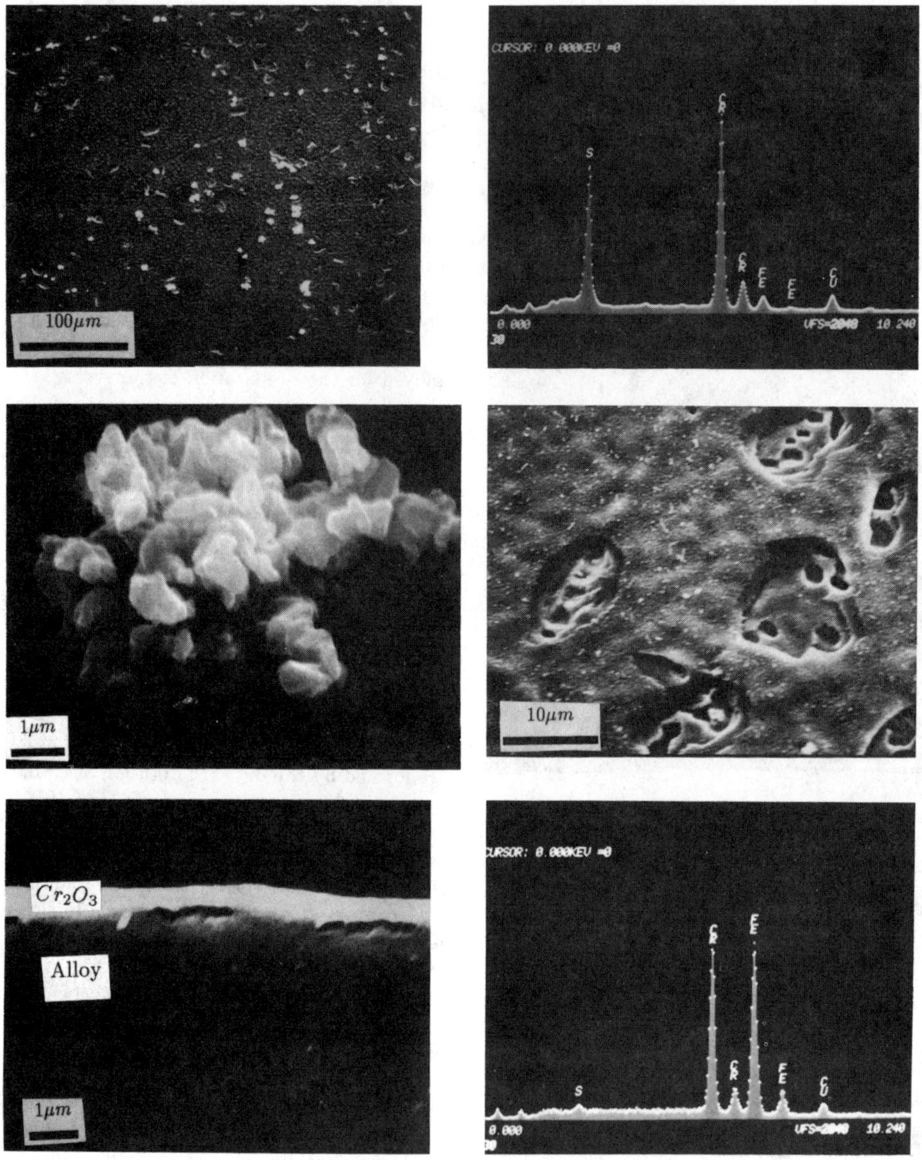

Figure 4: One hour preoxidation in $H_2 - H_2O$ ($P_{O_2} = 10^{-19}$) followed by 20 minute exposure to $H_2 - H_2O - H_2S(P_{O_2} = 10^{-19}, P_{S_2} = 10^{-6.5})$. (a) CrS_{1+x} particles grow on the top surface of oxide scale at preferential locations. (b) Higher magnification of a CrS_{1+x} particle. (c) Polished cross-section through Cr_2O_3 scale. (d) XEDS analysis for 4(b). Cu peak is from coating of the sample. (e) Enlargement of a region from 4(a) showing fine particles of CrS_{1+x} and spalling of patches during cooling which reveal the alloy/scale interface region. (f) XEDS for alloy/scale interface at the spallation sites in 4(e).

a further deterioration of the alloy substrate. For convenience of comparison to the results presented in Figures 4 and 5, the same one hour preoxidation in $P_{O_2} = 10^{-19} atm$ was effected, but at the end of the preoxidation run a 25-hour oxidation/sulfidation was performed at $P_{O_2} = 10^{-19} atm$ and $P_{S_2} = 10^{-6.5} atm$. The corresponding kinetic results are shown in curve C in Figure 2. Mixtures of Cr and Fe sulfides were observed in the external scale.

Figure 7a is a back-scattered electron (BSE) image of the overall polished cross section. Only a small portion of the scale remains in contact with the alloy substrate when examined under the microscope in cross-section. The sites where a thicker external scale forms correspond to deep penetration of internal precipitation (both internal oxidation and sulfidation). BSE images reveal the internal precipitation zone clearly even without chemical etching. Figure 7b is selected from a portion of Figure 7a where good contact of scale to substrate was maintained. The interest here is focused on the interface between the internal precipitation

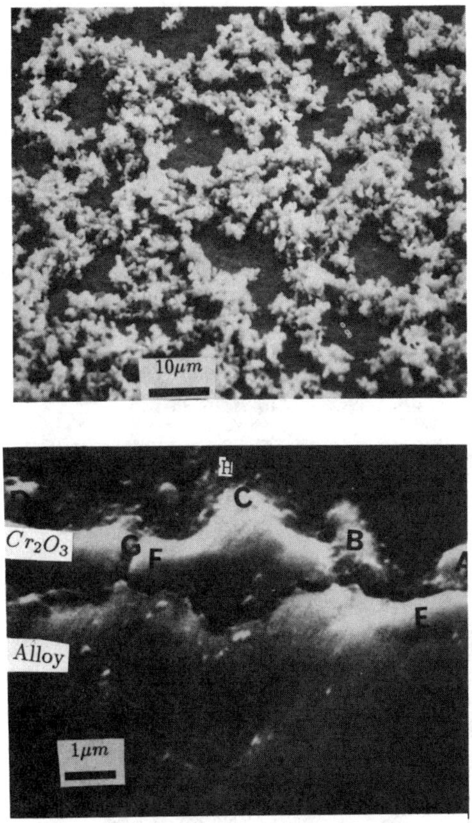

Figure 5: The same preoxidation treatment as for the sample shown in Figure 4, but subjected to 40 minute subsequent oxidation/sulfidation. (a) CrS_{1+x} particles on the surface of Cr_2O_3. (b) Polished transverse section through the oxide scale.

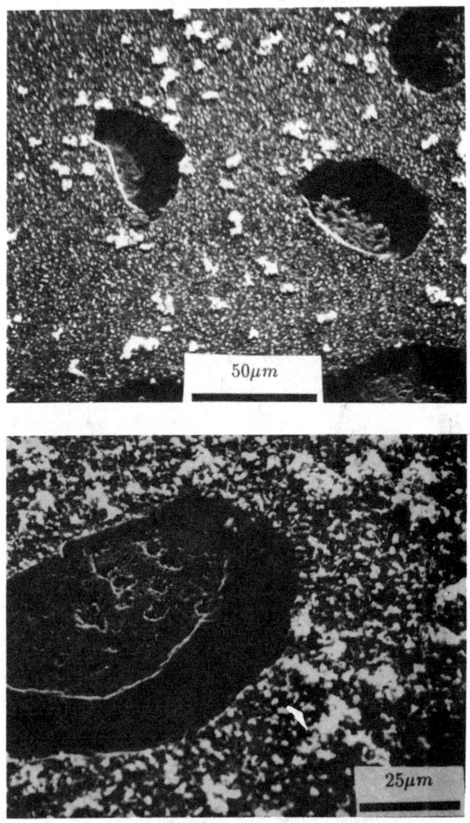

Figure 6: The same gas composition as for specimen in Figure 4, but (a) 8 hour preoxidation and 2 hour subsequent oxidation/sulfidation. (b) 10 hour preoxidation and 6 hour subsequent oxidation/sulfidation.

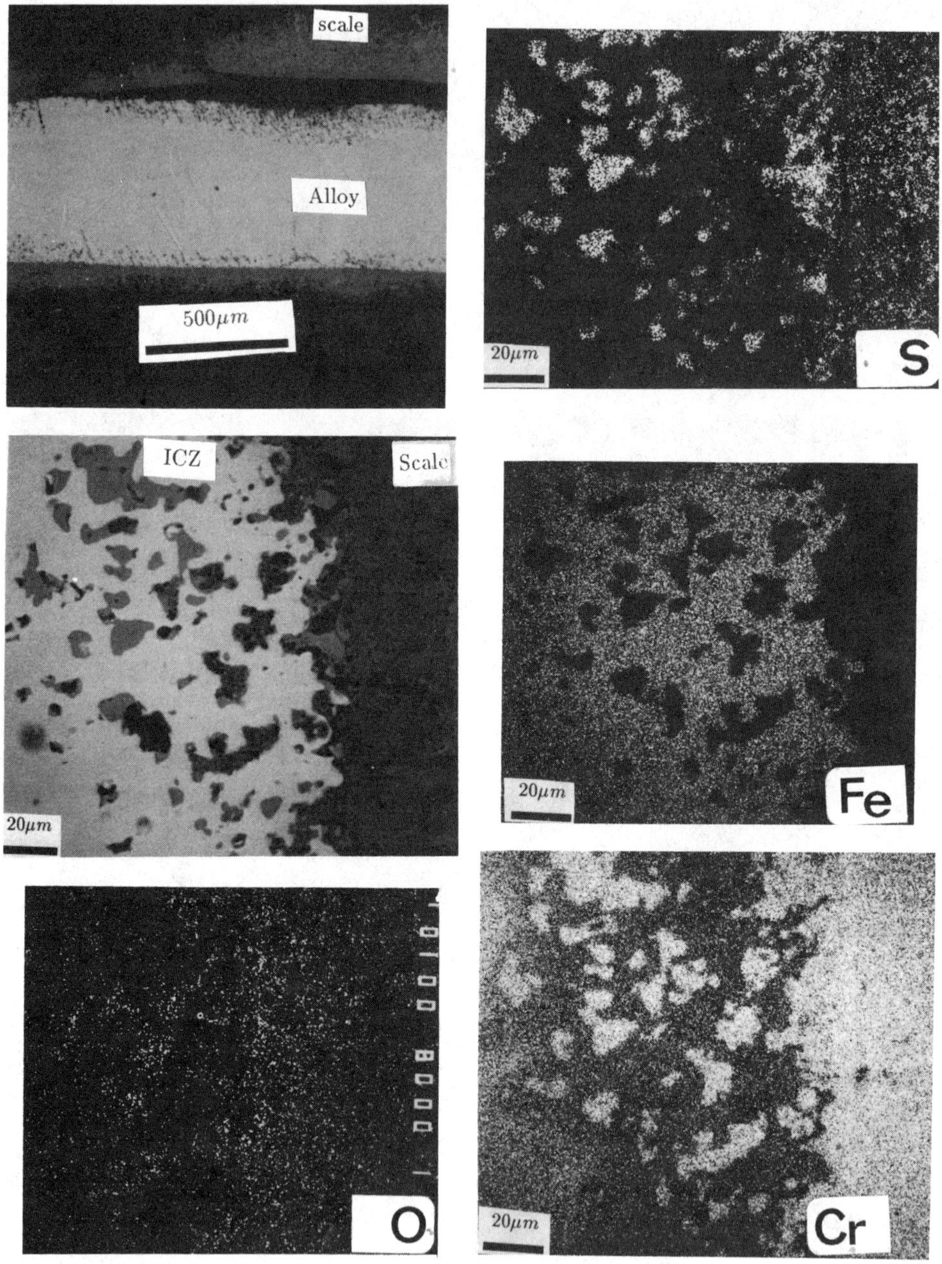

Figure 7: Same 1 hour preoxidation conditions as for specimen in Figure 4, 25 hour subsequent oxidation/sulfidation. (a) Overall transverse section. (b) Partial transverse section reveals internal mixed oxidation/sulfidation zone and inner layers of external scale. (c-f) Corresponding X-ray dot maps for O, S, Fe and Cr respectively. ICZ refers to internal corrosion zone.

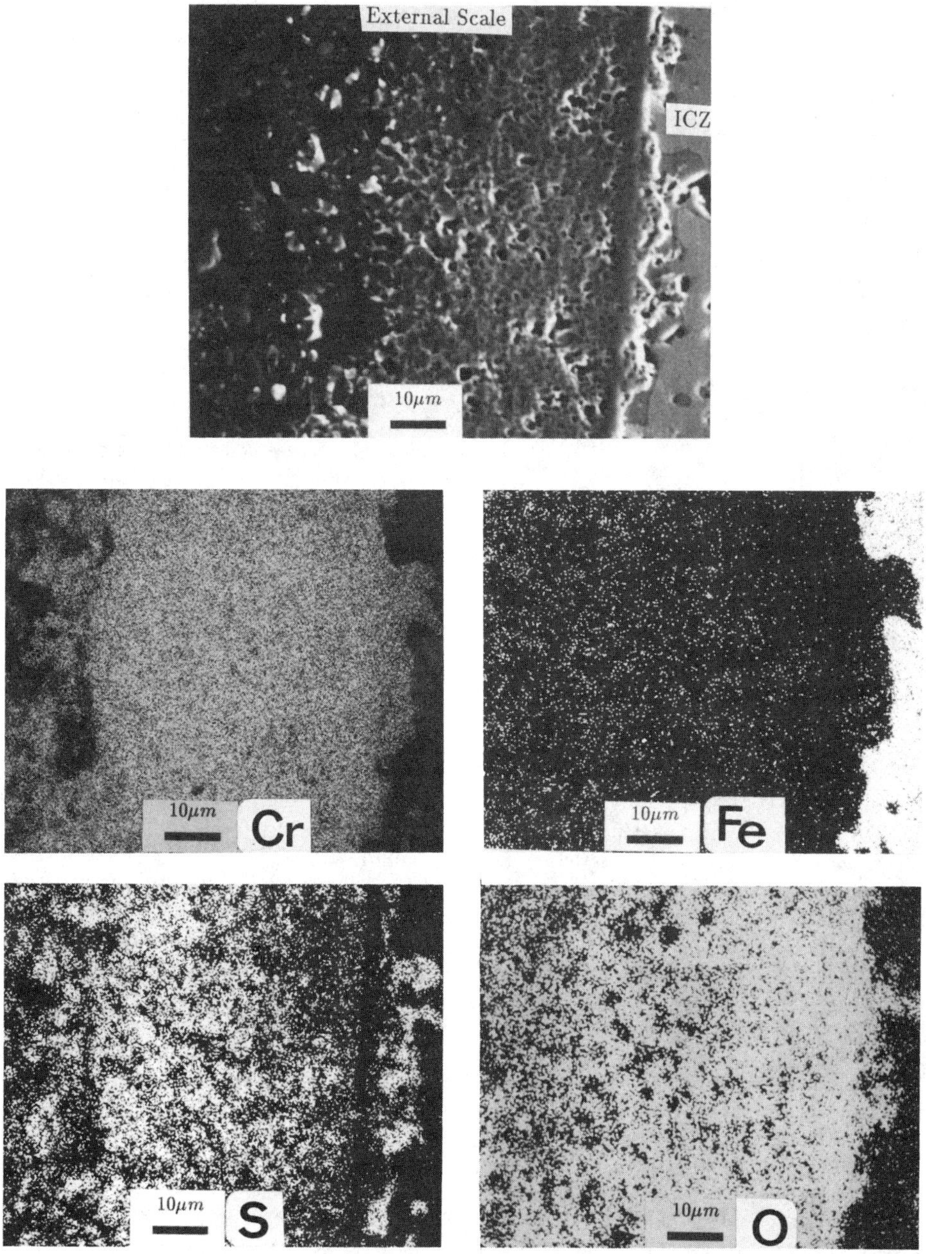

Figure 8: The same condition as for specimen illustrated in Figure 7. EPMA images of cross-section through the whole external scale and the vicinity of scale/alloy substrate interface. (a) BSE image. (b-e) X-ray dot maps in Cr, S, Fe and O respectively. ICZ refers to internal corrosion zone.

Table 2: Quantitative Analysis ($wt.\%$) at sites Indicated in Figure 9

Point	Fe	Cr	S	O
1	2.19	57.27	40.00	~ 0.01
2	87.26	12.45	~ 0.01	~ 0.01
3	87.64	11.74	~ 0.01	~ 0.01
4	6.73	52.64	23.67	16.96
5	89.10	9.59	~ 0.01	~ 0.01
6	5.66	54.9	28.61	10.83
7	5.60	55.3	28.80	10.60
8	57.94	40.95	~ 0.01	~ 0.01

zone (internal oxidation and sulfidation) and the inner external scale, where an advancing internal sulfidation zone is followed by an advancing internal oxidation/sulfidation zone. Figures 7c, 7d, 7e and 7f correspond to X-ray dot maps for O, S, Fe and Cr for the region illustrated in Figure 7b. A complementary band on the the S and the Cr or O maps indicates the presence of the Cr_2O_3 scale developed in the preoxidation stage, also observable in Figure 7a. Figure 7e shows the absence or negligible occurrence of Fe in the intermediate and inner external scales as well as in the internal oxides and sulfides.

Figure 8 characterizes the overall external scale of specimen #C in Table 1. Fe is found to occur in the outermost external scale, beneath which negligible Fe is detected. The outermost scale consists of CrS_{1+x}, Cr_2O_3 and $Fe_{1-x}S$ (hexagonal pyrrhotites of different polytypes) as determined from glancing angle X-ray diffraction methods discussed in section 4. The intermediate scale, consisting of CrS_{1+x} and Cr_2O_3, is not a solid solution but a mixture. The inner scale is Cr_2O_3 formed by preoxidation with some sites breaking down.

Quantitative point analyses were performed for both the external scale and internal oxidation/sulfidation zone. Table 2 shows the results for the points identified in Figure 9. The maximum depletion of Cr in the internal oxidation/sulfidation zone is to about $10wt.\%$.

We note that homogeneous internal oxidation to form Cr_2O_3 precipitates does not

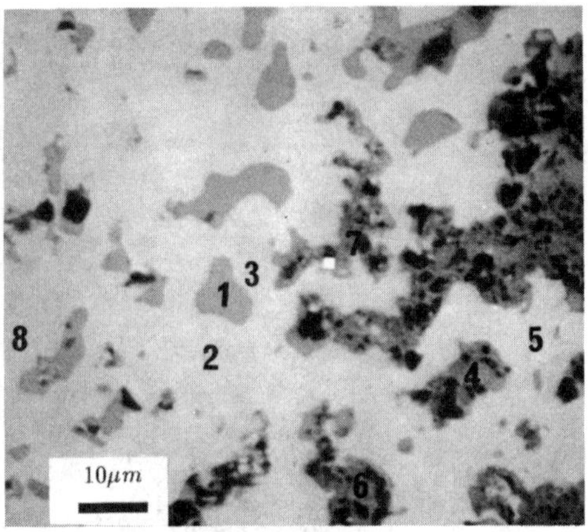

Figure 9: Same conditions as for specimen in Figure 7. BSE cross section image showing locations for quantitative analysis in the internal mixed oxidation/sulfidation zone. Results are given in Table 2.

occur. Rather, internal oxidation, following the advancing internal sulfidation, takes place preferentially at such sites where internal sulfide, CrS_{1+x}, has formed. This selectivity derives from two conditions: (1) easier nucleation and (2) relatively low Cr content favorable for internal oxidation of Cr. A composition gradient of Cr is found to exist in the alloy substrate between the particles of internal oxide/sulfide.

Figure 10 gives quantitative composition profiles for each element transversing the overall cross-section using electron microprobe analysis (EPMA). Since neither the external scale nor the internal oxidation/sulfidation zone grows uniformly, the measured composition profiles were determined for typical sites where well-defined external scales and internal corrosion zones were established.

3.4 Characterization by X-Ray Diffraction

Conventional X-ray diffraction (XRD) methods give an accurate result for structural

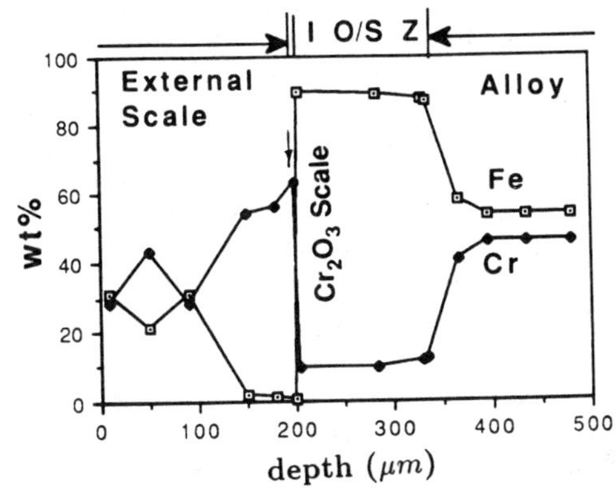

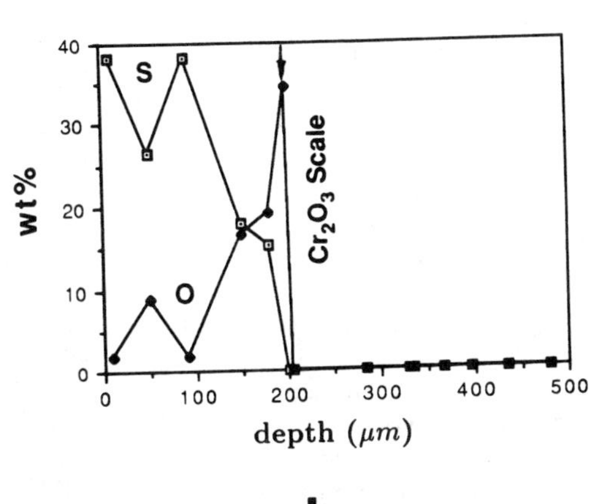

Figure 10: Same conditions as for specimen in Figure 7. Quantitative composition profiles along the transverse section for the external scales and for the alloy matrix. "I O/S Z" refers to internal oxidation/sulfidation zone.

analysis if the bulk sample is thick enough, but encounter limitations in analyzing the structure of a thin scale or scales formed on a substrate. X-ray photons penetrate into the substrate to produce diffraction patterns from as deep as $10\mu m$ or further into the substrate, in which case there may be no detectable diffraction information from an associated thin scale or film using conventional XRD techniques. By employing glancing angle X-ray diffraction techniques (GAXRD) instead, reducing the incident angle to as small as 1°, even a thin film 30nm thick will diffract sufficiently to be distinguished from substrate. For multilayer scales or films, the structure of each layer may be determined by fixing various incident glancing angles. Because of limitations on the length of this paper, only a brief description of X-ray results is given here.

For a sample preoxidized for 10 hours in H_2 - H_2O ($P_{O_2} = 10^{-19} atm$) and exposed a further 6 hours to H_2 - H_2S - H_2O ($P_{O_2} = 10^{-19} atm$, $P_{S_2} = 10^{-6.5} atm$), diffraction peaks were obtained from the alloy substrate and from the Cr_2O_3 scale by employing conventional bulk XRD methods. By reducing the incident angle to 1.5^o with the GAXRD method, interference from the substrate was eliminated, and diffracted peaks were identified from hexagonal CrS_{1+x} ($x = 0.14$), two polytypes of $Fe_{1-x}S$ ($x = 0.05$) and rhombohedral Cr_2O_3. Longer time exposure (1 hour preoxidation + more than 10 hours oxidation/sulfidation) resulted in an outer external scale consisting of $Fe_{1-x}S$, CrS_{1+x} and Cr_2O_3; an intermediate layer of CrS_{1+x} and Cr_2O_3, with a decreasing fraction of CrS_{1+x} and a corresponding increase of Cr_2O_3 along the direction into the alloy substrate; and an inner external scale of Cr_2O_3 which developed principally in the preoxidation period. Microscopic examinations indicate that this inner Cr_2O_3 scale was, in fact, often compromised at sites 10 - $20\mu m$ apart.

4 Discussion

Figure 1 illustrates a kinetic boundary which is located in a region where oxides are thermodynamically stable phases. Within the area between the kinetic and thermodynamic boundaries, reaction phases which are found are mixtures of both oxide and sulfide and not the pure oxides which are thermodynamically the more stable phases in equilibrium with the corresponding gas mixtures. In part, the presence of this kinetic boundary can be attributed to the relative ease of dissociation of H_2S compared to H_2O; at 1173K, for example, $\Delta G^{\circ}_{1173K}(H_2O) = -182kJ/mol$ whereas $\Delta G^{\circ}_{1173K}(H_2S) = -32kJ/mol$[17]. This difference derives, of course, from details of the molecular bonding states of H with O and S, which are beyond the purview of the present paper[18]. In part, however, the kinetic boundary is also attributable to details of microstructural development of the scale, a phenomenological account of which will serve to document the deleterious effect of sulfur on pre-existing Cr_2O_3 scales.

Initially, in the presence of sulfur in the gas mixtures, the outward migration rate of Cr appears to be enhanced, as evidenced by the kinetic results in Figure 2. Under conditions where Cr_2O_3 is stable thermodynamically, however, a direct chemical reaction between preformed Cr_2O_3 and H_2S or S_2 does not appear to occur. Instead, CrS_{1+x} particles form on the surface of the external Cr_2O_3 scale in the first instance, as shown in Figure 4. In terms of the kinetic results depicted in Figure 2, the preoxidation treatment delays the onset of catastrophic sulfidation, but it does not provide a significant improvement in subsequent resistance to H_2 - H_2S - H_2O environments. Failure of some oxide scales, grown on certain other metallic alloy systems, to sulfur-bearing atmosphere, has been reported by numerous authors[8-15].

The effect of alloy inhomogeneities — specifically alloy grain boundaries — on the scale development is suggested by the intriguing decoration of the oxide scale surface by CrS_{1+x} particles. The average grain size in the Cr_2O_3 scale was less than $1\mu m$, while that of the alloy substrate was greater than $10\mu m$, the same as the cells delineated by CrS_{1+x} particle decoration. While the evidence that the decoration replicates the position of the underlying alloy grain boundaries is circumstantial (or even coincidental), there are several mechanisms which could explain such a long-range effect of substrate boundaries through a thin (less than $1\mu m$) oxide scale: boundaries in the scale locating above boundaries in the alloy due to preferential oxide/substrate orientation relationships; localized strain in the oxide occurring above alloy grain boundaries due to intrusion of levering oxide into these boundaries; local buckling of the scale above alloy boundaries, etc. No ridges in the outer scale topography were observed which might have been coincided with decoration sites, but notably, the decoration effect did disappear with thicker initial oxide layers.

According to Benlyamani *et al*[16], the lattice diffusion of sulfur is 4 to 5 orders of magnitude lower than the grain boundary diffusion of S in pure Cr_2O_3 crystals. Since fine grains were usually obtained in the high temperature oxidation process, the resulting inward migration rate of sulfur through the scale could easily be three orders of magnitude higher than accounted for by pure lattice diffusion. Interactions of S with O, Cr and Fe may enhance the transport of each species respectively.

Stages in the transition from protective Cr_2O_3 scale to breakdown can be deduced from the microstructural developments chronicled in the cross-section micrographs, such as those illustrated in Figures 5, 7 and 8. A schematic representation of these stages is made in Figure 11, which includes depiction of the breakdown stage in Figures 11d and 11e. During initial exposure of the Cr_2O_3 scale to H_2 - H_2S - H_2O, the overall integrity of the scale is maintained but becomes compromised at local sites, leading to enhancement of Cr diffusion, formation of surface CrS_{1+x} particles, but also enhanced Fe transport through the oxide scale. Direct experimental observations show that breakdown and fast scaling rates are localized at sites, such as edges and boundaries. While, in the absence of S, outward migration of Fe is negligible through the Cr_2O_3 scale, introduction of S leads to appreciable diffusion of

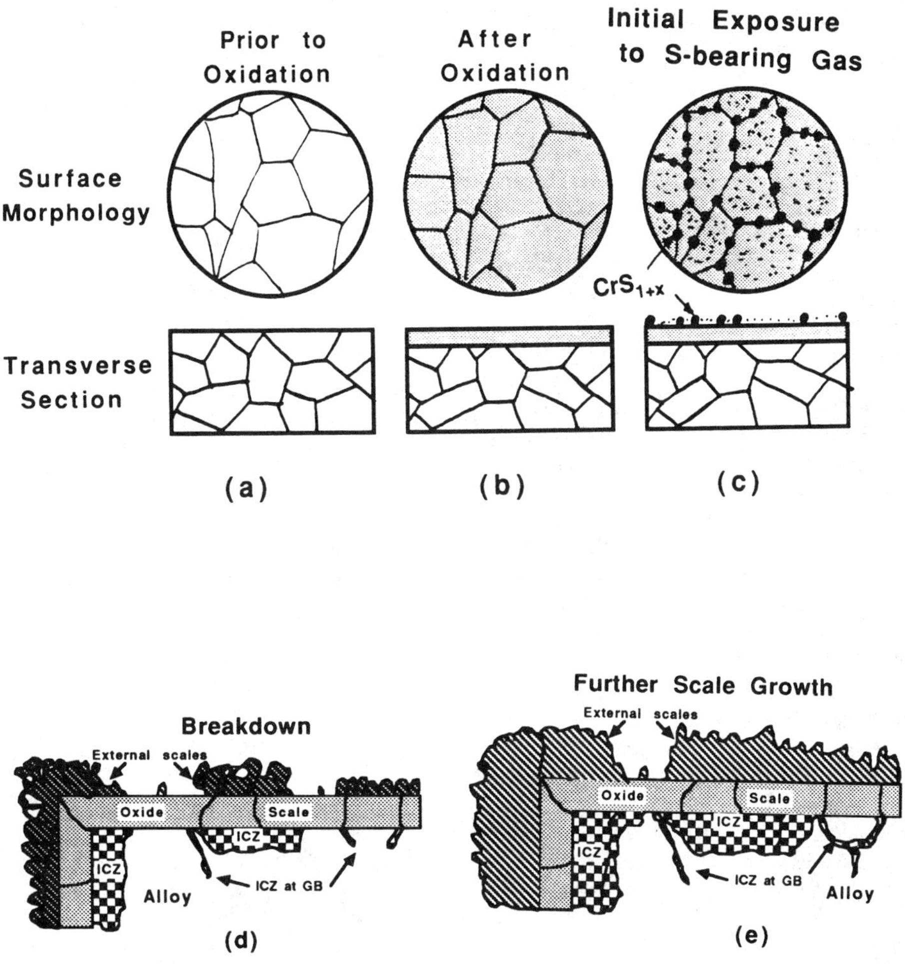

Figure 11: (a-c) Schematic representations of initial stages in the transition of Cr_2O_3 scale in H_2S-bearing gas mixtures from protective behavior to breakdown. (d-e) Schematic representations of the scale in eventual breakdown corrosion, as deduced from Figures 7a, 8a and 9. ICZ refers to internal corrosion zone; GB to grain boundary. Cr_2O_3 scale is about $2\mu m$ thick, thicker portions of external scale about $150\mu m$, ICZ about $200\mu m$ along alloy grain boundaries. The drawing is not to scale.

Fe, with two polytypes of $Fe_{1-x}S$ ($x = 0.05$) forming in the outer portion of the external scale . A gradient of sulfur level exists across the intermediate external scale — comprising a mixture of Cr_2O_3 and CrS_{1+x} — within which the sulfur activity is apparently not high enough to favor formation of $Fe_{1-x}S$ thermodynamically. Considerable competition between the reactions of Cr with O and with S occurs, however, even though Cr_2O_3 is the only stable compound in equilibrium with the corresponding gas mixtures. Penetration of sulfur through Cr_2O_3 scale at the same time results in formation of Cr sulfide beneath the scale and, subsequently, simultaneous internal oxidation and sulfidation.

Good contact seems to exist at the interfaces between CrS_{1+x} ($x = 0.14$) internal sulfide and the alloy, but voids are found at the interface between the mixed Cr_2O_3 + CrS_{1+x} layer and the alloy substrate. The voids are possibly filled with, or the result of, the accumulation of SO_2 or other insoluble gas molecules. Growth in the particle size of internal Cr sulfides was observed with increasing sulfidation/oxidation time, but across the whole of the internal corrosion zone, composition gradients were negligible; only local concentration gradients of Cr and Fe existed between the internal particles, independent of the position in the internal corrosion zone.

Nucleation of internal CrS_{1+x} occurs preferentially along alloy grain boundaries. This observation is schematically depicted in Figures 11d and 11e. The deep penetration of internal CrS_{1+x} along these grain boundaries also implies that fast diffusion of S occurs in this alloy. The steep gradients of Cr and Fe, illustrated in Figure 10, at the transition from the internal corrosion zone to the unreacted alloy matrix, indicate that internal sulfidation is dominated by the diffusion of sulfur. The first factor in the depletion of Cr is the advance of this internal sulfidation zone, the second the subsequent growth of internal oxides and sulfides. Because the high concentration of Cr in the original Fe-45Ce alloy initially inhibits the internal oxidation, substantial depletion of Cr by internal sulfidation enhances the formation of internal oxides which form first at the CrS_{1+x}/substrate interface.

5 Summary

1. Sulfur-free Cr_2O_3 scales formed by preoxidation treatments provide, in the initial stage, a measure of corrosion resistance to a subsequent exposure to $H_2 - H_2O - H_2S$. At sulfur partial pressure $\geq 10^{-6.5} atm$ at 1173K, this scale is eventually compromised and breaks down.

2. For $P_{S_2} \leq 10^{-8}$ at 1173K, the preformed oxide scale remains stable and protective, but long time exposure produces voids at the scale/substrate interface which may be filled with gaseous species.

3. Examination of preoxidized samples indicates the importance of both lattice diffusion and short-circuit diffusion of Cr through the Cr_2O_3 scale. The fastest diffusion paths for Cr are possibly related to the positions of grain boundaries of the substrate, particularly for short time preoxidation treatments.

4. S was early on detected at the alloy/scale interface at $P_{S_2} \geq 10^{-6.5}$ at 1173K. This implies that inward diffusion of S coupled with the outward diffusion of Cr through the oxide scale for the short time exposure before breakdown of the scale.

5. The Cr_2O_3 scale breaks down at localized sites, below which large amounts of internal CrS_{1+x} and Cr_2O_3 grow to form an internal corrosion zone (both internal oxidation and internal sulfidation) in the alloy.

6. Depletion of Cr in the internal corrosion zone by internal sulfidation enhances the formation of internal Cr_2O_3 and possibly SO_2 gas bubbles at CrS_{1+x}/substrate interfaces.

Acknowledgements

Support by the office of Basic Energy Sciences, U.S. Department of Energy (CZ, GJY) and subsequently by the Division of Materials Research, National Science Foundation (CZ, LWH) is gratefully acknowledged. LWH is indebted to the Master and Fellows of Balliol College, Oxford for their hospitality during a sabbatical visit. The authors would like to thank Ms. Elizabeth VanKirk for her efforts in coordinating communications and Mr. Bruce Pint for reading of the manuscript during that visit.

References

1. G. J. Yurek, M. H. LaBranche and Y. K. Kim, High Temperature Corrosion in Energy Systems, M. F. Rothman, ed. the Metallurgical Society of AIME, Warrendale, PA (1985) p. 295.
2. K. Przybylski, A. J. Garratt-Reed and G. J. Yurek, Comm. of American Ceramic Society 69, C-264 (1986).
3. G. J. Yurek and K. Przybylski, Mater. Sci. and Eng. 87, 125 (1987).
4. M. H. LaBranche, A. Garratt-Reed and G. J. Yurek, J. Electrochem. Soc. 130, 2405 (1983).
5. G. J. Yurek and M. H. LaBranche, Proc. Corr.- Eros.- Wear of Mater. in Fossil Energy Systems, Alan V. Levy ed. NACE, Berkeley, CA, (1982) p. 933.
6. M. H. LaBranche, Ph.D thesis, MIT, Cambridge, MA (1985).
7. Y. K. Kim, Ph.D thesis, MIT, Cambridge, MA (1988).
8. G. Romeo, H. S. Spacil and W. J. Pasko, J. Electrochem. Soc. 122, 1329 (1975).
9. P. A. Mari, J. M. Chaix and J. P. Larpin, Oxid. Met. 17, 315 (1982).
10. G. M. Kim and G. H. Meier, Proceedings of the Symposium on High Temperature Materials Chemistry - IV. Z. A. Munir, D. Cubicciotti and H. Tagawa, eds. The Electrochem. Soc. Inc. Pennington, NJ (1988) p. 79.
11. F. H. Stott, F. M. F. Chong and C. A. Stirling, International Congress of Corrosion. Toronto (1984) Vol. 2, p. 1.
12. M. C. Pope and N. Birks, Oxid. Metals 12, 173 (1978).
13. P. Singh and N. Birks, Oxid. Metals 13, 472 (1979).
14. P. Kofstad and G. Akesson, Oxid. Metals 12, 505 (1978).
15. P. Kofstad and G. Akesson, Oxid. Metals 13, 1 (1979).
16. M. Benlyamani, F. Ajersch and G. Kennedy, Proceedings of the Symposium on High Temperature Materials Chemistry - IV. Z. A. Munir, D. Cubicciotti and H. Tagawa, eds. The Electrochem. Soc. Inc. Pennington, NJ (1988) p. 314.
17. O. Kubaschewski and C. B. Alcock, Metallurgical Thermochemistry, 5th ed., Pergamon Press, Oxford, (1979) p. 380.
18. C. A. Coulson, Valence, Oxford University Press, (1961) Chap. VII, p. 155.

Oxidation and sulphidation of type 310S stainless steel at elevated temperatures*

J.S. Wolf
Department of Mechanical Engineering, Clemson University, Clemson, South Carolina 29634, U.S.A.

O.B. Cavin
High Temperature Materials Laboratory, Oak Ridge National Laboratory, Oak Ridge, Tennessee 37831-6064, U.S.A.

Abstract

Type 310S stainless steel in the form of flat or bent sheet was exposed to air and SO_2 at temperatures near 900°C. Protective scales, composed of sesquioxide and spinel layers, were formed in both atmospheres. Only specimens exposed in air produced uniform deformation. In air, the bent specimens exhibited angular closure while in SO_2 the angles opened with time at temperature. The interaction of sulfur with the oxide scales apparently alters their crystallinity, microstructure and stress state. The control of oxide grain size may be one approach in the development of alloys resistant to breakdown in sulfidizing atmospheres.

INTRODUCTION

Currently, there exist major economic forces which encourage the development of energy conversion technologies (1). One major problem in that development is the selection of materials for long-term application in the presence of both oxygen and sulfur-bearing gases at temperatures where they are subject to deterioration. An often-cited mode of deterioration is the loss of an initially protective quality of the scales formed upon alloys; a phenomenon termed "breakaway" corrosion (2). This is prevalent for many structural alloys which depend upon chromium-rich scales for their protection at elevated temperatures. While it has been the subject of many engineering and some scientific studies, a recent review of the breakaway phenomenon indicates that little is known about its origin; except that it may possibly be associated with stress development in the reactive scale layer/alloy substrate composite (2).

*Research sponsored by the U.S. Department of Energy, Assistant Secretary of Conservation and Renewable Energy, Office of Transportation Systems, as part of the High-Temperature Materials Laboratory User Program, under contract DE-AC05-84OR21400 with Martin Marietta Energy Systems, Inc.

It is well known that alloys which display outstanding corrosion resistance must develop a superior barrier film to isolate the metallic component from its environment. Of the methods for producing barrier layers on alloys, those with self-building alumina, chromia or silica scales seem to hold the most potential (3,4). Hence, to realize that potential, we must understand the details of the surface compounds formed and their evolution so that we may control the transport of matter through them.

We have chosen to probe that understanding by examining the scales formed on Type 310S stainless steel in both air and sulfidizing environments. Our particular objective was to examine the relation between: the structure of these scales, their growth kinetics and their propensity for stress development at temperatures near 900°C.

PROCEDURE

The general procedure used in this research is presented here; procedural details are more fully explained elsewhere (5).

Test Material - The material selected for this study was Type 310S stainless steel (310S), known to resist high-temperature oxidation due primarily to its high chromium content (6). The actual composition of this alloy is within the nominal analysis with one exception: the presence of Sw05 percent titanium; Table I.

Specimen Design - Two different specimen geometries were used: a flat specimen and an "ell"-shaped specimen; the latter is shown in Figure 1. Both specimens had a hole at one end for suspension purposes. The "ell" specimens were designed to sense oxidation- induced strain in two ways: a) by closure of the "ell" angle and b) by separation of the centers of the two fiducial gage holes. Both deformations may, in turn, be dependent upon details of the grain boundary network of the scale and the rapid material transport or local chemical reactions which occur in association with it (8,9).

TABLE I
CHEMICAL COMPOSITION OF TYPE 310S STAINLESS STEEL

ELEMENT	WEIGHT PERCENTAGE	
	Nominal (6)	Test alloy (7)
Chromium	24-26	25
Nickel	19-22	20
Manganese	2.0 max.	1.8
Silicon	1.50 max.	0.47
Carbon	0.08 max.	0.062
Titanium	0.00	0.05
Phosphorous	0.045 max.	0.044
Sulfur	0.03 max.	0.007
Iron	Balance	Balance

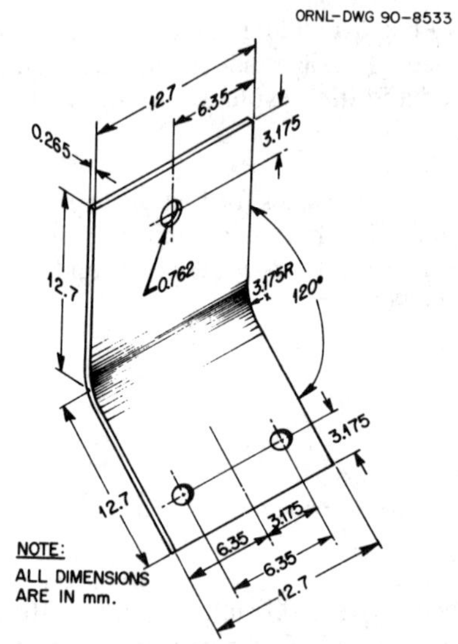

Figure 1.Strain-sensing "ell"-shaped specimen.

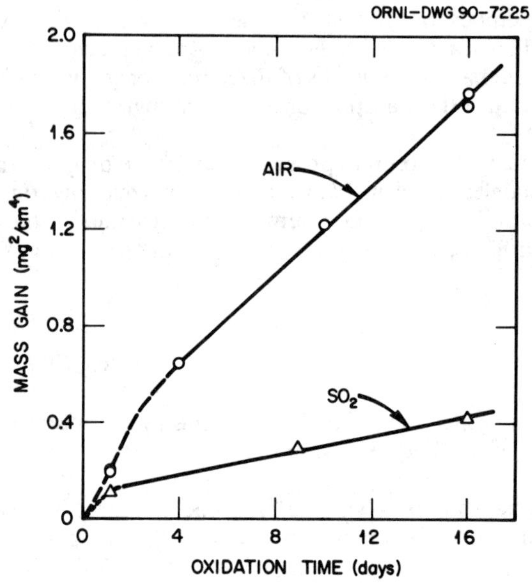

Figure 2. Square of specific mass gain for specimens exposed to air and SO2 at 900°C.

Specimen Preparation - All specimens were fabricated from one lot of 0.16 cm thick 310S stock. The material was cold-rolled, hydrogen annealed at 900°C for 15 minutes and was then finish-rolled to 0.0254 cm. Specimen blanks, nominally 2.54 by 1.27 cm, were sheared from this sheet, drilled, ground to a 600-grit finish, electropolished and cleaned. Subsequently, some blanks were formed into "ells" by bending the sheet about a mandrel. All specimens were then stress-relief annealed in hydrogen for 5 minutes at 900°C and re-electropolished. The mass and dimensions of each specimen were determined prior to testing. In addition, the bend angle and hole spacing for each annealed "ell" specimen were measured. Subsequent to exposure, the same critical dimensions and mass were determined. Part of the analysis below is based upon the differences between these pre- and post-exposure measurements.

Specimen Exposures - Tests were conducted in both air and a sulfidizing atmosphere. The sulfur-bearing gas was produced by the thermal dissociation of calcium sulfate which produces SO_2 as its primary gaseous dissociation product (5). Specimens were exposed in these environments at temperatures of 800 and 900°C for times ranging from one to 16 days; most of the specimens being exposed at 900°C. Exposures were conducted by hanging specimens in vertical furnaces.

Specimens to be sulfidized were mounted individually in quartz capsules fitted with a mounting hook and incorporating a small charge of $CaSO_4$ placed in the bottom of the capsule. Prior to sealing, these loaded capsules were baked out at 200°C for 16 hours while simultaneously being evacuated to a pressure of 10^{-7} torr. The evacuated capsules were sealed and hung in vertical furnaces.

Four tests were run in air using the "ell" specimens. Two of these were exposed at 800°C for one and 4 days and two at 900°C for one and 5 days to determine the magnitude of the deformations developed during oxidation. Specimens were run in SO_2 at 900°C for one, 9 and 16 days for the same purpose.

Tests were also run in air using flat specimens to obtain data concerning the oxidation rate, the scale structure and the compounds developed on 310S at 900°C. These tests were removed from the furnace after: 1, 4, 9 and 16 days. Tests run in SO_2 for the same purposes included three specimens exposed at 900°C for periods of 1, 9 and 16 days.

Analytical Methods - The methods used to analyze specimens included: 1) gravimetric analysis to determine the specific mass change, 2) dimensional analysis to determine the angular and linear deformations, 3) optical and scanning electron microscopy (SEM), the latter with associated energy dispersive x-ray spectroscopy (EDS), to determine the structure of scales and their approximate chemistries and 4) x-ray diffraction both at room and elevated temperatures to gain information as to the development and crystalline nature of the surface phases. Some of these methods are discussed below.

Two different orientations of the scale layers were analyzed using optical and SEM techniques: 1) the topography of the gas-scale interface as viewed normal to the surface and 2) the transverse section of scales using mounted and etched specimens. For both orientations, specimens were examined in a JEOL model JSM-35CF scanning microscope. Surface topographies were photographed and selected areas of the specimen surfaces were analyzed using an integral Tracor-Northern model TN 5500 X-Ray Analyzer. Specimen cross-sections of the scale and the adjacent alloy substrate were also photographed and analyzed. These EDS data were deconvoluted to correct for interferences between chromium, manganese and iron. For optical metallographic analyses, transverse sections of selected specimens were mounted, polished and immersion etched to reveal the grain structure of both the substrate and reaction products.

The crystal structures of the 310S substrate and the scales formed upon it were determined in air at both room and elevated temperatures by automated x-ray diffraction techniques. Examinations of sulfidized and most of the oxidized specimens were conducted at room temperature. For these analyses, a Scintag model "PAD V" diffractometer equipped with a liquid nitrogen-cooled intrinsic germanium detector was used. The output of this device was stored for analysis using a Data General Model MV-2000 computer which could access a master file containing the JCPDS diffraction standards. Because many of the scales examined evidenced both line broadening and specular reflections, portions of many diffraction patterns were deconvoluted via an interactive computer method. For elevated temperature analyses, a similarly-configured Scintag model "PAD X" diffractometer equipped with a Buehler high temperature furnace was used.

RESULTS AND DISCUSSION

The results of this investigation are presented and discussed in the following order: 1) thermochemical and kinetic considerations, 2) specimen deformations, 3) metallographic observations and 4) x-ray diffraction observations.

Thermochemical and Kinetic Considerations - For the 310S alloy, the affinities of the major constituent metals for oxygen and sulfur are essentially parallel; i.e., the affinities increase in the order: nickel, iron, chromium, manganese and silicon (10-14). The high reactivity of manganese has been observed here in its ubiquitous tendency to combine in the oxide reaction products. However, the extremely high mobility of manganese necessary for its participation in the reaction products is not explicable by its high reactivity alone.

Another important consideration is the potential for the formation of liquid phases in those cases where the sulfidizing atmosphere has been employed. It is well known that sulfides, as a class, melt at lower temperatures than their corresponding metallic oxides and, unlike the oxides, may be liquid at the test temperatures used here (15). For example, nickel sulfide and its eutectic with nickel represent cases which, if they had formed at 900°C, would be molten. For atmospheres richer in sulfur than those used here, it has been shown that internal sulfidation occurs at temperatures as low as 650°C in this steel (16). However, as discussed below, no evidence for the formation of a liquid phase was found.

For the tests conducted in SO_2 at 900°C, the partial pressures of the gases within the capsule have been estimated (5). At test initiation, the sulfur pressure is relatively low and the oxygen pressure is relatively high: i.e.: $(P_{S2}) = 5.00 \times 10^{-25}$ Atm. and $(P_{O2}) = 1.70 \times 10^{-6}$ Atm. As the tests progress, oxygen is preferentially fixed by the reactive specimen, decreasing its pressure, while that of the sulfur increases; eg.: $(P_{S2}) = 2.80 \times 10^{-11}$ Atm., and $(P_{O2}) = 5.60 \times 10^{-11}$ Atm.

For exposures in both air and SO_2, no evidence of catastrophic scale breakdown was found and only microscopic spalling was noted. The observed specific mass gains of those specimens exposed in air and SO_2 for times up to 16 days at 900°C are shown in Figure 2. These data indicate that, except for an initial transient, the kinetics are parabolic. The reaction in air exhibited a rate constant of 3.88×10^{-3} $mg^2cm^{-4}hr^{-1}$; that in SO_2 was 1.11×10^{-3} $mg^2cm^{-4}hr^{-1}$. As sulfur was not found to be overtly involved in compound formation, the low reactivity in SO_2 may be associated with the lower oxygen partial pressures involved in these tests.

Specimen Deformation - Deformation of "ell" specimens at 800 and 900°C were associated with the stresses generated during oxidation and/or sulfidation. These observed deformations are summarized in Table II. It is seen that the linear strain of specimens subjected to oxidation in air is much larger than that in specimens similarly exposed in SO_2. This occurred in spite of the fact that only oxide phases formed in both cases and, while a lesser amount of oxide was present on the sample exposed in the sulfidizing environment, the strains found there were less than they would have been had the deformation been proportional to the total amount of scale formation.

For air oxidation, a time dependent angular closure occurred despite gravitational effects which should have produced an opening of the bend angle. An extension of the gage length was also noted which increased with exposure time. The angle closure and strain rates were calculated on the basis that they too obeyed a parabolic relationship. These deformation rates are essentially constant, indicating that they are directly related to the oxidation process. Because the ratio of air oxidation rates to angle closure rates at 800 and 900°C are greater than 1:2, the mechanical deformation probably also reflects a decrease in the creep strength of this alloy with rising temperature.

TABLE II
DEFORMATION OF ELL-SHAPED SPECIMENS AT 800 AND 900°C

Temperature (°C)	Exposure Time (Day)	Angle Closure (Deg)	Closure Rate ($DegHr^{-1/2}$)	Linear Strain Strain (%)	Linear Strain Rate ($Hr^{-1/2}$)
		AIR ATMOSPHERE			
800	1	2.0	0.41	N/A	N/A
800	4	4.5	0.46	N/A	N/A
900	1	26.3	5.37	1.73	0.353
900	5	63.6	5.81	3.42	0.312
		SULFIDIZING ATMOSPHERE			
900	1	-4.0	-0.82	0.23	0.047
900	9	-6.0	-0.41	0.38	0.026
900	16	-9.5	-0.48	0.80	0.041

The deformation of specimens exposed to SO_2 at 800 and 900°C differed grossly from that in air. At 800°C, neither angle closure nor dilatation was observed; while at 900°C, the angles of the ells increased and the strain rates were an order of magnitude less than those found in air. Because these weak dilatations were apparently related to the parabolic reaction, it is presumed that the opening of the ells is primarily a creep-related phenomenon induced by gravity; whereas, the small linear strains are related to the reaction process.

Apparently, the cooperative tractive forces between the scale and its substrate are greatly diminished in the sulfidizing atmosphere. This essential lack of deformation for specimens exposed to SO_2, compared to exposures in air, is the major issue which requires explanation. It is believed that a primary cause of this anomalous behavior is associated with the phenomena which lead to the inclusion of metallic particles in the scale of the sulfidized specimens (discussed below); there being no evidence for similar formations in air. A second factor which may have diminished the deformation of the sulfidized specimens is the fact that they do not, at least initially, form compact scales; rather, open bladed structures as shown in Figure 3. The generation of lateral stresses is therefore not possible in the outer reaches of these scales. Conversely, the scales formed in air are compact and cooperative tractions arising from stress generation within them may be large.

Metallographic Observations - The structure of exposed specimens was examined by optical microscopy, SEM and EDS. These analyses indicate that the 310S steel contained sigma phase embedded in an austenite matrix (5). While the scales formed in air and sulfidizing atmospheres were both bilayered oxides, their structural details differed grossly.

The specimen exposed in air for 9 days at 900°C exhibited a thin scale layer, a zone of coarse austenite crystals (denuded zone) and sigma phase particles in fine austenite at the interior. Its transverse structure is shown in Figure 4. This scale exhibits two layers parallel to the alloy-oxide interface. The inner layer is composed of fine columnar grains having a typical plan dimension of approximately 0.25 micron and a length of 3 microns. Overlying it is a layer of coarse, equiaxial grains having a typical diameter of 2 microns. Smoothly -contoured voids both at the metal-oxide interface and upon the grain boundary network of the underlying alloy are also shown. From this observation, we may infer both that there was limited plasticity of the scale layers and that the rate of scale formation was probably slowed by transport across voids. Examination by EDS techniques indicated that chromium and manganese exist at high concentration in the scale layers and are depleted in the subsurface zone. Conversely, the nickel and silicon appear not to participate in the development of the external scale. This observation may reflect the higher vapor pressures of manganese and chromium; a portion of their transport probably taking place across the void spaces.

The surface structure of an air-formed scale developed after 1 day of exposure is shown in Figure 5. It is seen that the outermost portion of that scale is composed of faceted grains. Examination of specimens exposed in air for longer times indicated that the grains exhibited an octahedral symmetry, as would be expected of spinel formation. An EDS analysis for this specimen, Table III, indicated that the granular scale is rich in both chromium and manganese. Beneath this external scale, and as revealed at a micro-spalled area, was a filamentary network of silicon-rich material which was interpreted as an internal oxide (5).

TABLE III
EDS ANALYSES OF A 310S SPECIMEN AFTER 1 DAY IN AIR AT 900°C;
SUMMARY OF RELATIVE ELEMENTAL ABUNDANCE

Chemical Element	Outer Scale	Alloy Substrate
Iron	0	100
Chromium	0	40
Nickel	0	22
Manganese	39	0
Silicon	0	5.5

The air-formed scales developed after 9 days exposure at 900°C are more mature versions of those just discussed. Figure 6 illustrates a transverse section of such a scale clearly showing its two-layered structure. The outer layer exhibits relatively large blocky grains, while the inner layer was of much finer structure. Voids are again evident upon the grain boundary network of the alloy. Details of the scale structure are more clearly evident in Figure 4 where the inner layer appears to be composed of fine columns approximately 0.25 micron in width.

The results of an EDS analysis of this section are summarized in Table IV. The outer layer of the scale contains only the metals chromium and manganese, while the inner one contains essentially only chromium. It is noted that the denuded zone, beneath the external scale, is relatively poor in chromium as expected. This description of the scale is reinforced by data gathered from an EDS analysis done on the external surface of the scale. The chemistry of the inner layer was determined at an area where micro-spalling had occurred.

TABLE IV
EDS ANALYSES Of A 310S SPECIMEN AFTER 9 DAYS IN AIR AT 900°C;
SUMMARY OF RELATIVE ELEMENTAL ABUNDANCE

		A) Transverse Section		B) Normal Section	
Chemical Element	Denuded Zone	Inner Scale	Outer Scale	Inner Scale	Outer Scale
Iron	100	1	0	2	0
Chromium	54	100	100	100	100
Nickel	25	0.2	0	0.5	0
Manganese	0.6	0.2	52	3	42
Silicon	0.8	0.4	0.6	0.7	0

The topography of the scales formed in the SO_2 atmosphere differed markedly from those developed in air. Most striking was the formation of blade-like products at the gas-oxide interface. A photograph which depicts this structural aspect is shown in Figure 3 for a 1-day exposure at 900°C. For longer exposure times, the structure also exhibits blades; however, nodules of a second reaction product formed between them and the nodules dominate this structure; see Figure 7.

The scale chemistries after a 1 day exposure to SO_2 at 900°C are summarized in Table V. It is seen that the surface is chromium-rich and that it exhibits an appreciable concentration of manganese. These data suggest that manganese may initially be associated with blade formation and that, at later times, some of the manganese leaves the blade sites to develop the nodular product. The concept of manganese-rich nodules was reinforced by EDS examination of specimens exposed for longer times at 900°C. The relative abundance of manganese in the overall scales was found to increase strongly as the exposure times increased; the manganese concentration being half that of the chromium in the nodules per se after 9 days.

TABLE V
EDS ANALYSES OF 310S SPECIMENS SULFIDIZED FOR 1 DAY AT 900°C.
SUMMARY OF RELATIVE ELEMENTAL ABUNDANCE

Element	Integral Chemistry	Blade Chemistry
Iron	4	7
Chromium	100	100
Nickel	2	2
Manganese	22	15
Silicon	0.7	0.7
Sulfur	0	0

This type scale evidently remains protective for times of up to 400 hours, as indicated by the low parabolic rates in the sulfidizing atmosphere. Indeed, a parallel behavior is exhibited by this alloy in high-BTU coal gassification atmospheres where the scale remains protective for up to 3,000 hours before its breakdown occurs (17). That time, approximately 10 times longer than the tests here, is unfortunately also far shorter than the target design life of certain coal gassifier components. Thus, the focus of future research must shift to the determination of why and how such breakdown occurs.

To further investigate the scale structure developed in SO_2, a transverse section of the specimen exposed for 9 days at 900°C was prepared. It exhibits a two-layer scale, as is shown in Figure 8. Of special interest are the bright spheroidal particles of metallic character embedded in the inner layer of scale near the metal-oxide interface; a feature not observed in air-oxidized specimens. The results of an analysis of that transverse section are summarized in Table VI.

The particles appear to be rich in chromium; however, this is believed to be an artifact due to signal "spill-over" from the surrounding chromium-rich inner scale layer. While similar sulfide shapes and dispositions have been observed for nickel-base alloys exposed at similar temperatures (18), the data here clearly indicate that the particles are not sulfides. Rather, these features are identified as austenite particles which are believed to represent the remnants of alloy grains.

It is postulated that the austenite grains had been encircled, first by sulfide and later oxide; being encapsulated as the scaling proceeds. This process could involve the transport of sulfur through the oxide grain boundary

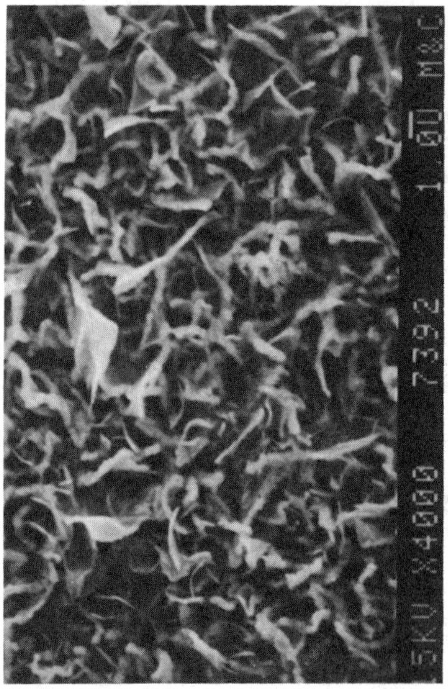

Figure 3. Blade-like surface structure of scale developed in SO_2 after one day at 900°C.

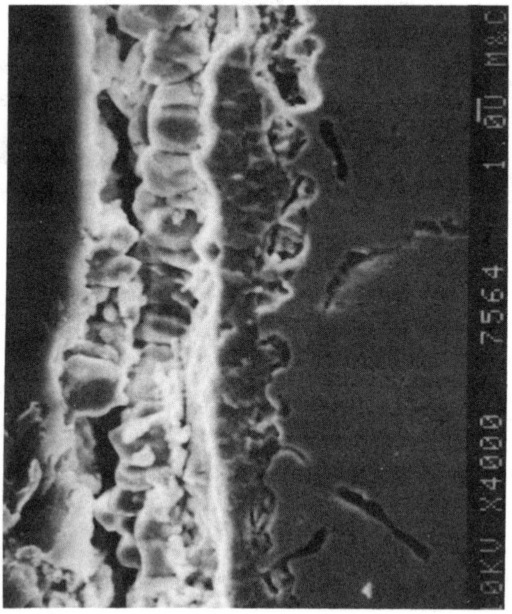

Figure 4. Transverse section of two-layer scale development in air after 9 days at 900°C.

Figure 5. Surface structure ot scale developed in air after one day at 900°C; note idiomorphic crystals.

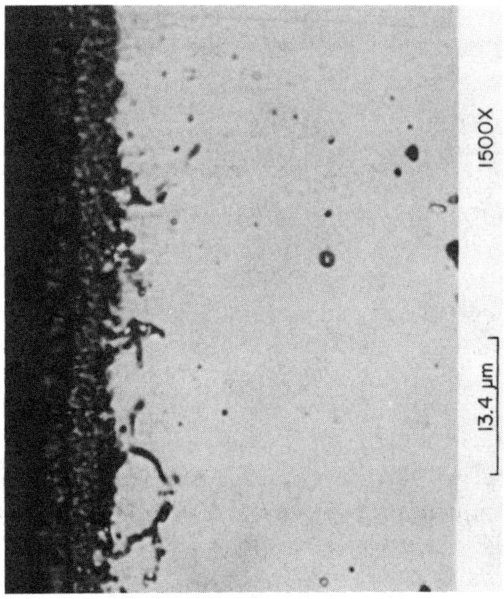

Figure 6. Transverse section of scale developed in air after 9 days at 900°C; note no entrained metal.

network. Given a large volume concentration of relatively small particles so trapped in the insulating scale layer, the local temperature may rise rapidly as their oxidation continues (19), thus enhancing the possibility of the breakdown of an otherwise protective scale layer. We propose that one portion of the breakdown mechanism of protective scale layers in sulfidizing atmospheres may involve such sulfur transport and reaction in the scale and alloy grain boundary networks.

TABLE VI
EDS ANALYSES OF A 310S SPECIMEN AFTER 9 DAYS IN THE SULFIDIZING ATMOSPHERE AT 900°C; SUMMARY OF RELATIVE ELEMENTAL ABUNDANCE

Chemical Element	Spheroidal Particle	Inner Layer	Outer Layer
Iron	30	2.1	0
Chromium	100	100	100
Nickel	9.01	0.8	0
Manganese	5.9	6.1	54
Silicon	2.0	1.1	1.9
Sulfur	0	0	0

The precise mechanism by which sulfur traverses the scale remains an open question. It has been shown that, for an Fe-25 Cr-10 Al alloy at 900°C, sulfur associated with the grain boundary networks of the oxides slows the chromia formation rate while increasing that of the alumina (20). As chromia may thicken by the transport of cations along that same path; it appears as if sulfur may serve as a temporary blocking agent. Ultimately, however, sulfur will proceed to the alloy-oxide interface, and the protective nature of the barrier will be lost. For the case of unalloyed iron which had been completely oxidized, and subsequently exposed to a sulfur-bearing gas, it was found that the sulfur was associated only with the grain boundary network of the scale and that it had completely penetrated the oxide body in a matter of minutes at 900°C (21). Further, in diffusion experiments involving chromia per se, it was found that the diffusion of sulfur via grain boundaries was very much larger than that through the bulk oxide (22). These observations indicate that the transport of sulfur via oxide grain boundaries is a phenomenon which must be considered.

Other mechanisms for the transport of sulfur through scales have been considered, including: its dissolution in the oxide followed by diffusive transport and its transport as a gas via undefined scale discontinuities (23). Obviously, further work will be required to define the effective transport paths of sulfur and, subsequently, eliminate them in scales formed in sulfidizing applications.

X-Ray Diffraction Observations - Both room temperature and 900°C investigations were conducted for the purpose of determining the crystalline nature of the surface compounds. The results of these investigations are summarized in Table VII. For one specimen simultaneously oxidized and x-rayed in air at 900°C for 5 hours, the primary reaction products were: a spinel oxide similar in structure to $MnFe_2O_4$ (the external scale layer) and a sesquioxide similar in structure to $Cr_{1.3}Fe_{0.7}O_3$ (the inner scale layer). These scales were evidently highly stressed initially, as the (311) interplanar spacing of the spinel increased while the (104) spacing of the sesquioxide decreased monotonically during the initial few hours of scale development (5).

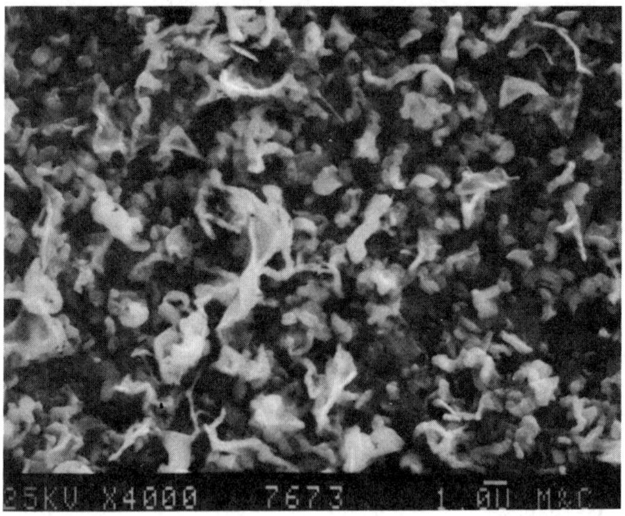

Figure 7. Surface structure of scale developed in SO_2 after 16 days at 900°C; note nodule development.

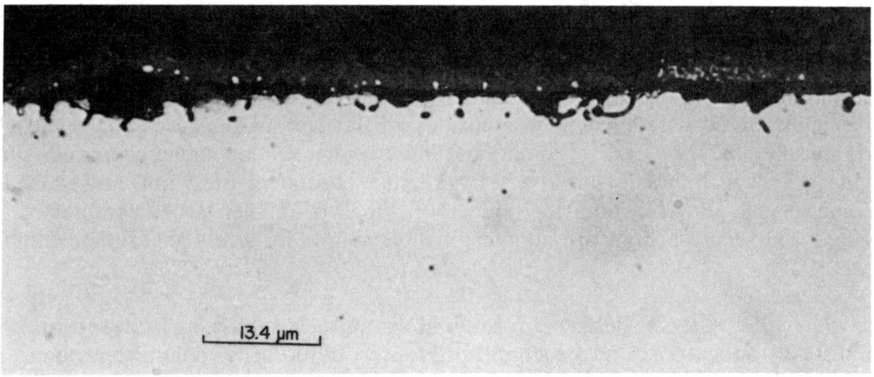

Figure 8. Transverse section of scale developed in SO_2 after 9 days at 900°C; note presence of entrained metal.

Inspection of the specimen oxidized in air for 1 day at 900°C by x-ray diffraction again shows the presence of both a spinel and a sesquioxide as surface compounds; however, no silicon-bearing compound was identified. This indicates that the silicon which had been found by EDS may have been either in an amorphous form or at a concentration below the limit of detectability by x-ray diffraction. While the spinel pattern was matched to $MnFe_2O_4$, the EDS analysis presented earlier indicates that it is more likely to be a compound of the general formulation $(Mn,Cr)_3O_4$. Similarly, the sesquioxide pattern was matched to $Cr_{1.3}Fe_{0.7}O_3$, while the EDS analysis indicated that it is more likely to be a distorted form of Cr_2O_3. In parallel with the data for shorter times, the compounds found after 16 days exposure in air at 900°C, are probably also an $(Mn,Cr)_3O_4$ type spinel product in the outer scale layer and a distorted Cr_2O_3 compound in the inner scale layer. Of special interest in this analysis is that two spinels, rather than one, have formed in this late stage of scale growth; see Table VII.

This result, coupled with the essentially textureless diffraction pattern as a whole, is suggestive of an incongruent solid state reaction occurring within the spinel phase as time progresses. It is possible that such reaction is encouraged by chemical gradients which occur within the scale and arising, fundamentally, from the fact that the material transport during the initial portion of oxidation may be grossly different from those which occur later. It is noted that apparently some strain does persist even in this outer layer and that it may be due to compositional gradients within that layer.

X-ray diffraction analysis of a specimen exposed in an SO_2 atmosphere for one day at 900°C indicated the very strong presence of Cr_2O_3 (undistorted) and less intense patterns representing the spinels $Mn_{1.5}Cr_{1.5}O_4$ and $MnFe_2O_4$. In view of the EDS data for this specimen, it is more probable that the latter spinel is again of the $(Mn,Cr)_3O_4$ type. For such exposure, the scale is again fundamentally bilayered; however, the underlying scale layer is a nearly undistorted form of Cr_2O_3, the lack of distortion perhaps arising as a result of the non-contiguous blade-like structures. The dominant component of the overlying spinel, which apparently nucleates upon the inner layer, has a somewhat smaller lattice parameter (.845 nm) and corresponds closely to a structure of the $Mn_{1.5}Cr_{1.5}O_4$ type. It appears that the chromia forms first as vertical blades and later grows laterally to become a continuous inner layer. During this process, nodular crystals of one or both spinels form between the blade tips which, in turn, overgrow the chromia at still later times.

From metallographic and x-ray examinations, it is clear that the scales formed in both environments are bilayered, having sesquioxides which were proximal to the alloy and which are chemically similar to Cr_2O_3. Upon these inner layers develop spinel chemically similar to $(Mn,Cr)_3O_4$ which initially has a very large lattice parameter (.855 nm) and which later undergoes a solid-state reaction to produce two spinels of smaller lattice parameters. It is suspected that such reactions are another possible source for stress generation within the scale.

Thus, while similar, the scale layers formed in air and the sulfidizing atmosphere exhibit some distinct topographic and structural differences; differences which are probably also reflected in the stress state in these scales. Another investigator has found that the 310S alloy also exhibits the "spinel over sesquioxide" scale sequence in coal gassification atmospheres (17). He has also identified the oxide phase in the grain boundary network of this alloy as a manganese-iron silicate, in agreement with the findings here regarding that filamentary network.

TABLE VII
SUMMARY OF ROOM TEMPERATURE X-RAY DIFFRACTION RESULTS FOR SPECIMENS EXPOSED AT 900°C.

Exposure Condition	Structure Type	Probable Composition	Lattice Parameter (nm)	Remark
Air, 5h	$MnFe_2O_4$	$(Mn,Cr)_3O_4$	0.855	*
	$Cr_{1.3}Fe_{0.7}O_3$	Cr_2O_3**		(110),(012)Tex
Air, 1d	$MnFe_2O_4$	$(Mn,Cr)_3O_4$	0.851	*
	$Cr_{1.3}Fe_{0.7}O_3$	Cr_2O_3**		(110)Tex
Air, 16d	$MnFe_2O_4$	$(Mn,Cr)_3O_4$	0.850	Two Spinels*
	$MnFe_2O_4$	$(Mn,Cr)_3O_4$	0.846	Two Spinels*
	$Cr_{1.3}Fe_{0.7}O_3$	Cr_2O_3**		(110)Tex
SO_2, 1d	Cr_2O_3	Cr_2O_3		(104)Tex
	$MnFe_{1.5}Cr_{1.5}O_4$	$MnFe_{1.5}Cr_{1.5}O_4$	0.845	Two spinels*
	$MnFe_2O_4$	$(Mn,Cr)_3O_4$	0.850	Two spinels*

(*) : Spinels exhibit essentially random texture and specular reflection.
(**): Probably a distorted form of chromia.

Extension of the Results - For the scales formed on 310S to remain protective in application, the oxygen pressure in the gas must exceed the equilibrium dissociation pressure of the dominant protective scale product, Cr_2O_3, by approximately 10^3 times (24). This heuristic limitation infers that the activity of sulfur must not reach the critical value for compound nucleation in the region of the alloy-oxide interface; there being no simplistic relationship between that activity and the activity of gases at the gas-oxide interface (25). Thus, the problem of preventing scale breakdown in the long term must focus upon the quality of the protective oxide barrier with respect to sulfur transport; a kinetics consideration most probably modulated by details of the scale structure. For the case of iron-base alloys, relatively little is known regarding the relationship between scale microstructure and reaction kinetics (26); however it is believed that fine chromia grain sizes lead to faster rates of reaction for alloys such as Fe-25Cr (27) - an alloy similar to 310S.

While only trace amounts of sulfur were detected in scales formed in the sulfidizing atmosphere, it is believed that sulfur does play a role in the evolution of the scale microstructure. It is well known that sulfur can permeate the protective oxide scales formed on many otherwise corrosion-resistant alloys where it subsequently becomes associated with the alloy grain boundary network and prevents the maintenance of the initial protective scale layer (28). As the zone beneath the scale is depleted in chromium, nickel sulfide formation may at later times be promoted which can lead to a breakaway reaction (29). While the presence of sulfur in the alloy grain boundaries was not confirmed, we believe that this transport route does play a role in the breakdown of initially protective scales.We have shown that in an SO_2 environment the inner chromia layer evolves from a network of finely-spaced blades which, presumably, grow laterally to produce a consolidated structure of large grain boundary area (fine grain size). We hypothesize that these boundaries may serve as short-circuit diffusion paths for the accelerated transport of both: a) manganese, in the early stages of scale formation and b) sulfur, in the latter stages of scale formation. Such transport would help to explain our observations regarding the development of the manganese-rich outer scale layer and the metallic particles encapsulated in the scale, as discussed above.

The fine grain size exhibited by the chromia scale layer is, on this basis, undesirable. It is therefore suggested that methods for developing coarser oxide grain sizes for more protective surfaces may be a fruitful area for future research.

CONCLUSIONS

Sheet specimens of Type 310S stainless steel of known dimension and form were exposed to air and a sulfidizing atmosphere at temperatures near 900°C. Based upon the analyses of these specimens by various techniques, the following conclusions are drawn:

1) The scales formed in air and a sulfidizing atmosphere are protective and non-spalling.
2) Only those scales formed in air provide the cooperative interfacial forces required to produce uniform dilatation and deformation of the substrate alloy.
3) The scales formed in both atmospheres are highly stressed at 900°C. The stresses appear to be long-range in air and more localized in the sulfidizing atmosphere.
4) The scales produced in the sulfidizing environment appear to mock those produced in actual coal gassification applications.
5) Two-layer scales are formed in both atmospheres; the inner layer being a sesquioxide and the outer layer being a spinel. However, the structural details of the scales formed in these two cases is grossly different.
6) Metallic particles, found embedded in the protective scales formed in the sulfidizing atmosphere only, may result from the grain-boundary transport of sulfur through the scale and may be a precursor to late-occurring scale breakdown.
7) The control of protective oxide grain size may prove to be a fruitful approach in the future development of sulfidation resistant alloys.

REFERENCES

(1) P.H.ABLESON, Am. Scientist, 75, 584 (1987).
(2) H.S.HSU, ORNL-6323 (Oct.1986).
(3) N.BIRKS, G.H.MEIER & F.S.PETTIT, J.Metals, 39, 28 (1987).
(4) R.H.KANE: Process Industries Corrosion, B.J.Moniz & W.I.Pollock Eds, NACE, Houston, TX (1986).
(5) J.S.WOLF, O.B.CAVIN and J.H.DEVAN, ORNL TM (In press).
(6) ASM Metals Reference Book, 2nd Ed., Am. Soc. for Metals, Metals Park, OH (1983).
(7) ORNL Analytical Chemistry Division; Analysis No. 880125-111 (1988).
(8) J.S.WOLF, Doct. Diss., Univ.of Fla., April 1965.
(9) J.S.WOLF, NASA TN D-5266, June, 1969.
(10) K.K.KELLY, Bureau of Mines, Bull. 406, 1937.
(11) J.P.COUGHLIN, Bureau of Mines, Bull. 542, 1954.
(12) S.R.SHATYNSKI, Oxidation of Metals, 11, 307 (1977).
(13) F.GESMUNDO, C.deASMUNDIS & C.BOTTINO, Oxidation of Metals, 14, 15 (1980).
(14) T.C.TIEARNEY & K.NATESAN, J. Matls. for Energy Syst., 1, 13 (1980).
(15) S.MROWEC & K.PRZYBYLSKI, High-Temperature Matls.and Processes, 6, 1 (1984).
(16) P.I.WILLIAMS, R.G.FAULKNER, L.W.PINDER & D.J.LEES, Corr.Sci., 27, 595 (1987).
(17) R.A.PERKINS, High Temperature Corrosion, NACE-6, 345 NACE, Houston, TX (1983).
(18) V.NAGARAJAN, R.D.SMITH and I.G.WRIGHT, Oxidation of Metals, 31, 123 (1989).
(19) J.S.WOLF, AFML-TR-74-265, April 1975.
(20) Y.-K.KIM, K.PRZYBLSKI & G.J.YUREK, Fundamental Aspects of High Temperature Corrosion II, p 259, Electrochem.Soc., Pennington, NJ (1986).
(21) F.C.YANG & D.P.WHITTLE, Corr.Sci., 23, 285 (1983).
(22) M.BENYLAMANI, F.AJERSCH and G.KENNEDY, J.Electrochem. Soc., 136, 843 (1989).
(23) P.SINGH & N.BIRKS, Oxidation of Metals, 19, 37 (1983).
(24) K.NATESAN, Corrosion, 41, 646 (1985).
(25) F.S.PETTIT, J.A.GOEBEL & G.W.GOWARD, Corr.Sci., 9, 903 (1969).
(26) K.ISHIGURO & T.HOMMA, High Temperature Corrosion, NACE-6, 28, NACE, Houston, TX (1983).
(27) M.J.GRAHAM, High Temperature Corrosion, NACE-6, 139, NACE, Houston, TX (1983).
(28) J.STRINGER, A.J.MINCHENER, D.M.LLOYD & H.R.HOY, Proc. of the Sixth Intnl. Conf. Fluidized Bed Combustion, 2, 433, USDOE (1980).
(29) R.C.LOBB & H.E.EVANS, Oxdation of Metals, 15, 371 (1981).

External and internal sulphidation of ferritic and austenitic alloys in H_2-H_2O-H_2S mixtures after preoxidation

H.J. Grabke and R.E. Lobnig
Max-Planck-Institut für Eisenforschung GmbH,
Düsseldorf, Federal Republic of Germany

Abstract

Chromium oxide layers on high temperature alloys may fail in oxidizing and sulfidizing atmospheres by external sulfidation, sulfide growth on top of the scale, or by internal sulfidation underneath the scale. The mechanisms of these processes have been investigated by exposure of ferritic Fe-20Cr and austenitic Fe-20Cr-12Ni alloys to sulfidizing/oxidizing H_2-H_2O-H_2S mixtures after preoxidation in H_2-H_2O at 973 and 1173 K. Chromium depletion upon oxidation of the austenitic alloy leads to higher Fe^{2+} content in the scale and extensive FeS formation on top of the scale above the alloy grains, on the ferritic alloy mainly Cr_3S_4 is formed above the alloy grain boundaries. The extent of internal sulfidation is minor, sulfur is transported to the alloy only by gaseous transport through cracks and pores, in the ferritic alloy sulfur ingress leads to precipitation of sulfides beneath the scale - in the austenitic alloy sulfur causes internal sulfidation along the grain boundaries. The external sulfidation appears to be more harmful and critical for the stability of a protective chromia scale.

1. Introduction

In the range of temperatures 600 to 1000°C in many applications at high temperature high alloy steels are used which rely for oxidation resistance on the formation of a slow growing, dense and well adherent Cr_2O_3-layer. In coal gasification atmospheres these high temperature alloys are exposed to an oxidizing and sulfidizing environment. Perkins (1 - 5) demonstrated that all commercial chromia-forming high temperature alloys failed and showed break-away corrosion in environment of relativily low oxygen activity and high sulfur activity even under conditions where Cr_2O_3 is the stable phase. This failure by rapid sulfide growth also occurs after preoxidation after an incubation time which depends on the thickness of the chromia layer, its mechanical integrity, on the alloy composition and on the oxygen and sulfur activities in the atmosphere. As yet it is not clear which mechanisms are decisive for the deterioration of the chromia layer (1 - 10). Two main mechanisms are in discussion:

- external sulfidation, i.e. growth of sulfides of the alloying elements Fe, Mn, Cr and Ni on top of the scale, correlated to outward diffusion of the resp. cations through the scale,
- internal sulfidation, i.e. growth of the stable sulfides of Cr, Mn, etc. beneath the scale, caused by transport of sulfur through the scale via diffusion of gas molecules (H_2S) through pores or cracks.

The external sulfidation was already observed by Perkins (1). The extent of external sulfidation is partially related to the amount of spinels present in the scale (11 - 14), inclusions of spinels offer fast diffusion paths for the outward diffusion of cations. But external sulfide formation should be possible in principle also on chromia scales, since Cr^{3+} and extraneous ions such as Fe^{2+}, Mn^{2+}, Ni^{2+} can diffuse through chromia. In this study the external sulfidation on chromia scales was investigated which were produced by preoxidation at low oxygen pressures where spinels could not be formed. Concerning the internal sulfidation it can be stated that sulfur is virtually insoluble in chromia, as was proved by radiotracer studies using ^{35}S with high sensitivity (15). Thus for internal sulfidation the sulfur must be tranported through pores or cracks of the chromia scale by a sulfur bearing gas, in this case H_2S. The investigations on internal sulfidation after preoxidation described therefore are a test on the imperviousness of the scale.

The studies were intended to give information on the severity of both mechanisms, external and internal sulfidation, and their role in the failure of carefully preoxidizid alloys.

2. External Sulfidation After Preoxidation

2.1 Preoxidation und sulfidation/oxidation conditions.

To investigate the external sulfidation after preoxidation samples of Fe-20Cr and Fe-20Cr-12Ni were oxidized in H_2-H_2O mixtures under conditions where Cr_2O_3 is the only oxidation product: at 973 K at 10^{-27} bar O_2 and 1173 K at 10^{-21} bar O_2, at both temperatures for 150 h. A scanning electron micrograph of a scale produced at 1173 K shows a continuous layer, by EDX no other elements than Cr were detected in the scale and XRD analysis confirmed that the layer consisted only of Cr_2O_3. The oxide scales also were analysed by secondary neutrals mass spectrometry (SNMS), see Fig. 1, the sputter profile of Fe-20Cr shows some Si and Fe in the scale, the sputter profile of Fe-20Cr-12Ni indicates a markedly higher Fe-content and also Si, both enriched near the surface. The higher Fe-content in the scale on the austenitic alloy is correlated to the higher Fe-activity below the scale, caused by Cr-depletion in the oxidation of the austenite. The diffusion of Cr is slower by a factor 100 in austenitic steels compared to ferritic steels, so that Cr-depletion in Fe-20Cr-12Ni leads to higher iron activity below the scale compared to Fe-20Cr.

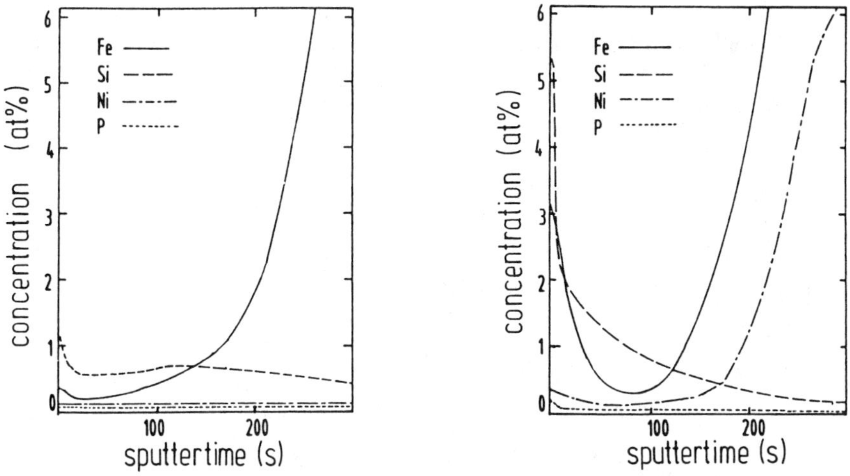

Fig. 1 SNMS-sputter profiles of samples oxidized at 973 K and $pO_2=10^{-27}$ bar
a) Fe-20Cr b) Fe-Cr-12Ni

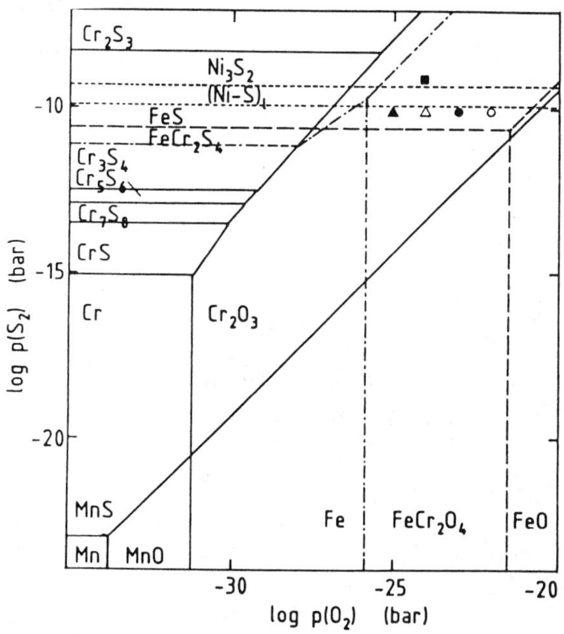

Fig. 2 Thermochemical stability diagram, superimposed stability fields of the systems Fe-S-O, Mn-S-O, Cr-S-O and Ni-S-O at 973 K
Atmospheres investigated are indicated by symbols
(1) ▲ , (2) △ , (3) ● , (4) ○ , (5) ■

After preoxidation the samples were exposed to sulfidizing-oxidizing conditions by changing the atmosphere at reaction temperature. The following conditions have been applied:

at 973 K	log pO_2	log pS_2
(1)	-25	-10
(2)	-24	-10
(3)	-23	-10
(4)	-22	-10
(5)	-24	- 9
at 1073 K		
(6)	-21.6	- 8.5

The conditions of exposure at 973 K are indicated in the thermochemical stability diagram Fig. 2, formation of FeS is possible, in atmosphere (5) also Ni-sulfide may be formed.

2.2 Investigations of ferritic alloys after sulfidation/oxidation

After 200 h exposure in atmosphere (1) at 973 K on Fe-20Cr large amounts of chromium sulfide Cr_3S_4 had formed on the chromia layer, especially above the grain boundaries of the alloy Fig. 3. In spite of the fact that the formation pressure of FeS was exceeded, after this experiment no FeS was observed. On the alloy Fe-20Cr-0,1Ce the formation of the external chromium sulfides was somewhat retarded.

In atmosphere (2) at an increased oxygen pressure also some chromium sulfide was detected after 200 h but its amount was markedly decreased. However, iron sulfide had grown on top of the chromia layer, especially above the grains of the alloys, the regions above the grain boundaries of the alloy were free of FeS crystals, see Fig. 4. The amount of sulfides is much less on the alloy with Ce. After exposure in this environment for 25 h sulfide formation was not yet detected, after 100 h external FeS growth was observed, but only above few alloy crystals.

Upon increasing the oxygen partial pressure further, in atmosphere (3) and (4) the external sulfide formation is suppressed more and more. After 200 h in atmosphere (3) only very small amounts of sulfides are detected on the chromia layer of the ferritic alloy. After 200 h exposure in atmosphere (4) the chromia layer is not affected at all, no sulfur could be detected in the layer.

At an increased sulfur pressure, in atmosphere (5), the phenomena are similar as described, however, FeS formation is favored in relation to chromium sulfides. At increased temperature, 1073 K, an exposure was performed in an environment corresponding to atmosphere (1) concerning the distances to the equilibrium-lines Fe-FeS and Cr-CrS. Also under these conditions the Cr_2O_3 layer on the ferritic alloys is rather protective, only small amounts of Cr_3S_4 were observed on Fe-20Cr, whereas on the oxide layer of Fe-20Cr-0.1Ce no sulfide was detected after 200 h exposure.

2.3 Investigations of austenitic alloys after sulfidation/oxidation

After 200 h exposure in atmophere (1) the alloy Fe-20Cr-12Ni showed thick

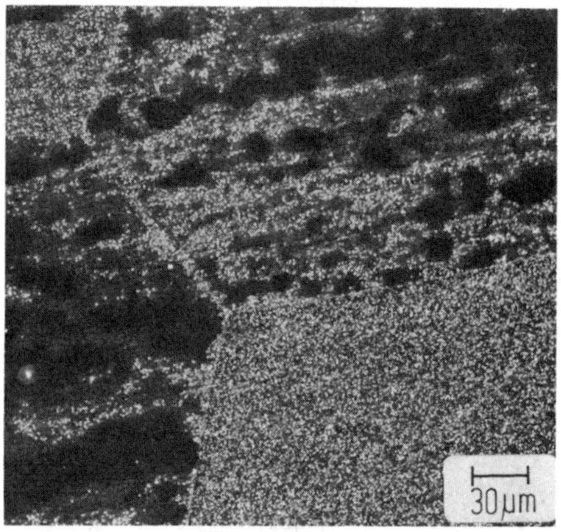

Fig. 3 SEM micrograph of the surface of Fe–20Cr–0.01Ce after exposure at 973 K $pO_2 = 10^{-25}$ bar, $pS_2 = 10^{-10}$ bar

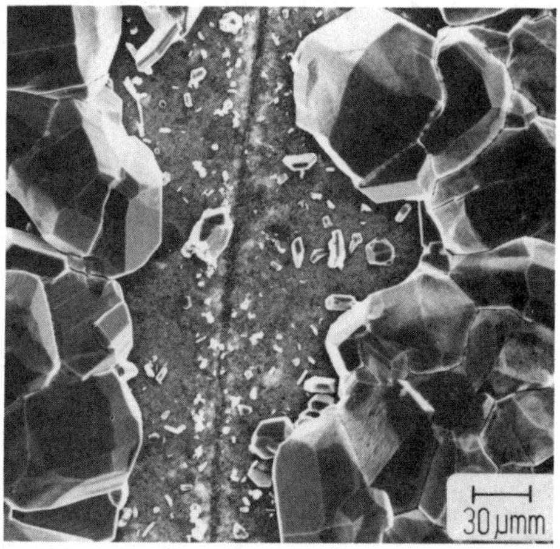

Fig. 4 SEM micrograph of the surface of Fe–20Cr–0.01Ce after exposure at 973 K, $pO_2 = 10^{-24}$ bar, $pS_2 = 10^{-10}$ bar

Fig. 5 SEM micrograph of the surface of Fe–20Cr–12Ni after exposure at 973 K $pO_2 = 10^{-25}$ bar, $pS_2 = 10^{-10}$ bar

layers of FeS, Fig. 5, only small areas were free of FeS and covered with chromium sulfides. For the alloy Fe-20Cr-12Ni-0.002Ce the sulfidation was somewhat retarded, after 100 h only small amounts of FeS were present on this alloy whereas the Ce-free alloy already showed a thick layer of FeS.

In atmosphere (2) at a somewhat higher oxygen pressure the amount of sulfides formed is less but the distribution is similar. Above the grain boundaries of the alloys again only small amounts of chromium sulfides are observed but no FeS.

For the austenitic alloys the oxygen pressure must be raised higher to suppress external sulfide formation, even in atmosphere (4) after 200 h some iron sulfides could be indentified.

At the elevated sulfur pressure, in atmosphere (5) two sulfide layers are observed, an outer layer of mixed sulfide (Fe,Ni)S and an inner layer of FeS. The exposure in atmosphere (6) at 1073 K led to strong sulfidation, the samples were covered completely with FeS, no Cr_2O_3 could be detected below this layer and internal sulfidation under formation of chromium sulfides growing into the alloy grain boundaries had started.

2.4 Thermogravimetric experiments

The kinetics of preoxidation in H_2-H_2O-mixtures and subsequent oxidation sulfidation in H_2-H_2O-H_2S mixtures were investigated in thermogravimetric experiments at 973, 1073 and 1173 K. The preoxidation was conducted for 150 h under conditions where only Cr_2O_3 is formed, the sulfidation/oxidation followed at the same temperature. The conditions are listed in the following table

T (K)	preoxidation log $p(O_2)$	sulfidation/oxidation log $p(S_2)$	 log $p(O_2)$
973	-27	-10	-25
1073	-24	- 8.5	-21.6
1173	-21	- 7.25	-18.9

At these values the distances are about equal to the equilibrium lines CrS-Cr_2O_3 and Fe-FeS in the thermochemical stability diagrams at the different temperatures.
The oxidation kinetics of the alloy Fe-20Cr can be described at all temperatures by the parabolic rate law, the values are obtained for the parabolic rate constants:

T (K)	$k_p (g^2\ cm^{-4}\ s^{-1})$
973	$1.7 \cdot 10^{-15}$
1073	$1.4 \cdot 10^{-14}$
1173	$1.1 \cdot 10^{-13}$

After sulfidation/oxidation at 973 K the Cr_2O_3 layer was equally covered with Cr_3S_4, at 1073 K the sulfidation/oxidation led to some Cr_3S_4 on the oxide layer above the alloy grain boundaries, at 1173 K no sulfides were detected.

At all temperatures the rate of mass increase is enhanced upon changing conditions from oxidizing to sulfidizing/oxidizing (Fig. 6), the change of rate is decreasing with increasing temperature. The decrease of rate change is caused by the increase in oxide thickness with increasing temperature of preoxidation. This was proved by an experiment in which the specimen was preoxidized at 1173 K which leads to a higher oxide thickness. The linear rate constants of sulfidation/oxidation were after preoxidation at 973 K: $k_1 = 1.1 \cdot 10^{-10}$ g $cm^{-2}s^{-1}$ and after preoxidation at 1173 K much lower: $k_1 = 1.4 \cdot 10^{-11}$ g $cm^{-2}s^{-1}$ - these values are inversely proportional to the resp. oxide thicknesses, which indicates rate control by the diffusion flux of Cr-cations through the layer.

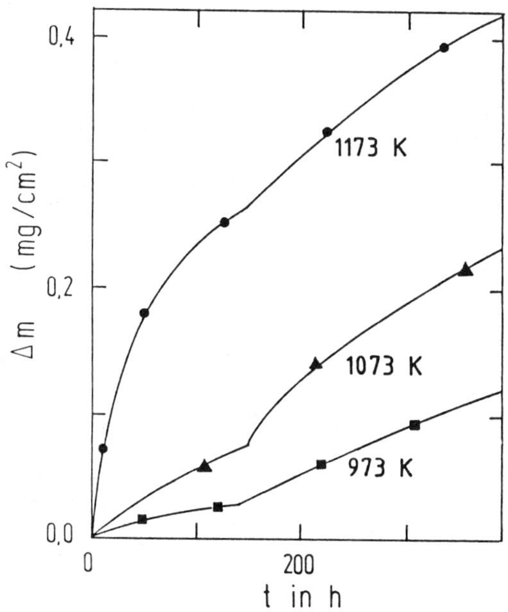

Fig. 6: Mass increase during preoxidation and during subsequent oxidation/ sulfidation of Fe-20Cr at 973 K 1073 K and 1173 K

The oxidation kinetics of the alloy Fe-20Cr-12Ni can be described at all temperatures by the parabolic rate law, the values are obtained for the parabolic rate constants:

T (K)	$k_p (g^2 cm^{-4} s^{-1})$
973	$2.6 \cdot 10^{-14}$
1073	$4.1 \cdot 10^{-14}$
1173	$1.1 \cdot 10^{-13}$

At the lower temperatures the Cr_2O_3 growth on the austenitic alloy is faster than on the ferritic alloy, which may be caused by the higher Fe-content of the scale (Fig. 1) At the low pO_2 in these investigations the iron cations will be divalent and will lead to formation of increased concentrations of interstitial chromium cations Cr_i^{3+} and increased cation diffusivities. The effect of the Fe impurity on the self-diffusivity decreases with increasing temperature.

At 973 K the sulfidation/oxidation of Fe-20Cr-12Ni is much faster than the sulfidation/oxidation of Fe-20Cr, see Fig. 7a) The high rate of mass increase is caused by FeS growth, in contrast to Fe-20Cr where growth of Cr_3S_4 was observed. Since the Cr_2O_3 layer stays on the alloy the sulfidation/oxidation rate is lower compared to the case without preoxidation. At 1073 K the difference in the sulfidation/oxidation behaviour after preoxidation for the two alloys is less, see Fig. 7b), and at 1173 K there is nearly no difference.

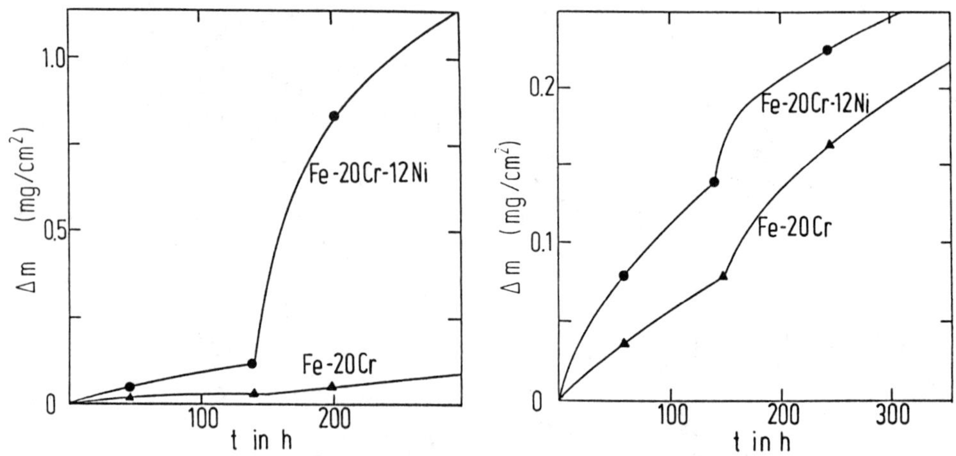

Fig. 7 Mass increase of Fe-20Cr and of Fe-20Cr-12Ni during pre-oxidation and during subsequent oxidation/sulfidation
a) at 973 K b) at 1073 K

3. Internal Sulfidation After Preoxidation

3.1 Conditions of preoxidation and sulfidation/oxidation, measurement of sulfur penetration.

To investigate the internal sulfidation after preoxidation samples of Fe-20Cr, Fe-20Cr-0.01Ce, Fe-20Cr-12Ni and Fe-20Cr-12Ni-0.002Ce were oxidized at 1173 K and $pO_2 = 10^{-20.5}$ bar 150 h. Subsequent sulfidation/oxidation was performed at the following conditions

	log pO_2	log pS_2
1173 K	-20.0	- 8.8
973 K	-26.0	-11.5

(The second experiment was conducted after an equilibration anneal of specimens preoxidized at 1173 K at $pO_2 = 10^{-26}$ and 973 K). The gas compositions were chosen in a range where Cr_2O_3 is the stable phase but no external FeS growth is possible. Only formation of chromium sulfides and cerium sulfide beneath the oxide scale was to be expected, in case that sulfur can penetrate the scale somehow.

For detection of the sulfur penetrated the most sensitive method was applied, measurement of the radioactive tracer ^{35}S which had been doped into the gas mixture. Concentration profiles of the sulfur coming from the gas atmosphere and penetrating through the oxide layer into the alloy were determined using the method of residual activity measurement after stepwise polishing the surface of the samples. With this method absolute values of the sulfur content could be obtained, which had entered the material from the gas phase with high sensitivity.

3.2 Results on sulfur penetration through chromia layers.

The chromia layer after 150 h preoxidation and 700 h sulfidation/oxidation looked still very dense and protective, no sulfur could be detected on this layer by EDX. The measurements of the radioactive tracer led to the data shown in Fig. 8 and given in the following table, which compiles data on the amount of sulfur uptake from surface to the inner of samples and the penetration depth.

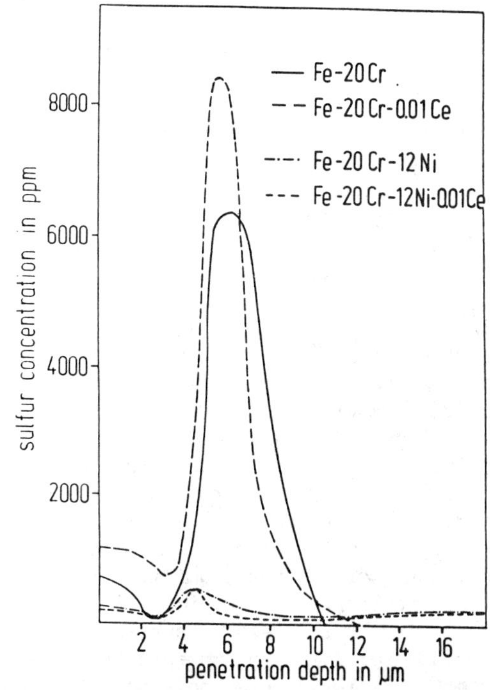

Fig. 8 Sulfur penetration profiles in preoxidized samples after oxidation/sulfidation at 1173 K for 700 h

	Fe-20Cr	Fe-20Cr-0.01Ce	Fe-20Cr-12Ni	Fe-20Cr-12Ni-0.002Ce
973 K				
μg	3.5	2.2	1.2	55
μm	57	54	1.1	62
1073 K				
μg	18.0	19.9	11.0	5.8
μm	352	344	350	338

Generally, the sulfur ingress is at 1173 K about 5 to 10 times higher than at 973 K, and the values for the austenitic alloys are about 2 to 3 times smaller than for the ferritic alloys. The additions of Ce clearly decrease the sulfur ingress.

The profiles of sulfur penetration at 1173 K, shown in Fig. 8, also demonstrate the higher sulfur uptake in the ferritic alloys. Furthermore, the penetration profiles indicate a sulfur enrichment at the oxide/metal interface, which can be correlated to the formation of sulfide precipitates at this interface. Precipitation is possible here of sulfides chromium and of cerium sulfides. In the interior of the samples the sulfur content is low, autoradiographs have shown that the sulfur diffuses inward preferably at grain boundaries, internal sulfidation along grain boundaries takes place.

4. Discussion

Upon exposure of an alloy covered with a chromia scale to a sulfidizing-oxidizing atmosphere external and internal sulfidation are possible, i.e. growth of sulfides on top of the scale and sulfide formation below the chromia scale.

For external sulfidation outward diffusion of cations must occur through the scale, the metal activity and sulfur activity on the surface of the scale must be high enough for sulfide formation. The chromium depletion below the oxide scale on the austenitic alloys leads to higher Fe-contents in the scale and to higher cation diffusivities. In the range of oxygen pressures investigated the disorder in Cr_2O_3 is governed by the formation of interstitial Cr_i^{3+} ions

$$Cr_2O_3 = 2Cr_i^{3+} + 6e^- + 3/2\ O_2$$

which is favored by introduction of Fe^{2+} ions on Cr lattice sites. Thus, oxidation rates result on the austenitic alloy which are higher than on the ferritic alloys - and also the external sulfide formation is much more extensive on the austenitic alloys. Extensive FeS growth was observed on the chromia scale of the austenitic alloys, mainly above the centre of the alloy grains. This observation is explained by the chromium depletion and increased iron activity both being more marked in the centre of the grains. On the ferritic alloys predominantly the sulfide Cr_3S_4 was formed, in spite of the fact that the reaction conditions were in the stability field of Cr_2O_3. This observation indicates a nonequilibrium situation, a much higher chromium activity and lower oxygen activity on the scale surface than expected in equilibrium with the gas atmosphere. The chromium sulfide preferentially grew above the grain boundaries of the alloy, where the supply of chromium from the bulk alloy is most rapid by grain boundary diffusion.

The rate of mass increase is enhanced immediately after the addition of the sulfidizing component to the gas mixture, this effect is valid for both kind of alloys. The increase in corrosion rate must be correlated to an increase in the outward diffusion flux of cations. Upon formation of external sulfides a decrease of oxygen activity may occur on the oxide layer which leads to an enhanced concentration of Cr_i^{3+} and faster cation diffusion. According to the defect equilibrium the diffusivity of Cr_i^{3+} increases with decreasing pO_2 in a range of low oxygen activities. Such change in interfacial oxygen activity, defect concentration and diffusivities in the chromia layer may explain the increase in corrosion rate in the sulfidizing/oxidizing environment, however, the abruptness of the effect is difficult to understand.

For internal sulfidation inward transport of sulfur through the scale must occur, since it was proved that the solubility of S in Cr_2O_3 is virtually nil /15/ the sulfur must be tranported by gaseous diffusion or surface diffusion through cracks or pores of the scale. The results indicate that the oxide scales on the ferritic alloys are much more porous and less protective than the scales on the austenitic alloys, since the overall sulfur uptake is much less for the austenitic alloys. In the ferritic alloys the sulfur is scavenged immediately below the oxide scale by precipitation of chromium sulfides and/or cerium sulfides. In the austenitic alloys sulfur penetration was observed to greater depth in the alloy, this may be correlated to the chromium depletion below the scale - the sulfur can penetrate deeper into the alloy before being precipitated, especially at the grain boundaries.

5. Conclusions

The scale on chromia forming high temperature alloys may fail in oxidizing-sulfidizing environments by external sulfidation, i.e. growth of sulfides on top of the scale, or by internal sulfidation, i.e. formation of sulfides below the scale. The morphologies of sulfidation after preoxidation and the mechanisms of these processes were investigated.
The ferritic alloy Fe-20Cr and the austenitic alloy Fe-20Cr-12Ni were preoxidized at 900°C in a H_2-H_2O mixture at low pO_2 to obtain a Cr_2O_3 layer. Analysis by SNMS showed a somewhat higher concentration of iron in the layer on Fe-20Cr-12Ni.
Subsequent exposure to oxidizing-sulfidizing H_2-H_2O-H_2S mixtures at 700°C and at a pS_2 at which FeS can be formed, leads to extensive growth of FeS on top of the Cr_2O_3 layer in case of the austenitic alloy and formation of small fringes of Cr_3S_4 on the chromia layer in the case of the ferritic alloy. The FeS in the first case is situated above the middle of the alloy grains and the Cr_3S_4 in the second case is situated above the alloy grain boundaries. The different behaviors are explained by differences in element distribution and diffusivities in the scale and in the alloy.
Exposure after preoxidation to oxidizing-sulfidizing H_2S-H_2-H_2O mixtures at 900°C and at a pS_2 at which only chromium sulfides can be formed, leads to internal sulfidation but only to a very small extent. The sulfur uptake was monitored by radioactive ^{35}S and is in the range of a few µg's after 700 h. Precipitates are formed below the scale, the sulfur uptake is somewhat higher for the ferritic alloy. Differences in sulfur penetration are determined by the scale porosity, since no sulfur is dissolved in Cr_2O_3 and must be transported by gaseous or surface diffusion through cracks and pores.
The most dangerous process in the failure of Cr_2O_3 scales seems to be the external sulfidation under formation of thick iron-, manganese- and/or nickel

sulfide deposits on the scale by outward diffusion of the resp. cations.

Acknowledgement

The study was supported by the Deutsche Forschungsgemeinschaft.

References

1. R. A. Perkins, in: Proc. Petten Int. Conf. on Alloy 800, 1978, Eds.: Betterridge et al., North Holland Publ. Company, p. 213
2. R. A. Perkins: Proceedings of the conference on corrosion/ errosion of coal conversion materials, Ed. A. V. Levy, NACE, Jan. 1979, p. 351
3. R. A. Perkins: in Proceedings of conference on 'Environmental degradation of high temperature materials', Spring Residential Conference, London, 1980, Chameleon Press, Vol. 2, 5/1
4. R. A. Perkins: in Proceedings of conference on 'Environmental degradation of high temperature materials', Spring Residential Conference, London, 1980, Chameleon Press, Vol. 2, 5/17
5. R. A. Perkins: in: 'High temperature corrosion', Ed. R. A. Rapp, 1981, NACE, Houston, Texas
6. F. H. Scott, F. M. F. Chong, C. A. Stirling: 'High Temperature Corrosion in Energy Systems', Ed. M. Rothman, Met. Soc. AIME 1985, p. 253
7. J. Stringer, Sulfidation as industrial problem and limiting factor, Vortrag, Workshop on 'The high temperature corrosion of alloys in sulfur containing environments', Petten 12/13., Dec. 1985
8. F. H. Scott, F. M. F. Chong, in: Proceedings of conference on: 'Corrosion resistant materials for coal conversion systems', Ed.: D. B. Meadowcraft and M. J. Manning, 1983, London, Applied Science Publishers, p. 491
9. D. J. Baxter, K. Natesan: Corros. Sci., 26, 1987, 153
10. D. J. Baxter, K. Natesan: Oxid. Met., 24, 1985, 331
11. R. E. Lobnig, P. Papaiacovou, H. P. Schmidt und H. J. Grabke, in 'The Role of Active Elements in the Oxidation Behaviour of High Temperature Metals and Alloys', Ed.: E. Lang, Elsevier Applied Science, 1989, p. 317 - 333
12. H. J. Grabke, R. E. Lobnig, P. Papaiacovou, in: 'Selected Topics in High Temperature Chemistry', Ed. O. Johannesen, A. G. Andersen, Elsevier Applied Science, 1989, p. 263 - 289
13. R. E. Lobnig, H. P. Schmidt and H. J. Grabke, Materials Science and Engineering, A. 120 (1989) 123 - 128
14. R. E. Lobnig, H. J. Grabke, Corrosion Science, in press.
15. R. E. Lobnig 'Mechanismen der Oxidation and Sulfidierung der Modellegierungen Fe-20Cr und Fe-20Cr-12Ni' Thesis Universität Dortmund,1987

Kinetics of internal and external oxide formation for 316L and 310 stainless steels between 800°C and 1200°C

S. Gendron and F. Ajersch
Département de génie métallurgie, École Polytechnique de Montréal, Montréal, Québec, Canada, H3C 3A7

ABSTRACT

The oxidation of 316L and 310 steels was studied in CO-CO_2 gas mixtures in the temperature range from 800° and 1200°C using a microbalance. The parabolic rate constants (k_p) were found to be practically independent of the oxygen partial pressure in the range of the gas mixtures employed (Po_2 = 1.06 x 10^{-3} to 2.32 x 10^{-17} atm). A variation of alloy grain size from 30 to 100 μm was found to have no effect on the oxidation kinetics. The mean value of k_p for 316L was found to be 5 x $10^{-9} g^2 cm^{-4} s^{-1}$ at 1000°C with an activation energy of 58±5 Kcal mol^{-1} compared to 2 x $10^{-12} g^2 m^{-4} s^{-1}$ with an average activation energy of 62±5 Kcal mol^{-1} for 310 alloys. The 316 alloy (17%Cr) does not contain enough chromium for the formation of a compact protective layer of Cr_2O_3 whereas the 310 alloy (25%Cr) forms a protective external scale in addition to the internal oxide. Both morphological and quantitative chemical analysis of the various oxide and metal phases were characterized by SEM and EDX analysis. Rate controlling mechanisms for the oxidation of the two alloys are proposed.

INTRODUCTION

The performance of metallic materials at high temperatures and in corrosive environments is crucial to the long term performance of materials used in energy conversion processes, in combustion systems as well as in aerospace applications. The oxidation behavior of commercial stainless steels in oxidizing, sulfurizing and carburizing atmospheres still needs to be characterized more precisely due to the varying rate controlling factors which influence the corrosion-degradation process. The diffusion of metal atoms in the alloy, metal cations or oxygen anions in the oxide-layer, the formation of internal or external oxide or spinel layers, the fissurisation and spalling of the oxide layers, and the formation of protective or healing oxides, externally or internally, all add to the complexity of the characterisation of the process, and can become very specific to alloy composition, atmosphere and temperature. The measurement and determination of the reaction rates and oxidation mechanisms for these alloys becomes essential in the choice of materials for specific applications.

A typical example of the complexity of the various interactions that influence the degradation phenomena was demonstrated in the studies of Gilmour et al (1), (2) where different types of stainless steel boiler tubes were exposed to the gases of a fluid bed coal combustion reactor. The sample tubes were exposed to an extremely corrosive hot gas containing carbon, oxygen, sulfur and nitrogen as well abrasive particles of solid or liquid combustion products (CaO or $CaSO_4$). Their results showed that 310 and 347 stainless steels performed condiserably better than the 316 alloy. They were unable to interpret their results on the basis of metallographic examination of the reaction products. The product morphology was very complex due to the formation of oxides, spinels, sulfides and also of sulfates with preferential attack at grain boundaries. This work initiated studies (3) (4) of a more fundamental nature in determining the diffusion characteristics in polycrystalline and monocrystalline chrome oxide in order to evaluate its relative resistance to cation or anion transport in sulfur and oxygen containing atmospheres, and subsequently to the behavior of 310 and 316L alloys in similar controlled atmospheres. In the present study, the oxidation characteristics of these alloys are investigated under various oxygen potentials in the range from 800°C to 1200°C.

EXPERIMENTAL

Table I presents the composition of 310 and 316L alloy as specified by the ASTM norms and which were also analysed by X-ray fluorescence (EDX) using a JEOL 840 scanning electron microscope. The samples were prepared from flat stock of 1.25 cm thickness and sectioned to produce coupons 1 cm square and 1 mm thick weighing approximately 2.0g. These samples were then mechanically polished to a finish with 0.05 alumina powder and a small hole (1.5 mm) was drilled near the outer edge of one of the sides in order to suspend the coupon in a SETARAM microbalance by means of a platinum wire. The average grain size of the 316L alloy was about 50 μm whereas two different grain sizes of 310 alloy was used, one of 30 to 40 μm and the other of about 100 μm.

TABLE I: COMPOSITION OF ALLOYS

Alloy	Method	Analysis (weight %)					
		Fe	Cr	Ni	C	Mu	Si
316L	SEM EDX	Balance	16.2	12.0	*	1.7	0.1
	ASTM norms	Balance	19-21	10-12	<0.03	<2.0	<1.5
310(fine grain)	SEM EDX	Balance	25.0	20.4	*	1.7	0.1
310(med. grain)	SEM EDX	Balance	23.5	20.8	*	1.7	0.4
	ASTM norms	Balance	24-26	19-22	<0.25	<2.0	<1.5

* not detected

Oxidation experiments were carried out at three temperatures (800°, 1000° and 1200°C) which cover the upper range of industrial applications and which tests the material at its limits. Since mass changes at higer temperatures are larger, the experiments also result in more accurate values of the oxidation rate constant measured by the microbalance and allow for a more detailed morphological analysis.

In order to be able to fix oxygen partial pressures, a series of different CO-CO_2 gas mixtures were used. The values were calculated using the F.A.C.T. system (5) for thermochemical computation of complex gas mixtures. A minimum gas flow rate of $25 cm^3 min^{-1}$ was chosen in order to avoid control of oxidation by gas diffusion though the gas boundary layer. Since the internal diameter of the reaction tube is 2.24 cm, this flow rate represents an average linear velocity of about 1.0 x 10^{-2}m s^{-1}.

The advantage of using a platinum wire to suspend the sample was that any spalling of the scale would easily be detected by sudden weight decreases. The temperature was measured by placing a Pt-Pt10%Rh thermocouple just below the sample and was maintained constant to ±2°C during the period of the experiment.

RESULTS

Thermogravimetric experiments were carried out on 40 samples of the two alloys (310 and 316L) at temperatures of 800, 1000 and 1200°C and the results are presented in Table II. Some of the samples indicated a small change of slope at various points during the oxidation process and parabolic oxidation rate constants (k_p) were calculated for each rate change. In most cases, the change of slope was not very significant and could possibly indicate a change of oxidation mechanism. A much more significant change in k_p was observed for different alloy composition and increasing temperature. Some tests were repeated up to 4 times with a total variation of not more than 15% in k_p. Individual differences in the results were attributed to experimental variations as well as to the variation of the alloy microstructure for long duration exposures.

Table III summarizes the average values of the rate constants as a function of temperature and oxygen partial pressure for each of the two alloys. The average values of k_p were plotted as a function of temperature to obtain activation energies, ΔE, for each of the alloys as shown in Figure (1). The figure clearly shows the negligible effect of oxygen partial pressure but of a significant effect of alloy composition. Oxidation rates of 3 to 4 orders of magnitude higher were obtained for 316L alloy when compared to the 310 alloy. For example, at 1000°C the average value of k_p for 316L is about 5.0 x 10^{-9} when the value for 310 alloy is about 2.0 x 10^{-12} $g^2 cm^{-4} s^{-1}$ with corresponding activation energies of 58±5 $kcal(mol)^{-1}$ and 61±5 $kcal(mol)^{-1}$ and 64±5 $kcal(mol)^{-1}$ for the two grain sizes studied.

Oxidation of 316L alloy

Two oxide layers were observed for the oxidation of 316L alloy for the range of conditions of this study. The atmospheres were kept in the range of oxygen pressures where the external oxide is always within the stability of the Fe_3O_4 phase region. Micrographs at 1200°C show that the external layer of oxide is relatively compact with some large pores occuring mostly near the interface of the external oxide layer and the internal oxide layer which also has a number of small pores. Some fissurisation can also be observed in the outer layer. Figure (2) shows that the internal oxide layer is compact with a smooth transition from the base metal to the external oxide layer. It should be noted that the thickness of the internal layer is about five times the average grain size for exposure at 1200°C for 22 hours, and that the average grain size has actually increased from 100 to nearly 200 μm during the oxidaton period. Futher examination of the internal layer shows agglomeration of a metallic phase rich in nickel, as well as a large number of small pores, as shown in Figure (3).

Table IV shows the scale thickness of the internal and external oxide layer for different periods of time and various oxygen pressures, and it can be observed that the internal oxide is two to five times the thickness of the external layer. The relative growth rate of the internal and the external

TABLE II: SUMMARY OF OXIDATION EXPERIMENTS

Sample No.	Alloy	T (°C)	%CO	k_{PI} ($g^2/cm^4.s$)	k_{PII} ($g^2cm/^4.s$)	k_{PIII} ($g^2/cm^4.s$)	k_P(ave.) ($g^2/cm^4.s$)
1	316L	1000	9.62				5.0x10^{-9}
2	316L	1000	9.62				2.2x10^{-9}
3	310(**)	1000	0.968				-
4	310(*)	1000	9.62				-
5	304	1000	0.968	3.4x10^{-9}	9.1x10^{-10}		9.1x10^{-10}
6	310(*)	1000	0.968				1.4x10^{-12}
7	316L	1000	0.968				-
8	310(**)	1000	9.62				1.3x10^{-12}
9	310(**)	1000	0.0091	2.8x10^{-12}	3.8x10^{-12}		3.3x10^{-12}
10	310(**)	1000	0.102				1.0x10^{-12}
11	310(**)	1000	0.102				1.4x10^{-12}
12	316L	1000	1.29				3.3x10^{-9}
13	316L	1000	1.29				-
14	316L	1000	1.30	5.0x10^{-9}	7.8x10^{-9}	1.2x10^{-8}	8.3x10^{-9}
15	310(*)	1000	1.30				-
16	310(*)	1200	9.22				3.4x10^{-11}
17	316L	800	10.2	7.7x10^{-11}	5.0x10^{-11}		6.4x10^{-11}
18	316L	1200	9.22	2.2x10^{-7}	2.6x10^{-7}		2.4x10^{-7}
19	316L	1200	9.22	2.2x10^{-7}	2.7x10^{-7}		2.4x10^{-7}
20	316L	1000	10.2				5.6x10^{-9}
21	316L	1000	10.2				4.4x10^{-9}
22	310(*)	1200	10.2				4.8x10^{-9}
23	310(**)	1000	0.986				9.7x10^{-12}
24	310(**)	1200	0.986	4.3x10^{-11}	1.1x10^{-10}	6.2x10^{-11}	7.2x10^{-11}
25	310(**)	1200	0.986	9.3x10^{-11}	6.1x10^{-11}	1.0x10^{-10}	8.5x10^{-11}
26	310(**)	1200	0.086	6.2x10^{-11}	1.0x10^{-10}		8.1x10^{-11}
27	310(*)	1200	0.086	5.5x10^{-11}	7.6x10^{-11}		6.6x10^{-11}
28	310(*)	1000	0.086	1.5x10^{-12}	8.0x10^{-11}		1.2x10^{-12}
29	316L	1000	0.086	2.3x10^{-8}	1.5x10^{-8}	1.1x10^{-8}	1.6x10^{-8}
30	316L	1200	0.086	3.0x10^{-7}			3.0x10^{-7}
31	316L	1200	0.086	3.0x10^{-7}			3.0x10^{-7}
32	316L	1200	0.086	2.7x10^{-7}	3.6x10^{-7}		3.2x10^{-7}
33	316L	1200	0.0086	2.4x10^{-7}	3.0x10^{-7}		2.7x10^{-7}
34	316L	1000	0.0086	3.4x10^{-3}	1.3x10^{-8}		8.2x10^{-9}
35	310(*)	1000	0.0086	5.1x10^{-11}	6.5x10^{-11}		1.8x10^{-12}
36	310(*)	1200	0.0086	5.1x10^{-11}	6.5x10^{-11}		5.8x10^{-11}
37	310(**)	1200	0.0086				2.5x10^{-11}
38	316L	1200	80.				3.1x10^{-9}
39	316L	1200	9.96				2.0x10^{-7}
40	316L	1200	9.96				2.2x10^{-7}

*: fine grain: 30-40 μm **:medium grain: 100 μm

oxides is not easy to quantify from these results due to morphological irregularities of the external oxide as evidenced by fissures, detachment from the external layer and general porosity. For example, the external layer is of a nearly constant thickness between 100 to 200 μm for exposures from 5 to 22 hours at 1200°C using 10% $CO-CO_2$ gas mixtures, whereas the internal layer increases from an average of 430 to 800 μm for the same period. It is assumed that, due to the large pores and fissures of the external layer, it offers little resistance to the transfer of oxygen from the reactive gas to the internal layer.

TABLE III: AVERAGE OXIDATION RATE CONSTANTS

Alloy	T (°C)	P_{O_2} atm	ΔE (kcal/mol)	K_P ($g^2/cm^4.s$)
316L	800	$3.01x10^{-17}$	58±5	$6.4x10^{-11}$
	1000	$6.27x10^{-13}$ $7.59x10^{-11}$ $7.73x10^{-9}$ $8.00x10^{-7}$		$5.0x10^{-9}$ $7.0x10^{-9}$ $1.6x10^{-8}$ $8.2x10^{-9}$
	1200	$8.60x10^{-10}$ $1.06x10^{-5}$ $1.06x10^{-3}$		$2.4x10^{-7}$ $3.1x10^{-7}$ $2.7x10^{-7}$
310 medium grains (100μm)	1000	$6.27x10^{-13}$ $7.59x10^{-11}$ $7.73x10^{-9}$ $8.00x10^{-7}$	61±5	$1.3x10^{-12}$ $9.7x10^{-12}$ $1.2x10^{-12}$ $3.3x10^{-12}$
	1200	$8.60x10^{-10}$ $1.04x10^{-7}$ $1.06x10^{-5}$ $1.06x10^{-3}$		$3.4x10^{-11}$ $8.5x10^{-11}$ $8.3x10^{-11}$ $2.5x10^{-11}$
310 fine grains (30-40μm)	1000	$7.59x10^{-11}$ $7.73x10^{-9}$ $8.00x10^{-7}$	64±5	$1.4x10^{-12}$ $1.1x10^{-12}$ $1.8x10^{-12}$
	1200	$8.60x10^{-10}$ $1.04x10^{-7}$ $1.06x10^{-5}$ $1.06x10^{-3}$		$4.8x10^{-12}$ $7.1x10^{-11}$ $6.5x10^{-11}$ $5.8x10^{-11}$

Analysis of the reaction products was done by X-ray diffraction and EDX analysis. The X-ray diffraction patterns of the external oxide correspond to the spinels of Fe_3O_4, $FeCr_2O_4$, $NiFe_2O_4$ and $NiCr_2O_4$. These spinels result in very similar patterns because of their common basic structure and cubic parameter "a". The analysis of the internal and external layer resulted in similar cristallographic analysis but with a different chemical composition.

The concentration profiles of the cations or metal atoms in the alloy layer were determined by X-ray fluorescence (EDX) analysis. The profiles at 800, 1000 and 1200°C shown in Figures (4) (5) and (6) indicate that the external layer is an oxide rich in iron, corresponding to 95 to 99% Fe_3O_4 with some chrome or possibly nickel in solution. The stability of this phase can be confirmed by the alloy phase diagram shown in Figure (7). The binary potential-composition diagrams for this ternary system were calculated using the POTCOMP program of the F.A.C.T. system. In order to represent the ternary region as a function of oxygen pressure, approximation were made to draw the boundaries of the ternary phases as shown by the dotted lines in the diagram.

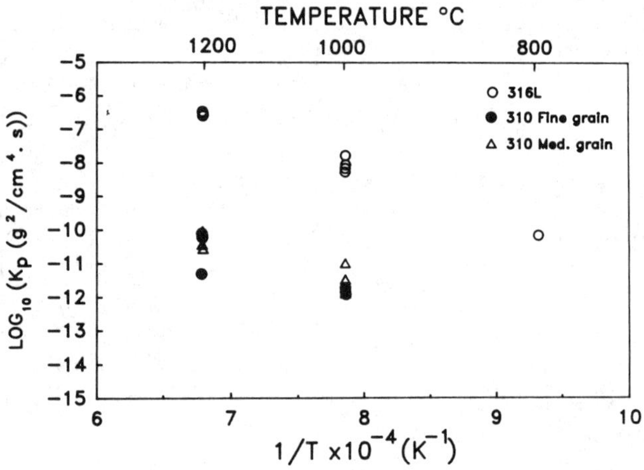

Fig. (1): Variation of the parabolic oxidation rate constant k_p as a function of temperature. Activation energies are given in Table III.

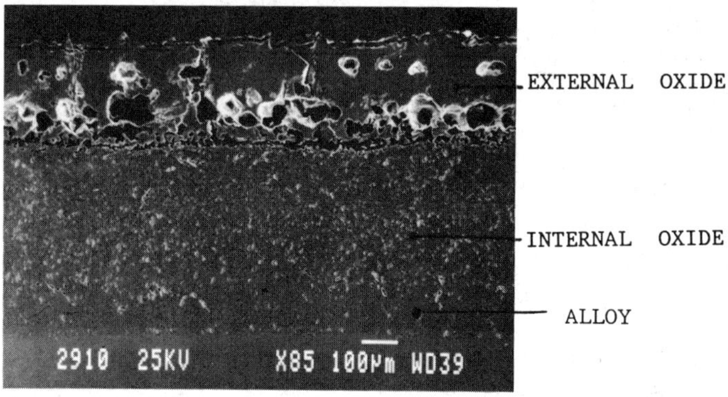

Fig. (2): Micrograph of internal and external oxide layers of 316L alloy oxidized at 1200°C

(P_{O_2} = 9.60 x 10^{-10} atm, time = 5 hrs.)

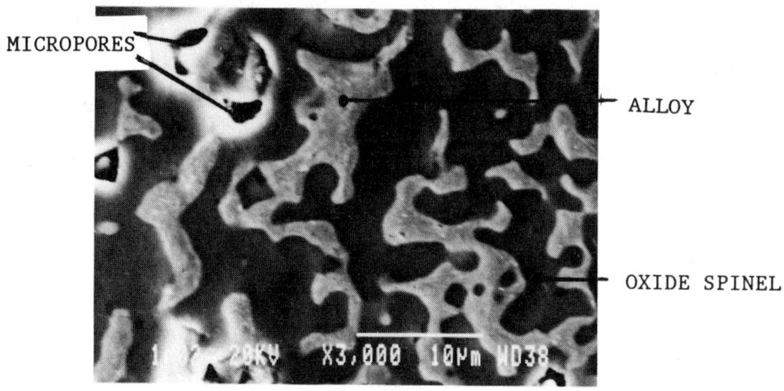

Fig. (3): Micrograph of internal oxide of 316 alloy oxidized at 1200°C

(P_{O_2} = 8.60 x 10^{-10} atm, time = 5 hrs.)

TABLE IV: RELATIVE SCALE THICKNESS OF EXTERNAL AND INTERNAL OXIDE LAYER AND INTERNAL OXIDE LAYER FOR 316L ALLOY.

Temp °C	P_{O_2} atm	time hours	internal layer μm	external layer μm	ratio
1000	8.00×10^{-7}	60	265	180	1.5
1200	1.06×10^{-3}	18	1000	200	5.0
800	3.01×10^{-17}	130	20	20	1.0
1000	6.27×10^{-13}	35	190	85	2.2
1200	8.60×10^{-10}	5	525	300	1.8
1200	8.60×10^{-10}	5	430	220	2.0
1200	8.60×10^{-10}	22	800	250	3.2

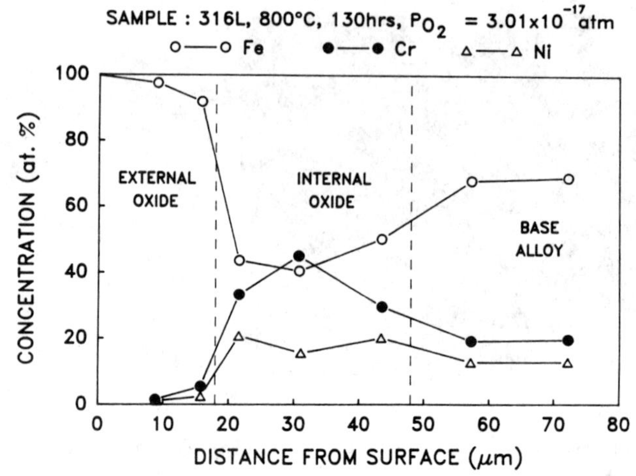

Fig. (4): Concentration profiles of cations across external and internal oxide layers of 316L alloy at 800°C.

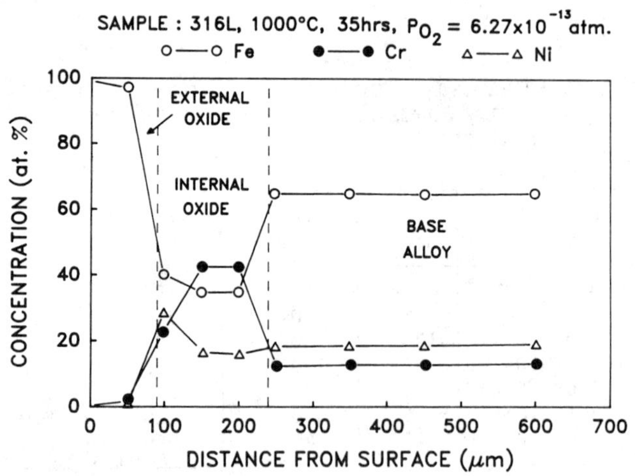

Fig. (5): Concentration profiles of cations across external and internal oxide layers of 316L alloy at 1000°C.

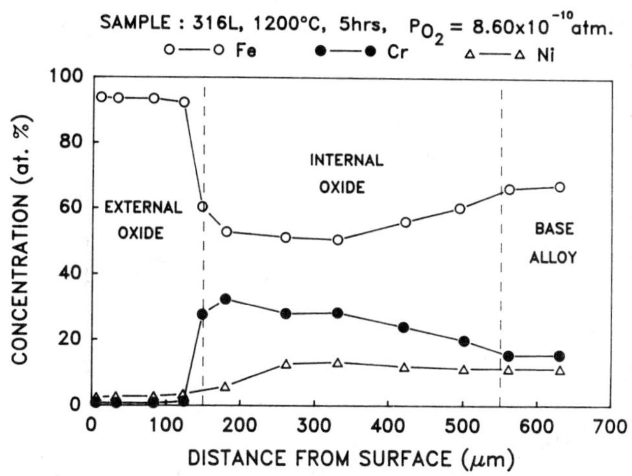

Fig. (6): Concentration profiles of cations across external and internal oxide layers of 316L alloy at 1200°C.

At 800°C the variation of the iron content in the external layer is far less abrupt then at 1200°C. The iron content in the internal layer decreases to a minimum value of about 40% at 800°C, 35% at 1000°C and 55% at 1200°C and returns to about 70% in the metal substrate. The average concentration of chrome in the external layer is essentially zero at the gas-oxide interface but increases up to 30 to 40% in the internal oxide region. Due to the different relative rates of migration of iron and chrome, the nickel content in the internal layer increases with respect to the base alloy. The variation of the composition in the internal layer is difficult to quantify because of the interdiffusion of the elements, the formation of pores, and also because of its two phase structure of oxide spinel and metallic phase. A point analysis of the spinel phase results in 1 to 5 atom % nickel, 30 to 40% chrome, the rest being iron, whereas the metallic particles are very rich in nickel, typically 55% Ni, 40%Fe and 5% chrome identified as the γ alloy phase. The equilibrium between the oxide and the metal phases can be corroborated by the ternary phase diagram at 1000°C shown in Figure (7).

An element distribution map (Figure 8) of the internal oxide also shows that chrome oxide is concentrated at the original metal grain boundaries whereas the interior of the grain contains more iron and nickel. As was indicated previously, the nickel is found mostly in the metallic phase and the amount appears to increase with the depth of the internal oxide, eventually reaching the alloy composition.

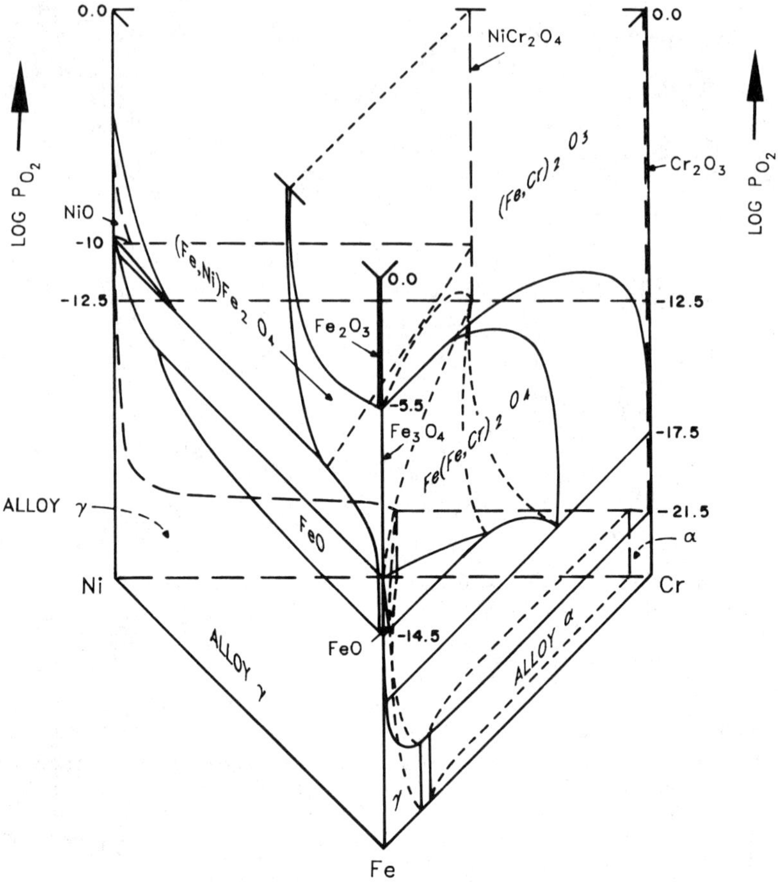

Fig. (7): Ternary oxygen potential-composition phase diagram of the Ni-Cr-Fe-O system at 1000°C.

Oxidation of 310 alloy

In contrast to the previous case, the 310 alloy does not form an internal oxide layer. The external oxide layer is compact and continuous but can have an undulating interface between the metal and the alloy. A porosity of about 15% is determined on the basis of the comparison of the experimentally measured value of k_p using the weight change vs time curve, and the calculated value determined from the average measured layer thickness and the theoretical densities of the oxide (Cr_2O_3) formed. A photomicrograph of this morphology is shown in Figure (9) where the sample was exposed to an oxygen pressure of 7.59×10^{-11} atom at 1000°C for a period of 65 hours. Large fissures between the oxide and the base metal were observed due to the differences in density and coefficient of thermal expansion between the two phases. Since no abrupt changes were observed during the experiment, it is assumed that the fissurisation occurs on quenching the sample.

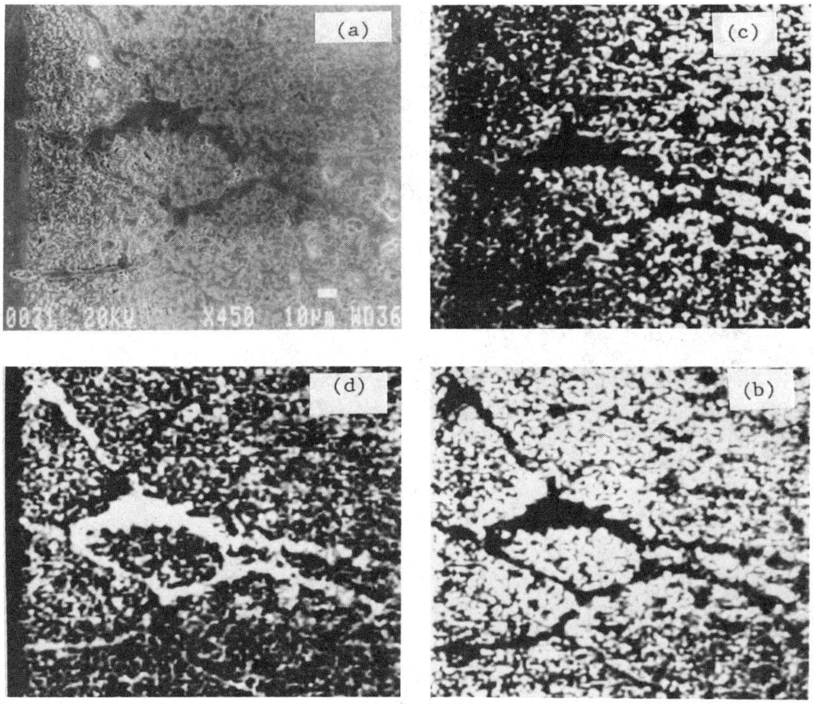

Fig. (8): Element distribution map of internal oxide layer of 316L alloy oxidized at 1200°C

P_{O_2}= 8.60 x 10^{-10} atm, time = 5 hrs.)

(a) Secondary electron map
(b) Fe distribution
(c) Ni distribution
(d) Cr distribution

The slow rate of oxidation of these alloys makes it difficult to perform the same type of analysis that was done as for the 316L alloy. Long duration test needed to be carried out and during this period the average grain size of the metal substrate increases from an average of 40 μm to 100 μm or from 100 μm to about 200 μm for the two alloys tested. Insufficient amounts of oxide layer did not permit analysis by X-ray diffraction. However EDX analysis clearly indicated that the external layer is essentially Cr_2O_3. For example, sample 6 (1000°C, p_{O_2} = 7.59 x 10^{-11}) produced an oxide with a cation concentration of 93% Cr, 4% Fe and 3% Mn whereas sample 9 (1000°C, P_{O_2} = 8.0 x 10^{-7}) showed a layer of 86% Cr, 7% Fe and 6% Mn. These values clearly indicate the formation of the sesqui-oxide $(CrFeMn)_2O_3$. Again it can be seen from the ternary phase diagram that the solubility of iron increases with increasing partial pressure of oxygen.

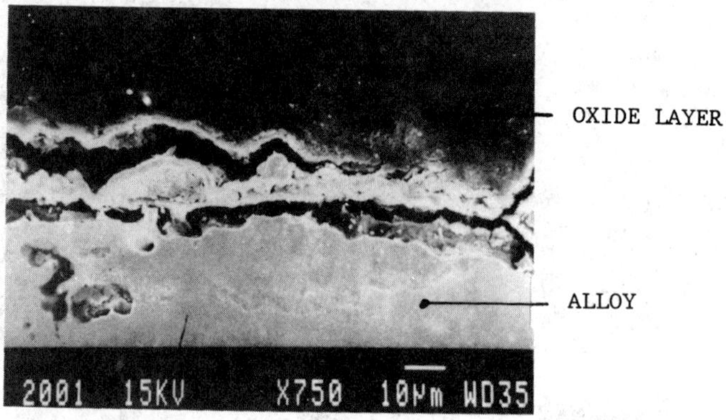

Fig. (9): Micrograph of external oxide layer of 310 alloy oxidized at 1200°C

(P_{O_2} = 8.60 x 10^{-10} atm, time = 5 hrs.)

DISCUSSION

On the basis of the findings in the literature and the experimental results obtained in this study, an oxidation mechanism for the 316L alloy is proposed and illustrated schematically in Figure (10). The external spinel layer grows principally by the diffusion of Fe^{2+} ions to the gas metal interface. This iron comes from the internal oxide zone which becomes depleted in iron and consequently enriched in chrome and nickel, as was shown in the concentration profile figures. Smith (6) has clearly shown that the diffusion coefficient of iron in the internal oxide is much higher than in the external layer and that the overall migration of iron is faster than that of chrome. This rapid diffusion of iron explains the formation of pores, particularly at the base of the external oxide layer and at the grain boundaries of the internal layer.

The nickel concentration in the metallic phase varies only slightly and it can be assumed that it does not really take part in the oxidating reaction except as an oxygen carrier because of its relatively large solubility for oxygen when compared to that of iron and chrome. As was previously shown, the internal oxidation is much more important than the external oxidation when the two layers are compared. It is surprising that Smith (7) had obtained a good correlation between his calculated and experimental values of k_p with elevated values of oxygen pressure for a temperature of 600°C. This was not found in the present study and can only be attributed to the effect of larger pores in both the external and internal layer at higher temperatures which permits easy access of oxygen to the internal oxide.

As a result, the rate determining step is therefore considered to be the diffusion of the cation species in the internal oxide or the diffusion of oxygen in the metallic region which has been depleted of iron and chrome. The activation energy for this mechanism was found to be 58±5 kcal(mol)$^{-1}$, a value which is higher than what would be expected for pore diffusion, or oxide grain boundary diffusion processes.

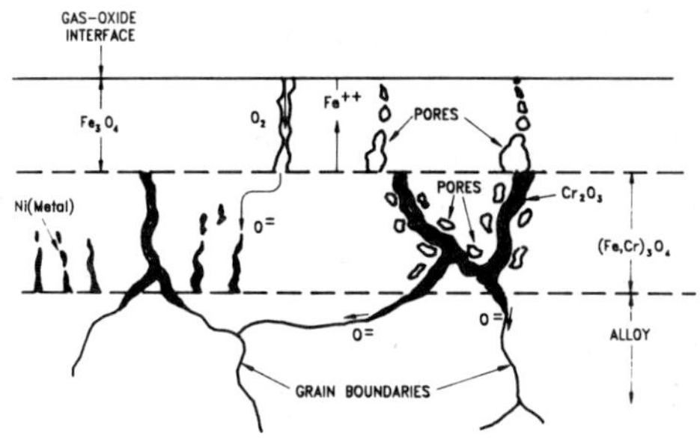

Fig. (10): Schematic representation of the high temperature oxidation mechanism of 316L alloy.

An oxide layer rich in chrome is also produced at the original metal grain boudaries. But, since the amount of chrome in the alloy is not sufficiently high, and its grain size is relatively large, this layer cannot completely surround the grains, and the internal oxidation proceeds far beyond the first layer of metal grains. The Hales and Smith model (8), (9) proposes the formation of a "healing" layer of Cr_2O_3 enveloping the metal grains and thus slowing down the internal oxidation process. This healing process is dependent on grain size. Calculations using this model would predict a healing layer to be formed in 5 minutes at 1200°C and in 3 hours at 1000°C for alloy grain sizes of 50μm. This behavior was not observed for the 316L alloy.

The morphology and the chemical analysis of the oxide scale is much simpler for the 310 alloy but the quantification of the ionic species which actually diffuse is still unresolved. The results indicate that the value of k_p is independent of the oxygen partial pressure supporting the findings of Hindam and Whittle (10) and of Skeldon et al (11) who propose that the chrome oxide is an n-type semiconductor with diffusion of oxygen or interstitial chrome in the lattice.

If, on the other hand, the rate determining step is assumed to be the diffusion of chrome in the alloy phase rather than in the oxide, the oxide can have p-type characteristics as proposed by Cadiou and Paidassi (12). The manganese and iron can also dissolve in chrome oxide according to two possible mechanisms:

(i) If Cr_2O_3 is considered to be an n-type oxide,

$$MO + e' + \frac{1}{2} O_2 = M'_{Cr} + 2O_O$$

and $$MO = M'_{Cr} + Cr^{\bullet} + \frac{1}{2} O_2$$

where $M = Fe^{2+}$ or Mn^{2+}

In this case the diffusion of the cations and consequently the value of k_p will increase with the amount of M dissolved.

(ii) If Cr_2O_3 is considered to be a p-type oxide,

$$MO + \frac{1}{2} O_2 = M'_{Cr} + h^{\bullet} + 2O_0$$

$$2MO + V''_{Cr} = 2M'_{Cr} + 2O_0$$

and $M = Fe^{2+}$ or M^{2+}

Here the diffusion of the cations and k_p will decrease with the amount of M dissolved in the Cr_2O_3. When the experimental values of k_p for the 310 alloy are compared to the values for the oxidation of pure chrome (10), which can vary from 10^{-11} to 10^{-12} g $cm^{-4}s^{-1}$ at 1000°C, we can conclude that the lower values of k_p observed support the hypothesis that the oxide is a p-type semi-conductor.

CONCLUSION

The oxidation behavior of stainless steels at elevated temperatures involves complex reaction mechanisms. The findings of this study indicate that the transition from internal catastrophic oxidation to a protective external layer occurs at about 20 atom % chrome in the alloy. The 316L alloy (17% Cr) does not contain a sufficient amount and forms an external layer rich in Fe_3O_4 with some chrome and nickel in solution and an internal oxide layer composed of spinels of chrome, iron and nickel. On the other hand, the 310 alloy (25% Cr) does form a protective chrome oxide layer resulting in oxidation rate constants three to four orders of magnitude smaller for the same environmental conditions.

The variation of oxygen pressure appears to have very little effect on the oxidation rates and consequently the values of k_p were found to be dependent on temperature and alloy composition only.

The oxidation mechanism proposed for the 316L alloy is the diffusion of oxygen in the internal oxide or in the chrome depleted zone of the alloy. The nickel concentration appears to have no effect on the oxidation kinetics even though it has a much larger solubility of oxygen in the metallic phase.

The oxidation phenomena for the 310 alloy are less evident even though only one oxide layer is formed. Both diffusion of chrome in the alloy phase or in the oxide phase are possible rate controlling mechanisms. The doping effect of manganese and iron in the Cr_2O_3 formed reduces the kinetics with respect to pure chrome. If the oxygen pressure is increased, the solubility of iron in the Cr_2O_3 also increases, which can lead to the formation of a hematite layer and consequently reduce its resistance to oxidation.

ACKNOLEDGEMENTS

The authors would like to acknowledge the Natural Sciences and Engineering Research Council of Canada for their financial support and to Prof. A.D. Pelton for the helpful discussions relating to the phase diagram determination.

REFERENCES

(1) J.B. Gilmour, A.W. Lui, D.C. Briggs, R.F. Knight and H. Hindam, Proceedings of NACE Corrosion '87 conference, San Francisco (1987).

(2) H. Hindam, J.B. Gilmour, R.H. Packwood, D.C. Briggs and A.W. Lui, Proceedings of NACE Corrosion '87 Conference, San Francisco (1987).

(3) M. Benlyamani, F. Ajersch and G. Kennedy, J. Electrochem. Soc., <u>136</u> (3), 343 (1989).

(4) M. Benlyamani, F. Ajersch and G. Kennedy, Oxid. of Metals, <u>29</u> 3/4, 203 (1988)

(5) A. D. Pelton, C.W. Bale and W. T. Thompson, F*A*C*T Users Guide, 2nd Edition, Ecole Polytechnique-McGill University, Montréal (1985).

(6) A.F. Smith, Corrosion Science, <u>21</u>, 9, 517 (1981).

(7) A.F. Smith, Corrosion Science, <u>22</u>, 9, 857 (1982).

(8) R. Hales, Werkstoffe and Korrosion, <u>29</u>, 393 (1978).

(9) A.F. Smith, M.O. Tucker and R. Hales, Oxid. of Metals, <u>17</u>, 5/6, 329 (1982).

(10) H. Hindam and D.P. Whittle, J. Electrochem. Soc. <u>130</u>, 7, 1519 (1983).

(11) M. Skeldon, J.M. Calvert, D.G. Lees, Oxid. of Metals, <u>28</u>, 1/2, 109 (1987).

(12) L. Cadiou and J. Paidassi, Man, Sci, Rev. Met., <u>LXVI</u>, 3, (1969).

Zirconium alloy oxidation — why do we still understand so little about it?

B. Cox
Centre for Nuclear Engineering, University of Toronto, Toronto, Ontario, Canada, M5S 1A4

Abstract

Much early work on zirconium alloy oxidation centred on the behaviour at high temperatures ($\geq$ 600°C), where the ratios of the lattice diffusion coefficients and the line diffusion coefficients for oxygen are small in both the oxide and the metal. At these temperatures the oxide is reasonably well behaved and remains coherent to considerable thicknesses. However, zirconium alloys are not used at these temperatures, but typically operate in the range 300-350°C in water in Nuclear Reactors. At such temperatures the diffusion of oxygen in the oxide occurs almost entirely at crystallite boundaries. Thus it is particularly important both to know and to understand the evolution of the microstructure of the oxide. In the metal phase the diffusion of oxygen in the lattice is very small, however, the line diffusion still remains significant so that grain boundary diffusion (localised at specific sites on some grain boundaries) can still perturb the protective nature of the oxide. It is also necessary to understand the breakdown processes in the oxide, interaction with species in the reactor water, and the effect of irradiation on all processes. The extent of the multifacetted problem is illustrated from recent work and the extensive gaps in our understanding identified.

1. Introduction

When Walt Smeltzer was working on the high temperature oxidation of zirconium alloys(1) things seemed relatively straightforward. Most experimental work was being done at temperatures where the oxidation kinetics appeared to be parabolic, and provided that alloying additions with significant solubility in zirconium were considered, and their expected valences in the ZrO_2 film were employed(2) things seemed to obey the Wagner-Hauffe rules(3). This period may be regarded as the "Age of Innocence" of zirconium oxidation studies, about the only serious arguments were over whether the oxidation kinetics were really parabolic, or were actually cubic, or changed from one to the other with increasing oxidation time.

However, zirconium alloys are not used in practical equipment at such high temperatures. Their commonest use at present is as fuel cladding alloys or other structural components in nuclear reactors cooled by high temperature water (either light or heavy). Typical operating temperatures in such systems are close to 300°C. When the kinetics of oxidation at such temperatures were studied in detail it was found that, although averaging results from large numbers of samples (Figure 1) can give results close to cubic kinetics (followed by a "breakaway" to linear kinetics), the examination of the detailed kinetics of individual specimens revealed a much more complicated situation (Figure 2). The kinetics were found to be much closer to a quartic than a cubic, and at least one well defined inflection (plateau) was found in the curves. It was also found that a significant fraction of the electrochemical potential equivalent to the free energy of the reaction persisted across the oxide film during the pretransition oxidation period (Figure 3)[4] Studies of the oxide microstructure revealed a microcrystalline thin film, possibly with an amorphous layer at the oxide metal interface, with evidence for crystallite growth and the development of a fibre texture as the oxide thickened, and with oxygen (diffusing primarily along crystallite boundaries) as the migrating species.

Also, as the oxide thickens it becomes cracked and porous and a transition to an approximately linear rate law ensues (Figure 1). This transition may be smooth and continuous when only fine porosity develops (Figure 4), as at high temperature and low pressure, or discontinuous and cyclic when episodes of grosser cracking recur[5].

Thus, although the Age of Innocence had now passed it appeared that, in the new Age of Reason, the oxide film growth on the alloys should be readily explicable if the following information were available.

- Accurate measurements of the lattice and grain boundary diffusivities of oxygen in ZrO_2 films.
- A detailed knowledge of the evolution of the oxide microstructure with increasing thickness.
- An understanding of electronic conduction through the oxide film, and how preferential conduction sites evolved with increasing oxide thickness.

Similarly the various oxide breakdown processes should yield to:

- A knowledge of the localised diffusion of oxygen along metal grain boundaries that contributes to oxide breakdown under inadequately oxidising conditions.
- Details of the mechanisms by which cracks and pores developed in the oxide film at the rate transition when excess oxidant was present.

In this period workers set about measuring the necessary quantities[5] and fitting together a comprehensive picture of the oxidation process. The results of this work for an alloy of the Zircaloy type (i.e. one containing intermetallic particles of Zr/Fe, Cr, Ni phases) are illustrated in Figure 5. Diffusion data showed that crystallite boundary transport was overwhelmingly easier than bulk diffusion, so that the kinetics could be seen in terms of changes in the volume fraction of boundary area as the oxide evolved, modified by changes in the electron transport process with time.

In recent years we seem to have passed from the Age of Reason to the Age of Disillusion because further work has continued to show that the overall process is even more complicated than was thought earlier. The remainder of this paper will be devoted to an attempt to explain the origins of this.

2. Surprises in the out-of-reactor observations

Because of the essentially pH independent nature of the oxidation process (Figures 6 and 7), excluding specific ion effects (e.g. Li^+); the lack of any significant effect of polarising specimens (Figure 8); the similarity of oxidation behaviour in a wide variety of media; and the identity of the oxidation rates in the water and steam phases at the same temperature, it was believed that the pretransition oxidation kinetics of zirconium alloys at reactor temperatures were controlled essentially entirely by solid state processes within the oxide film. Furthermore, this oxide film was believed to be non-porous and close to the theoretical density for ZrO_2. The primary diffusing species was oxygen, with diffusion coefficients for Zr in ZrO_2 below the detection level at these temperatures.

It was found however, at an early date[6], that some anodic oxide films apparently had an array of fine closed porosity within them. Such observations could easily be dismissed as an effect of electrolyte incorporation, since it did not appear to be universal. However, when apparently closed porosity was found in low temperature thermal oxides, and furthermore when these pores were highly crystallographic in nature, (Figure 9), it became far more difficult to explain[7]. It is very difficult to see how such large crystallographic features could develop by diffusional growth in a system where one component (Zr^{4+}) appears to be essentially immobile. A clue to the mechanism may be present in the observation that these features have only been seen so far in the presence of moisture (the environment was moist oxygen, and a repeat in carefully dried oxygen has not shown the same effects). Thus, a hydrothermal recrystallisation process may be responsible for the large crystallographic features observed. That the porosity was apparently closed was shown by the examination of both faces of the fracture (artificially produced in the oxide). We cannot completely exclude the possibility of finer porosity, not visible at resolutions achievable on the Scanning Electron Microscope, linking these crystallographic voids to the surface, however.

Hence, the simple picture of a completely diffusion controlled growth process in a compact oxide film, of close to theoretical density may have to be modified even for pretransition films.

Other recent results may require even more profound changes in our understanding of the corrosion process. It was again discovered early that porous anodic oxide films could be formed in some electrolytes[8] and that these pore structures were sometimes quite regular (Figure 10) like the cell structure formed in anodic aluminum oxide films. Such effects were, however, regarded as peculiarities of a few, typically strongly acidic electrolytes (e.g. H_3PO_4). Attempts to understand the specific deleterious effects of LiOH were focussed on incorporation of Li^+ in the ZrO_2 lattice and its effect on the oxygen diffusion coefficient. However, the initial effect of concentrated LiOH is not to change the pretransition kinetics but to change the way that porous post-transition oxide develops[9,10]. Evidence is now accumulating that even pretransition oxide films may be porous in pH 12 LiOH, and that these films may also resemble porous anodic alumina films[11]. Thus, the possibility that ZrO_2 films grow (at least in part) by a local dissolution and reprecipitation process becomes a strong contender for the mechanism by which porosity develops in these films. The conclusions from such an hypothesis would be that Li^+ does not modify the diffusion kinetics in pre-transition films, but attacks the oxide locally from the outside, reducing the rate of oxidation. The material dissolved out of the pores created locally would then reprecipitate as a surface layer of micro-porous oxide.

With such modifications to the picture the growth of ZrO_2 films in aqueous media begins to resemble much more closely the mechanism for the formation of thick passive films on (e.g.) stainless steels.

3. Surprises in the in-reactor observations

The unexpected long-term acceleration of the oxidation rate of Zircaloy-2 pressure tubes in CANDU reactors (Figure 11) was one of the contributing causes in the sudden rupture of the P2G16 pressure tube in August 1983[12]. The hypothesis proposed at the time to explain this accelerating oxidation was that in thick porous oxide films the access of hydrogen (from the bulk water) will become restricted and permit radiolysis of water within the pores of the oxide. The generation of oxidising radiolytic species in this circumstance will lead to an acceleration in the oxidation rate. Hydrogen is added to the bulk water in CANDUs and PWRs to inhibit radiolysis, and failure to maintain sufficient hydrogen in solution

(>1 cc/kg) is known to lead to increased rates of oxidation. A similar acceleration in the oxidation rate of Zircaloy-4 clad fuel in PWRs at high burnup[13], that cannot be readily explained by the thermal-hydraulics of the system and the temperature drop through the thick oxide on a heat transfer surface, has also been explained in the same fashion[14].

There is, however, a fallacy in this hypothesis[15]. In thick porous oxide films it has been long believed that the basic reduction of water occurs close to the metal/oxide interface where a thin protective barrier oxide persists. Under these conditions the cathodic half cell reaction (discharge of a proton) will occur within the porous film. Provided that less than 100% of this cathodic hydrogen is absorbed by the metal there will always be ample hydrogen to saturate the water in the pores, irrespective of whether sufficient has been added to the bulk water to suppress radiolysis or not. Since the "thick film hypothesis" very effectively explains the observations[14], the only way it could be salvaged mechanistically would be if in some way the cathodic half cell reaction could continue to operate at the oxide/bulk water interface even on porous oxides. This seems improbable, but may not be impossible given the effect of the high radiation fields in rendering the oxide film electrically conducting. It is clear that we still understand too few of the basic processes occurring under irradiation.

A further result of the increased conductivity of both oxide and water under irradiation may be the appearance of effects apparently caused by galvanic coupling in-reactor, in circumstances where they would not be observed in the laboratory (Figure 12). It is thought now that much of the effect of the radiation chemistry of water on the corrosion process may function merely through changes in the equivalent redox potential (ERP) of the water, and the susceptibility to galvanic effects caused by irradiation induced conductivity of both oxide and water.

The situation has, however, led to a reappraisal of the whole mechanism of in-reactor corrosion of zirconium alloys. It was found at an early stage in studies of this phenomenon that in-reactor weight gains (apparent oxide thicknesses) tended to be significantly <u>less</u> than expected from laboratory studies in the pretransition oxidation period. An example of this phenomenon is seen in curve II of Figure 7, but such observations have been very common. There was a tendency at the time to dismiss such observations as weighing errors, because of the low weight changes being measured, the low sensitivity of many balances in hot cells, and the fact that the initial (unirradiated) and final (irradiated) weights were often measured on different balances. However, such weight changes are almost invariably low, and seldom show a true statistical scatter about the expected value. Examinations of the oxides on such samples (Figure 13) has shown that they also contain arrays of pores[16]. Thus, again the possibility of a dissolution and reprecipitation process during the formation of thin oxide films appears to be present. This has led to a suggestion that the LiOH present in most PWRs, and its concentration in thick oxide films (even on non-heat transfer surfaces) is the primary cause.

Analyses of these thick oxide films have seldom shown evidence for significant Li concentrations (Figure 14); the surprise was the high fluoride content of the oxides on the pressure tubes. Reactor water is usually below 0.05 ppm fluoride, and the large concentration factor that is evident is surprising, even though fluoride is known to be strongly adsorbed by ZrO_2. No other source of this quantity of fluoride is known, the production in reactor by the nuclear reaction $^{18}O(n\gamma)^{19}F$ being too small to account for this amount. Fluoride at a peak concentration of 100 ppm in the oxide film is getting to the level where fairly severe immediate effects of fluoride from residual etchant are known to occur. Measurements of the concentrations of fluoride in the oxide on such occasions have given values as low as 100 ppm fluoride[17], although the results of such analyses are very variable and other workers[18] have argued that much higher fluoride concentrations are necessary before any effect is seen.

Nevertheless, the amounts measured in the oxide film on CANDU pressure tubes are high enough to cause concern. If the accumulation of fluoride in the oxide were the cause of acceleration of the oxidation rate at long times then this might be independent of whether the water pH was regulated with NH_4^+ or Li^+. This could then account for the similarity in long-term corrosion behaviour between Pickering #1/2 Zircaloy 2 pressure tubes and Hanford N-reactor tubes. This similarly could not be explained on a LiOH concentration mechanism as the N-reactor uses NH_4OH for pH control.

4. Conclusions

In recent years we have changed from a conviction that the oxidation of zirconium alloys was controlled entirely by solid state processes within the protective ZrO_2 film (or the residual barrier layer oxide after the bulk of the film becomes porous) to a realization that there are processes going on in high temperature water that are unique to that environment. Such processes as dissolution and reprecipitation of ZrO_2 as a method of converting impervious barrier oxide into porous thick oxide films are not now regarded as untenable, even though evidence supporting their existence remains almost entirely circumstantial. There also appears to be accumulating evidence for galvanic effects in-reactor that are not visible in the laboratory, perhaps because irradiation enhances the conductivity of both the water and the oxide.

Thus, although the growth of the barrier layer film is still thought to be understandable in terms of solid state processes such as oxygen diffusion along crystallite boundaries, the evolution of the oxide microstructure driven by the high compressive stresses and the localised electron transport processes, the breakdown of the oxide film has taken on a whole new aspect. Gross mechanical processes (such as cracking) and microscopical pore development (linked with the evolution of the oxide microstructure), while they are clearly still occurring, may be circumvented by direct chemical processes such as hydrothermal recrystallisation and oxide dissolution and reprecipitation.

The simple answer to the question posed by the title is that the zirconium alloy corrosion process in high temperature aqueous solutions in-reactor has proved to be very much more complicated than was thought originally.

References

1. W. W. Smeltzer et al., Papers on Zirconium Oxidation:- Can. Met. Quart., 1, 41, (1962), Acta Met., 12, 409, (1964), and 13, 55, (1965), J. Electrochem. Soc., 111, 1221, (1964), 112, 554, (1965), and 115, 927, (1968); Z. Metall., 71, 470, (1980).
2. H. A. Porte, J. G. Schnizlein, R. C. Vogel and D. F. Fischer, J. Electrochem. Soc., 107, 506, (1960).
3. K. Hauffe, Oxidation of Metals, Plenum, N.Y., 1965.
4. B. Cox, J. Nucl. Mat., 31, 48, (1969).
5. B. Cox, Advances in Corrosion Science and Technology, Vol. 5, Eds. Fontana and Staehle, p. 173, Plenum, N.Y., 1976.
6. R. A. Ploc and M. A. Miller, J. Nucl. Mat., 64, 71, (1977).
7. R. A. Ploc, J. Nucl. Mat., 82, 411, (1979).
8. B. Cox, J. Electrochem. Soc., 117, 654, (1970).
9. E. Hillner and J. Chirigos, U.S. Report, WAPD-TM-307 (Aug. 1962).
10. F. Garzarolli, J. Pohlmeyer, S. Trapp-Pritsching and H. G. Weidinger, Paper presented at I.A.E.A. Technical Committee Meeting on Zirconium Alloy Corrosion, Portland, Or., Sept. 1989.

11. N. Ramasubramanian, N. Preocanin and V. C. Ling, ASTM-STP-1023, p.187, (1989).
12. E. G. Price, Paper presented at Spring Meeting, Can. Electrical Assoc. Montreal, March 1985, AECL-8755.
13. B. Cox, EPRI Report NP-4287 (Oct. 1985).
14. A. B. Johnson, Jr., EPRI Report NP-5132 (April 1987).
15. B. Cox, Paper presented at I.A.E.A. Technical Committee Meeting on Zirconium Alloy Corrosion, Portland, Or., Sept. 1989.
16. B. Cox and V. Fidleris, ASTM-STP-1023, p. 245, (1989).
17. S. Kass, (in discussion), ASTM-STP-368, p. 40, (1964).
18. J. Rynasiewitz, Reactor Technology Report #22 (Chemistry), U.S. Report KAPL-2000-19 (1962).

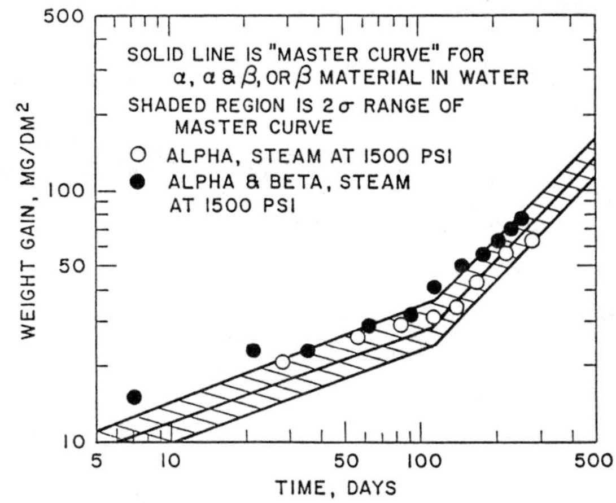

FIG. 1. Zircaloy-2 oxidation in steam and water at 360°C.

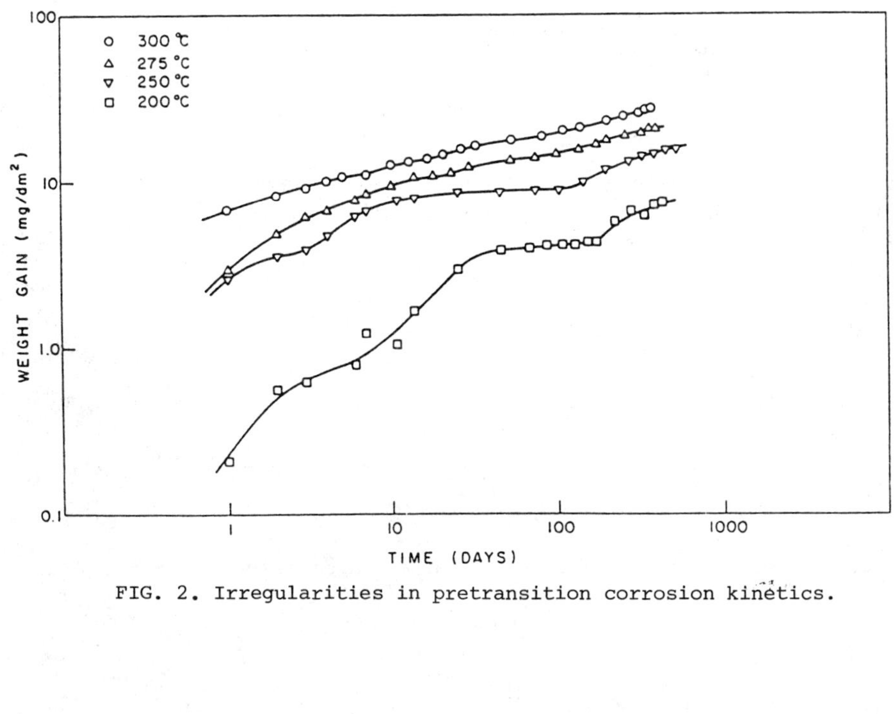

FIG. 2. Irregularities in pretransition corrosion kinetics.

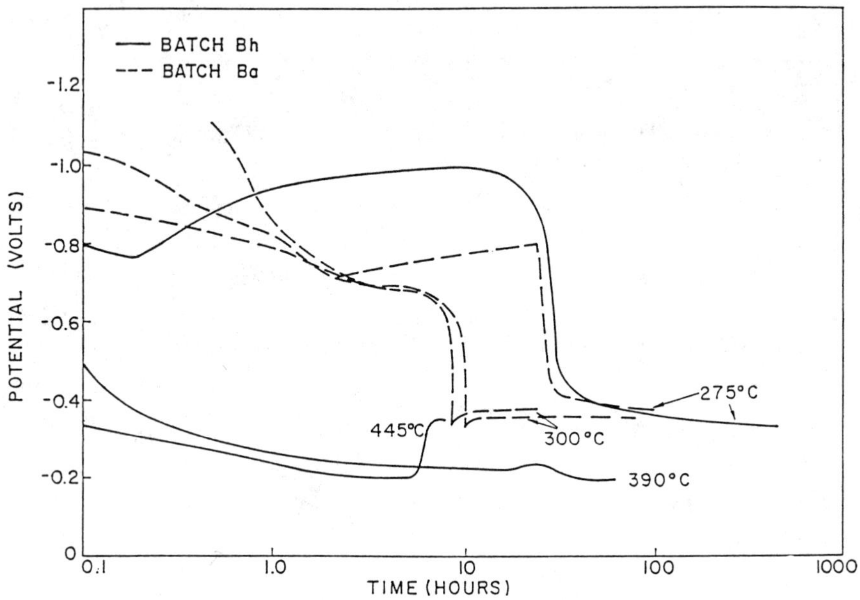

FIG. 3. Potential versus time curves for Zircaloy-2 in fused salt.

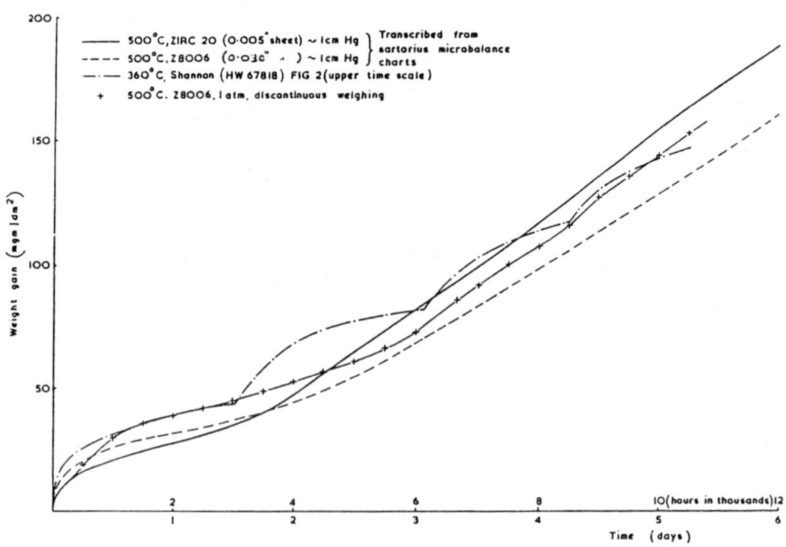

FIG. 4. The shape of the Zircaloy-2 oxidation curve at 360° and 500°C.

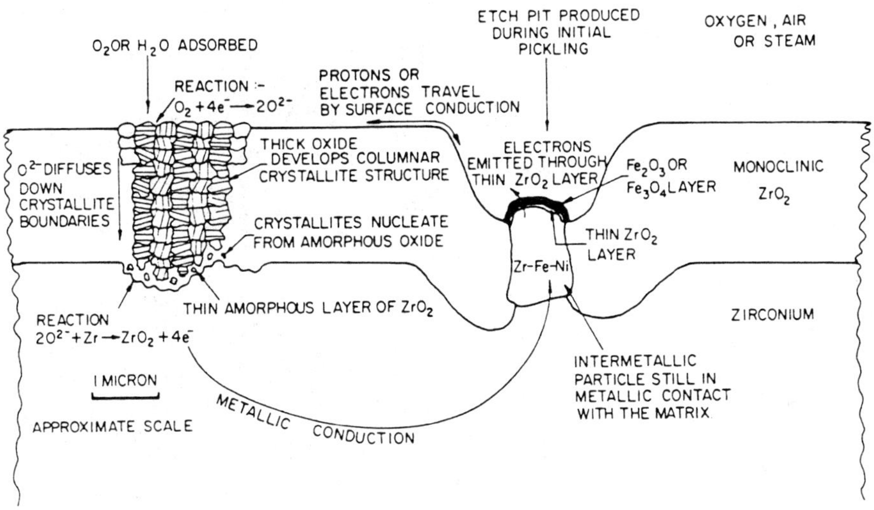

FIG. 5. Model of the oxide film and corrosion processes of Zircaloy-2.

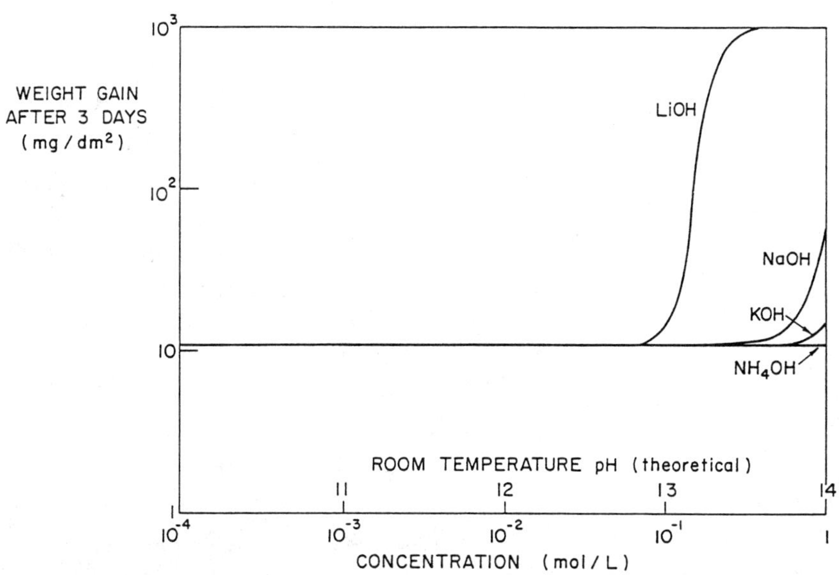

FIG. 6. Corrosion of zirconium alloys in alkaline solutions at 360°C[10]

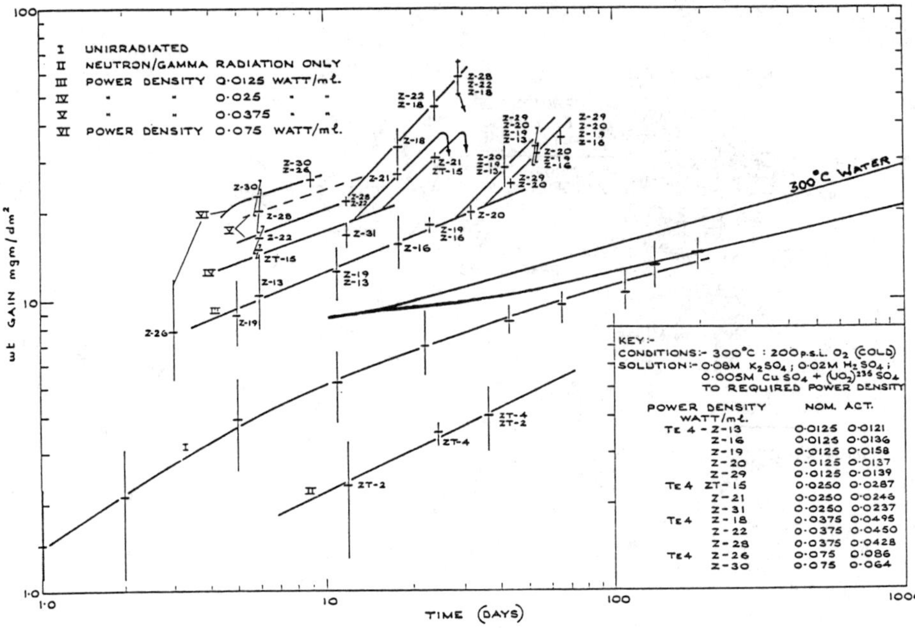

FIG. 7. Irradiation corrosion of Zircaloy 2 at 300°C.

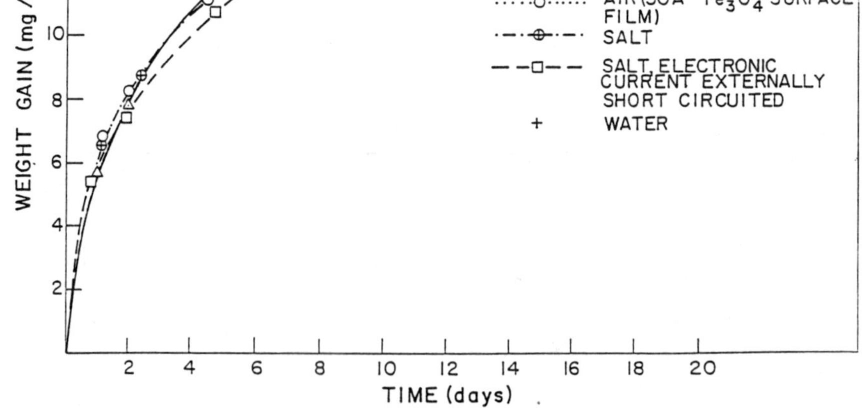

FIG. 8. Oxidation of Zircaloy-2 in various environments

FIG. 9. Crystallographic pores in a 8μm thick oxide film formed on zirconium in moist O_2 at 300°C.

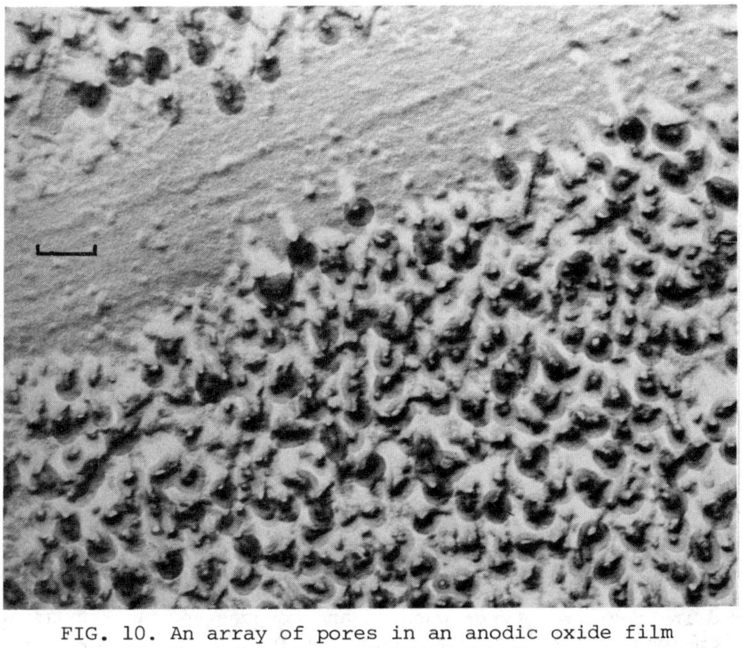

FIG. 10. An array of pores in an anodic oxide film formed in H_3PO_4 on Zircaloy-2. 0.1μm bar.

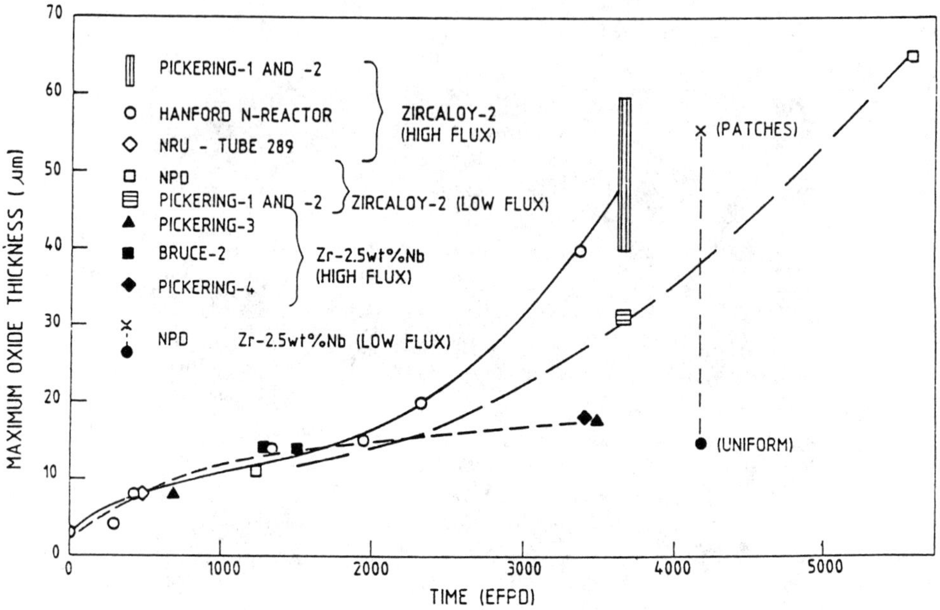

FIG. 11. Accelerating oxidation of Zircaloy-2 pressure tubes in-reactor.

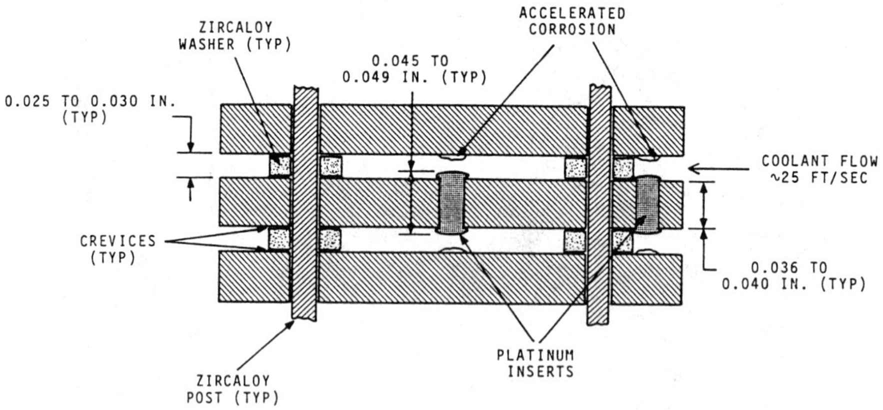

FIG. 12. Typical configuration of specimens mounted on racks, including a coupon containing platinum inserts. (Note: Accelerated oxidation occurred in crevices between washers and coupons and on coupons adjacent to coupons containing platinum inserts at locations opposite the inserts.)

FIG. 13. Pores developed in a thin oxide film formed in irradiated $UO_2SO_4/CuSO_4/H_2SO_4$ (see figure 7).

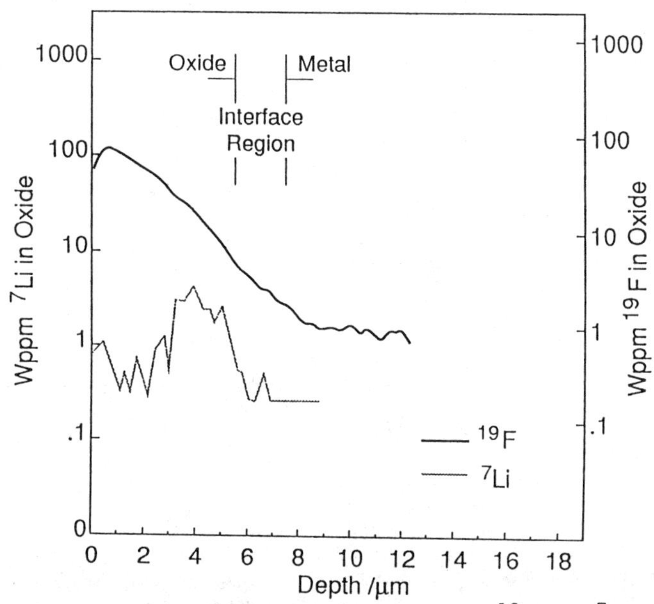

FIG. 14. In-depth concentration profiles for ^{19}F and ^{7}Li from ID Oxide of a CANDU Pressure Tube (Tube P3 J09).

Effect of microstructure on the corrosion of Zr-2.5Nb alloy

V.F. Urbanic and R.W. Gilbert
Atomic Energy of Canada Limited, Chalk River Nuclear Laboratories, Chalk River, Ontario, Canada, K0J 1J0

Abstract

Zr-2.5Nb alloy is used as a pressure tube material in CANDU reactors. Tests in high temperature water and steam have shown that corrosion of Zr-2.5Nb is very sensitive to changes in microstructure. Highest corrosion rates are generally seen on martensitic structures in β-quenched material, however, lower rates are obtainable with Widmanstatten structures produced by cooling more slowly from the β region. Corrosion resistance can be substantially improved by ageing β-quenched structures at temperatures below the monotectoid temperature of 610°C. Further improvements can be realized by quenching from the two-phase $\alpha+\beta$ phase field and introducing cold work between the quenching and ageing treatments. The presence of equilibrium α-Zr and the precipitation of β-Nb from the supersaturated α prime martensitic structure both appear to contribute to lowering the corrosion rate. Improvements in corrosion resistance are also observed when extruded and cold worked pressure tubes are thermally aged at 400-500°C. The improvements seem to correlate with the eventual decomposition of the β-Zr phase to the more stable α-Zr plus β-Nb phases.

The quenched, cold-worked and aged material as well as the extruded and cold worked material both undergo microstructural changes during fast neutron irradiation at reactor operating temperatures. Both show a fine distribution of β-Nb precipitates within the α-Zr grains. Their absence before irradiation and the inability to precipitate such particles by thermal ageing alone suggest that they are induced by the fast neutron irradiation. Additional precipitation in the α-prime matrix of the quenched, cold worked and aged material is also seen following irradiation at reactor temperatures. Re-exposure of irradiated material in out-reactor tests has demonstrated that the microstructural changes which evolve during irradiation, for both material conditions, increase the corrosion resistance of Zr-2.5Nb.

Introduction

The CANDU-PHW reactor is a pressure tube reactor that is fuelled with natural uranium oxide and is both cooled and moderated with heavy water. The fuel is contained within 0.5 m long fuel elements assembled into bundles which in turn reside inside the zirconium alloy pressure tubes. The temperature of the heavy water on the inside of the pressure tubes ranges from about 250°C at the inlet to about 300°C at the outlet end. Pressure tubes in the demonstration, prototype and first commercial CANDU-PHW nuclear power reactors were made of Zircaloy-2, whereas in all subsequent reactors the tubes have been made from Zr-2.5Nb alloy. This alloy was developed because its higher strength allowed a thinner tube to be used which is a benefit in terms of neutron economy. Pressure tubes in the quenched, cold-worked and aged condition were developed in the 1960s for a prototype CANDU boiling water reactor whereas those in the extruded and cold-worked condition are used in the commercial CANDU-PHW reactors. Zr-2.5Nb is a two-phase alloy, and its microstructures can be much more complex than those associated with the Zircaloys. Since pressure tubes have a design lifetime of 30 years, an understanding of how microstructure affects their corrosion behaviour is essential. During the development of Zr-2.5Nb as a pressure tube material, the effect of microstructure on its corrosion behaviour was one of the principal areas of investigation. It is the purpose of this paper to review some of this earlier work and to present some new results with respect to microstructural changes that appear to evolve during irradiation in high temperature water and the implications of these changes on corrosion behaviour.

Microstructure

The phases that can exist in the Zr-Nb alloy system are shown in the equilibrium phase diagram in Figure 1 (1). In pure zirconium, hexagonal close packed α-Zr is stable at temperatures up to 862°C. Above 862°C the stable phase is body centered cubic β-Zr. Niobium additions stabilize the β-Zr to lower temperatures such that β-Zr containing as much as about 20 wt% Nb can exist at the eutectoid temperature of about 610°C. At this temperature the solubility of niobium in α-Zr is found to be at a maximum of about 0.6 wt%. Below 610°C the equilibrium phases that can exist are α-Zr and β-Nb. β-Nb has a body centered cubic structure and contains about 85 wt% Nb. Commercial Zr-Nb alloys however contain oxygen as an impurity which is an α-Zr stabilizer. As a result, the α-Zr/β-Zr transformation in commercial Zr-2.5Nb occurs at somewhat higher temperatures than that shown in the equilibrium phase diagram.

Zr-2.5Nb alloy can exhibit a variety of microstructures depending on its fabrication history. When cooled slowly from the β-Zr phase field, needles of α-Zr first nucleate at the β-Zr grain boundaries and grow during cooling to develop a Widmanstatten structure typical of that shown in Figure 2. When water quenched from the β-Zr phase field, the β-Zr undergoes a diffusionless shear transformation to hexagonal close packed α-prime. A martensitic type structure containing fine needles of α-prime Zr is produced (Figure 3). In this condition, all the niobium is contained as a supersaturated, substitutional solid solution in the martensite needles.

When Zr-2.5Nb is heated into the two-phase α-Zr plus β-Zr phase field, both α-Zr and β-Zr are present. The relative proportions of the two phases present and the concentration of niobium in each at any given solution temperature can be determined from the equilibrium phase diagram. When the solution temperaature is relatively high in the two-phase field, water quenching produces a microstructure consisting of islands of equilibrium α-Zr in a matrix of martensitic α-prime (Figure 4). Thermal ageing treatments, with or without intermediate cold-working, can be performed to alter the mechanical properties of quenched structures. From mainly practical considerations, the optimum temperatures for thermal ageing are in the range 450°C to 550°C. The resolution of optical microscopy however is generally too low to reveal the more subtle effects of cold work and ageing on the microstructure; therefore, examination of thin foils by transmission electron microscopy is commonly used. Figure 5 is a transmission electron micrograph of the martensite matrix from a Zr-2.5Nb alloy which had been water quenched from the $\alpha+\beta$ phase field (870°C), cold worked 17% and subsequently aged at 500°C for 24 hours. This treatment is typical of that developed for CANDU-BLW reactor pressure tubes and is generally referred to as the heat treated condition. During quenching, the stresses produced by the martensitic transformation result in the formation of microtwins in the α-prime. Cold working introduces dislocation tangles which relax during ageing to form networks. Ageing of the supersaturated matrix results in the formation of Nb-rich precipitates. Coarse precipitates are evident on the twin boundaries whereas a finer precipitate is present throughout the martensite needles. Since the ageing process is both temperature and time dependent, the size, distribution and composition of the precipitates can vary depending on these ageing parameters. Detailed studies of this kind have been done to characterise the nature of these precipitates in β-quenched and aged Zr-2.5Nb structures (2,3). Ageing treatments below the eutectoid temperature eventually lead to the formation of precipitates of equilibrium β-Nb in the martensite regions, but there is no evidence for any precipitation of β-Nb or Nb-rich phases in the equilibrium α-Zr islands as a consequence of thermal ageing. Various aspects of thermal ageing treatments on both the microstructure and mechanical properties of quenched Zr-2.5Nb have already been described in several other papers (4,5,6,7).

Air cooling from the two-phase, α-Zr plus β-Zr region, results in structures that are quite different from those produced by water quenching from the same temperature. Under these circumstances, metastable β-Zr can be retained down to room temperature. Zr-2.5Nb alloy CANDU-PHW reactor pressure tubes which are extruded at 820 to 850°C, cold worked 20 to 30% and stress relieved in steam at 400°C for 24 hours are a good example of this. At the extrusion temperature, the tube microstructure consists of about 80% β-Zr and 20% α-Zr. During air cooling following extrusion, the β-Zr transforms to α-Zr by growing on the existing elongated α-grains produced during the extrusion, while the β-Zr becomes progressively enriched in niobium. In annealed and air-cooled material the situation is the same except that the α-grains would tend to be equiaxed and not elongated. Hence extruded tubes contain α-grains up to about 10 microns long, about 1 micron wide and about 0.5 microns thick and a grain boundary network of metastable β-Zr that contains 18 wt% to 20 wt% Nb. Cold drawing causes additional elongation of the α-grains and introduces a high density of dislocations to the microstructure. The microstructure typical of a cold worked tube is shown in Figure 6a. Stress relieving pressure tubes at 400°C reduces the dislocation density without causing any recrystallization to occur. In addition, stress relieving at temperatures less than 610°C will also cause the metastable β-phase to eventually decompose to an equilibrium structure consisting of α-Zr and β-Nb (8). At temperatures above about 500°C, the decomposition proceeds according to the following:

$$\beta\text{-Zr} \longrightarrow \text{Zr} + \beta\text{-enriched} \longrightarrow \text{Zr} + \beta\text{-Nb}$$

whereas at temperatures below about 500°C, it proceeds according to:

$$\beta\text{-Zr} \longrightarrow \text{omega} + \beta\text{-enriched} \longrightarrow \alpha\text{-Zr} + \beta\text{-Nb}$$

Heating a cold-worked pressure tube at 400°C for 72 hours results in some transformation to omega phase, and the coalescence of the thin β-Zr networks into more globular precipitates tend to produce rough α/β boundaries (Figure 6b). After heating at 500°C for 10 hours, the transformation has progressed even further with the β-Zr network showing evidence of decomposition to α-Zr and chains of discrete Nb-rich particles (Figure 6c). This decomposition can occur at temperatures as low as 300°C, typical of operating reactors. However on the basis of available time/temperature/transformation data for 12 to 25 wt% Nb alloys (8,9), the time required for complete decomposition could be several years. Nevertheless it should be remembered that since Zr-2.5Nb is a two-phase alloy, structural changes can occur during exposure to high temperature water and steam, and that in the case of long-term corrosion testing in particular, one should exercise caution in this regard when trying to interpret kinetic data.

Out-Reactor Corrosion

The out-reactor corrosion behaviour of Zr-2.5Nb is sensitive to microstructure. Figure 7 shows the difference in behaviour as a result of structural differences achieved by cooling from the β-phase at different rates (10). In tests performed in both high temperature water and steam, corrosion of a martensitic structure (Figure 7a) produced by water quenching is considerably higher than that for a Widmanstatten structure (Figure 7b) produced by rapid cooling. In these tests, oxide spalling also occurred on the martensite structures in the 316°C water test.

In water quenched material that has been subsequently aged, the effect of the quench temperature is important. Figure 8 compares the corrosion behaviour of both β-quenched and $\alpha+\beta$ quenched Zr-2.5Nb alloy which had been aged at 535°C for 6 hours prior to testing. At all three test temperatures, the corrosion behaviour of material solution treated at 825°C (two-phase region) is superior to that which had been solution treated at 950°C (β-Zr region). Further improvements in the corrosion resistance can be realized by introducing cold work between the quenching and ageing steps. Figure 9a shows the effect of up to 30% cold work on the corrosion response of material initially quenched from 880°C in the $\alpha+\beta$ region, cold worked and subsequently aged at 500°C for six hours. It can be readily seen that the corrosion resistance increases with increasing amounts of intermediate cold work prior to ageing.

The degree to which a quenched and cold-worked structure ages is a function of both the ageing temperature and ageing time. Increasing each individually or both at the same time increases the extent to which the second phase Nb-rich particles precipitate from the supersaturated martensitic matrix. Figure 9b shows the effect of increasing the ageing time at 500°C from 6 hours to 24 hours on the corrosion of material initially quenched from 880°C in the $\alpha+\beta$ region and cold worked 30%. The longer ageing times are associated with lower corrosion rates. In fact the corrosion response of structures developed in Zr-2.5Nb through these thermo-mechanical treatments was equal to or better than that of Zircaloy-2 in these same tests. The effect of increasing the ageing temperature for a given ageing time produces a similar improvement in corrosion resistance. These effects of ageing time and temperature on the corrosion resistance of quenched or quenched and cold worked Zr-2.5Nb have been documented by others as well (11,12,13,14).

A two-phase microstructure consisting of α-Zr and metastable β-Zr also shows an improvement in corrosion resistance when heated to temperatures below the eutectoid temperature of 610°C. As described earlier, the cold-worked Zr-2.5Nb pressure tubes used in CANDU-PHW reactors have a structure in which the metastable β-Zr can be decomposed by a heat treatment in this temperature range. Figure 10 shows that the corrosion of cold-worked pressure tubes in 316°C water can be reduced by heat treatments at 400°C for 72 hours and 500°C for 10 hours (15). The improvements in corrosion response after the 400°C and 500°C treatments seem to correlate with the progressive stages of β-Zr decomposition as illustrated by the microstructures in Figure 6. The behaviour of Zircaloy-2 in the same test is shown for comparison and illustrates how corrosion resistance comparable to Zircaloy can be achieved with suitably heat treated Zr-2.5Nb alloy. Figure 11 clearly shows that higher oxidation rates are associated with the metastable β-Zr phase in pressure tube structures. Here the oxide interface has been exposed by etching away the underlying metal of an oxidized sample of as-extruded pressure tube, revealing the presence of well defined, penetrating oxide ridges coincident with the filaments of metastable β-Zr.

Effect of Neutron Irradiation

Neutron irradiation of quenched, cold worked and aged pressure tube material at about 300°C produces defect clusters that collapse to form small dislocation loops which appear as "black dots", Figure 12a. The strains produced by these loops should be similar to those produced by dislocations introduced during cold working and thus may affect corrosion resistance in a similar manner. The contrast arising from the small loops formed at this irradiation temperature is very similar to that arising from small precipitates, however there are indications that excess niobium is being precipitated during the irradiation. In the α-prime phase of material irradiated to a fluence of 4.4×10^{25} n/m^2 ,after about 14 full power years in a power reactor, the amount and coarseness of niobium precipitate is much greater than in the starting material, Figure 12b. In the equilibrium α there is some indication of radiation enhanced precipitation as is shown by the presence of precipitates in material heated at 500°C to remove the irradiation damage, Figure 12c. Heating unirradiated material for 1000h at temperatures up to 500°C produces no such precipitates. Enhanced precipitation from the α-phase has been shown to occur in annealed Zr-2.5Nb irradiated at 570K (16). In this case the niobium content of the α-Zr would be about 0.6 wt% while the equilibrium α-Zr in quenched material would contain less.

Neutron irradiation of extruded and cold worked pressure tube material at about 300°C produces a dislocation loop structure similar to that seen in quenched materials, Figure 13a. Suppressing the loop contrast by imaging using a basal reflection reveals Nb-rich precipitates in the elongated α-grains, Figure 13b. During extended service, additional decomposition of the β-Zr stringers will occur but even at a fluence of 6.8 x 10^{25} n/m^2 after about 10 full power years of irradiation, the β-Zr stringers remain continuous with some indication of larger precipitates forming at the α/β boundaries, Figure 13c. X-ray diffraction shows that the omega phase and β-Zr, enriched to 50 to 60% Nb, are present but there appears to be no indication of β-Nb in sufficient size or quantity to be detected by x-rays.

Post irradiation corrosion experiments have provided some insight as to the role these modified structures might play during corrosion in reactor. In these experiments, unirradiated and irradiated specimens are oxidized out reactor and their behaviour compared. Results on quenched, cold-worked and aged Zr-2.5Nb alloy have been previously reported (12). In that work, specimens which were quenched from 870°C, cold worked 15% and then aged at various temperatures and times were irradiated in the G7 loop of the Engineering Test Reactor at Idaho Falls to a fluence of 1 x 10^{25} n/m^2 or exposed in an out-reactor autoclave at the Chalk River Nuclear Laboratories (CRNL). Both sets of specimens were exposed to 280°C water for up to 90 days. After the exposures some of the irradiated and unirradiated specimens had their oxide removed and their surfaces prepared in the same way they had been initially, while other irradiated and unirradiated specimens were left with their oxide intact. All four types of specimens were then re-exposed in an out-reactor autoclave at the same test temperature as before, and their weight gains compared with those from their initial in- and out-reactor exposures. Figure 14 shows the results that were obtained for specimens aged at 450°C for 100 minutes. Firstly, the initial exposures for up to 90 days show that considerably less corrosion occurs in reactor than out reactor. Secondly, the unirradiated and irradiated specimens with their oxides intact continued to corrode as if they were still out reactor and in reactor respectively. Thirdly, specimens with their oxides removed behaved essentially the same as specimens originally exposed in the out-reactor and in-reactor test. Similar observations were recorded for specimens which had been aged at different temperatures and for different times. These results implied that irradiation-induced changes in the microstructure of the quenched, cold worked and aged material were responsible for the low in-reactor corrosion observed in the initial 90 day test.

Another experiment was done using specimens from an extruded and cold-worked pressure tube which had been irradiated in the Pickering NGS to a fluence of 6.8 x 10^{25} n/m^2 after 9.5 effective full power years. Corrosion coupons prepared from the irradiated tube have been exposed to 300°C moist air for up to 204 days along with unirradiated archive specimens from the same tube. The results in Figure 15 show that the irradiated specimens are oxidizing at a significantly lower rate than their unirradiated counterparts. Like the earlier experiments with heat-treated material, these results with cold-worked material indicate that a more corrosion resistant microstructure has evolved during the course of this pressure tube's irradiation in a power reactor.

Discussion

The corrosion of Zr-2.5Nb alloy is very sensitive to the microstructure which can be altered significantly by various thermo-mechanical treatments. This should not be too surprising since the microstructure can consist of several different phases, each containing various concentrations of niobium and possessing either close packed hexagonal or body centered cubic structures. Increasing the niobium content of Zr-Nb binary alloys over the range 1 to 5 wt% generally causes an increase in corrosion in high temperature water (17,18). In steam, the corrosion rate is seen to increase over the range 1 to 20 wt% Nb, particularly in the early stages of oxidation when the exposure times are relatively short (19). However the situation becomes complicated by the fact that the alloy rankings can change with exposure time and temperature. Since Zr-Nb alloys are multiphase, heat treatable alloys, it is not unreasonable to suspect that structural changes, which could influence the kinetic behaviour, might even occur at the test temperature.

The difference in the corrosion behaviour between martensitic and Widmanstatten structures is most likely related to the niobium content in solid solution. In martensite, the supersaturated α-prime needles contain at least 2.5 wt% Nb if produced by β-quenching, and even more if produced by quenching from the $\alpha+\beta$ region, whereas in Widmanstatten structures, the niobium content of the α-Zr is much less and close to the equilibrium concentration of 0.6 wt%. Early Soviet (18) and Canadian (20) results both showed that corrosion rates of binary Zr-Nb alloys were dependent on the amount of niobium in solid solution in α-Zr with minimum rates being associated with concentrations of niobium close to the equilibrium solubility at the corrosion temperature. The crossover of the 360°C and 400°C weight gain curves in Figure 7a may in fact be a consequence of a changing microstructure due to a larger reduction in the niobium supersaturation from ageing reactions at the higher test temperature.

The solution temperature effects seen in Figure 8 are most likely due to the difference in corrosion behaviour of α-prime and α-Zr with the more corrosion resistant α-Zr being present in material solution treated and quenched from the two phase region. Some support for this can be gleaned from tests on Zr-2.5Nb in 300°C air (21,14) where it has been observed that the oxidation rate of quenched specimens decreases with increasing volume fraction of equilibrium α-Zr in an α-prime matrix structure.

The effects of cold work and ageing on the corrosion resistance of quenched Zr-2.5Nb structures has been the subject of several investigations (11,12,13, 14). The improvements in corrosion resistance with ageing have generally been associated with the precipitation of small β-Nb particles from the supersaturated α-prime phase, and thereby a reduction in the niobium concentration of the matrix structure. The benefits of higher ageing temperatures and longer ageing times are thus related to the additional precipitation arising from such treatments. There is some evidence that excessive ageing may in fact lead to an eventual loss of some of the initial benefits of such treatments (22), however the evidence for this is still rather weak. The benefits of intermediate cold working between the quenching and ageing steps have been ascribed to additional precipitation being promoted at defects introduced by the deformation processes (12).

Changes to a more equilibrium microstructure are also believed to be responsible for the improvements seen when cold-worked, pressure-tube structures are thermally aged. The morphological features seen at the oxide-metal interface demonstrate that the Nb-rich, metastable, β-Zr oxidizes faster than equilibrium α-Zr. Thermal treatments would tend to decompose the β-Zr, eventually producing the more corrosion resistant equilibrium α-Zr plus β-niobium structure. In achieving this end, incremental improvements in corrosion resistance would be expected with progressive β-Zr decomposition, as is observed in out-reactor tests.

During the course of irradiation, the microstructure of Zr-2.5Nb pressure tubes in both the heat treated and the cold-worked conditions changes. From post-irradiation corrosion testing, these changes coincide with an improvement in their corrosion resistance. On the basis of the response of specimens re-exposed with and without oxide, it can been argued that radiation induced changes to the alloy substrate rather than to the oxide layer or to the local coolant chemistry during irradiation was responsible for the improvement seen with heat-treated material (12). The additional precipitation of β-Nb seen in both the α-prime and α-Zr phases of irradiated pressure tube material is a change which is consistent with a reduction in corrosion rate. Although additional precipitation in the α-prime may occur from the combined effects of neutron irradiation and thermal ageing at the irradiation temperature, post-irradiation annealing experiments have demonstrated that radiation induced precipitation can also occur within the equilibrium α-Zr phase (16). Both processes are believed to be responsible for the observed irradiation effect on the corrosion of heat-treated material, an effect which becomes apparent at fluences as low as 1×10^{25} n/m^2.

In irradiated cold-worked pressure tube material, both the decomposition of β-Zr and precipitation of β-Nb within the α-Zr are observed. β-Zr decomposition can occur as a result of thermal ageing at the irradiation temperature, with perhaps some enhancement from neutron irradiation. However, the extent of β-phase decomposition in the irradiated pressure tube structure is considerably less than that which led to the significant reduction in corrosion of thermally aged pressure tubes tested in 316°C water. On the other hand, the presence of small β-Nb precipitates within the elongated α-Zr grains are truly a consequence of neutron irradiation since they cannot be made to precipitate in unirradiated material from thermal treatments alone (16). Irradiation defects in the metal most likely induce precipitation by either increasing the mobility of niobium in α-Zr and/or providing sites for easy nucleation. Irrespective of the actual precipitation mechanism, the net result is a corresponding reduction in the niobium concentration of the α-Zr matrix and a fine distribution of second phase particles. Although both β-Zr decomposition and β-Nb precipitation should individually contribute to increasing the corrosion resistance, processes associated with irradiation-induced precipitation within the α-Zr grains are believed to be the major contributor to the irradiation effect seen in these tests with cold-worked pressure tube material.

Summary

The corrosion resistance of Zr-2.5Nb alloy is sensitive to metallurgical microstructure. Highest corrosion rates are generally observed on martensitic structures produced by water quenching from the β-Zr phase field. Thermal ageing of martensitic structures below 610°C precipitates second phase β-Nb particles resulting in an improvement in corrosion resistance. Additional improvements can be obtained by quenching from the α+β phase region and by introducing cold work between the quenching and ageing steps. The corrosion resistance of a two-phase, α-Zr plus β-Zr structure, produced by air cooling from the α-Zr plus β-Zr region is improved if the metastable β-Zr is thermally decomposed to the more equilibrium α-Zr plus β-Nb phases.

The quenched, cold worked and aged material as well as the extruded and cold worked material both undergo microstructural changes during fast neutron irradiation at reactor operating temperatures. Both show a fine distribution of irradiation-induced β-Nb precipitates within the α-Zr grains. Additional precipitation in the α-prime matrix of heat-treated material also occurs during irradiation. The observed precipitation and the associated reduction in niobium concentration of the matrix structures are believed to be largely responsible for the increase in corrosion resistance of irradiated material detected in post-irradiation corrosion tests.

References

1. C.E. Lundi and R.H. Cox, Proceedings of the USAEC Symposium on Zirconium Development, Rep. GEAP-4089, Vol. 1 (1962).

2. S. Banerjee, S.J. Vijayakar and R. Krishnan, "Precipitation in Zirconium-Niobium Martensites", J. Nucl. Mater. 62 (1976).

3. C.P. O and G.C. Weatherly, "The Precipitation Behaviour of a Zr-2.5 Wt Pct Nb Alloy", Metallurgical Transactions A, 19A (1988).

4. C.D. Williams and R.W. Gilbert, "Tempered Structures of a Zr-2.5wt%Nb Alloy", J. Nucl. Mater. 18 (1966).

5. S.A. Aldridge and B.A. Cheadle, "Age Hardening of Zr-2.5wt%Nb Slowly Cooled From the (α+β) Phase Field", J. Nucl. Mater. 42 (1972).

6. C.D. Williams and C.E. Ells, "The Influence of Niobium in Irradiation Strengthening of Dilute Zr-Nb Alloys", Phil. Mag. 18, 154 (1968).

7. C.D. Williams and R.W. Gilbert, "On Structural Factors Influencing the Strength of Martensitic Δ' in Quenched Zr-2.5% Nb", Trans. JIM 9 (1968).

8. B.A. Cheadle and S.A. Aldridge, "The Transformation and Age Hardening Behaviour of Zr-19wt%Nb", J. Nucl. Mater. 47 (1973).

9. R.F. Hehemann, "Transformations in Zirconium-Niobium Alloys", Can. Met. Quart. 11 (1972).

10. S.B. Dalgaard, "The Corrosion Resistance of Zr-Nb and Zr-Nb-Sn Alloys in High Temperature Water and Steam - Part II", AECL-1308, (1961).

11. J.E. LeSurf, "The Corrosion Behaviour of 2.5Nb Zirconium Alloy", Applications-Related Phenomena for Zirconium and its Alloys, ASTM STP 458, (1969).

12. V.F. Urbanic, J.E. LeSurf, and A.B. Johnson, Jr., "Effect of Ageing and Irradiation on the Corrosion of Zr-2.5 Wt% Nb", Corrosion 31, 1 (1975).

13. D.G. Lees, "The Effect of Production Route, Ageing Time and Oxygen Content on the Properties of Zr-2.5wt% Nb", Corrosion Science 5 (1965).

14. B. Cox, "Long Term Oxidation of Zr-2.5wt% Nb Alloy", AECL-5610, (1976).

15. V.F. Urbanic, et al., "Oxidation and Deuterium Uptake of Zr-2.5Nb Pressure Tubes in CANDU-PHW Reactors", Zirconium in the Nuclear Industry: Eighth International Symposium, ASTM STP 1023, (1989).

16. C.E. Coleman, R.W. Gilbert, G.J.C. Carpenter and G.C. Weatherly, "Precipitation in Zr-2.5 wt% Nb During Neutron Irradiation", (Proceedings of a Symposium on Phase Stability During Irradiation, Met. Soc. of AIME), Also as AECL 7444, (1980).

17. AMBARTSUMYAN et al, "Mechanical Properties and Corrosion Resistance of Zirconium and its Alloys in Water, Steam and Gases at High Temperatures", (Proceedings of the Second International Conference on the Peaceful Uses of Atomic Energy, Vol. 5), (1958).

18. O.S. Ivanov and V.K. Grigorovich, "Structure and Properties of Zirconium Alloys", ibid., Vol. 5, (1958).

19. B. Cox, M.J. Davies and T. Johnston, "The Oxidation and Corrosion of Zirconium and its Alloys", AERE-R-3257, (1960).

20. C.E. Ells, S.B. Dalgaard, W. Evans and W.R. Thomas, "Development of Zirconium-Niobium Alloys", (Proceedings of the Third International Conference on the Peaceful Uses of Atomic Energy, Vol. 9), (1964).

21. M.G. Cowgill and W.W. Smeltzer, "The Oxidation Properties of a Zirconium-2.7 w/o Niobium Alloy in the Temperature Range 300-500°C", J. Electrochem. Soc. 114, 11 (1967).

22. L.V. Ramanathan, I. Costa and W.A. Monteiro, "Influence of Heat Treatment on Structure and Oxidation Behaviour of Zr-2.5Nb", J. of Testing and Evaluation, JTEVA 17, 3 (1989).

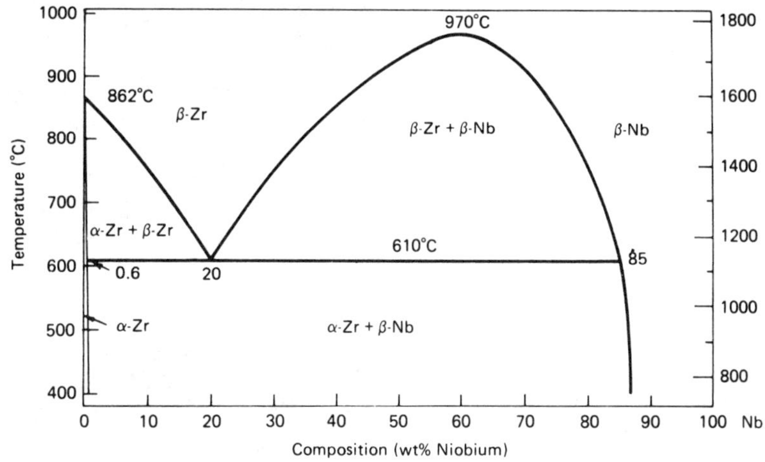

FIG. 1. Zirconium-Niobium phase diagram after Lundin and Cox (1).

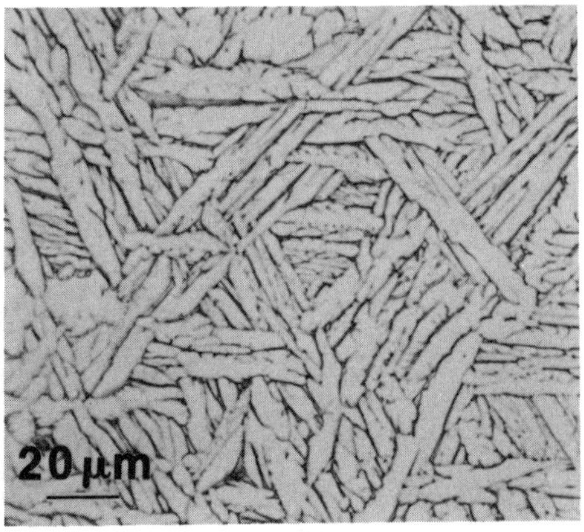

FIG. 2. Widmanstatten structure formed by slowly cooling Zr-2.5Nb from the β region.

FIG. 3. Martensitic structure formed by water quenching Zr-2.5Nb from the β region.

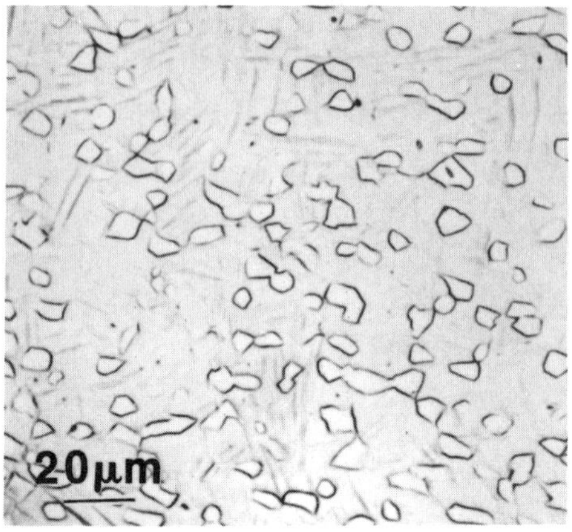

FIG. 4. Islands of equilibrium α in a martensitic α′ matrix formed by water quenching Zr-2.5Nb from the α+β region.

FIG. 5. Martensite matrix structure of Zr-2.5Nb water quenched from 870°C, cold worked 17% and aged at 500°C for 24 h.

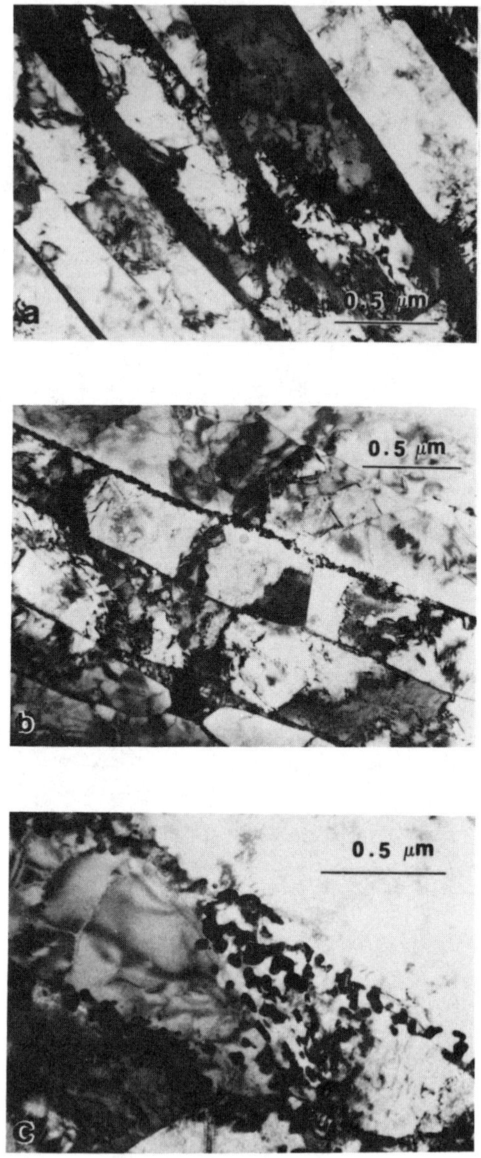

FIG. 6. Effect of thermal treatment on the microstructure of cold-worked Zr-2.5Nb pressure tube;
(a) as-received, (b) heated at 400°C for 72 h, (c) heated at 500°C for 10 h.

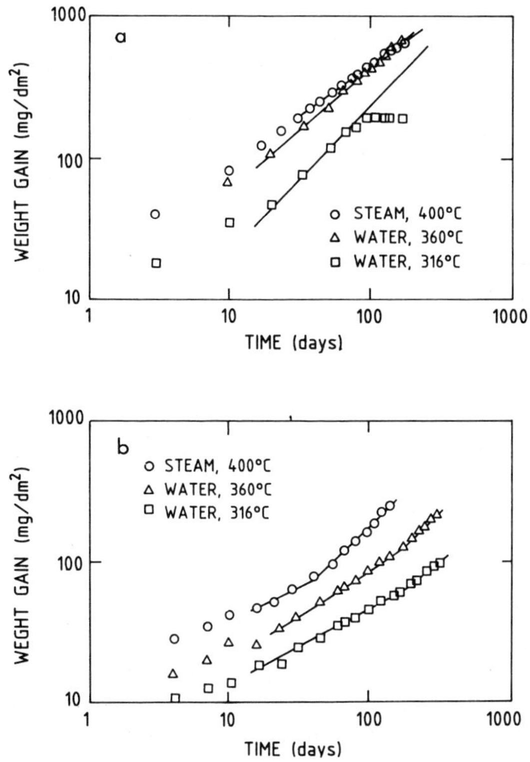

FIG. 7. Out-reactor corrosion of Zr-2.5Nb;
(a) water quenched from the β region (martensite),
(b) rapidly cooled from the β region (Widmanstatten).

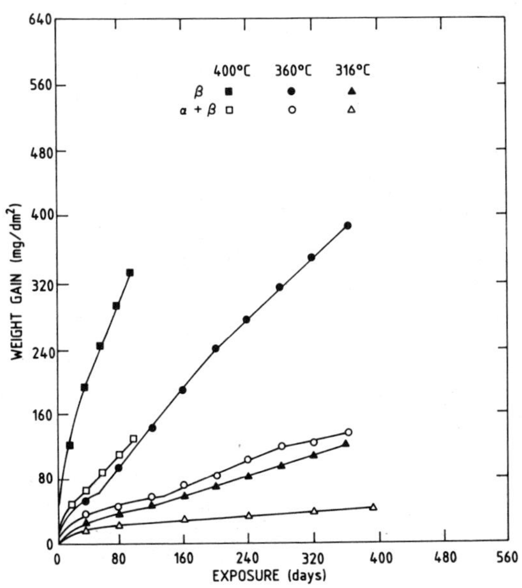

Fig. 8. Out-reactor corrosion of Zr-2.5Nb water quenched from the β region (950°C) and the $\alpha+\beta$ region (825°C), and subsequently aged at 535°C for 6 h.

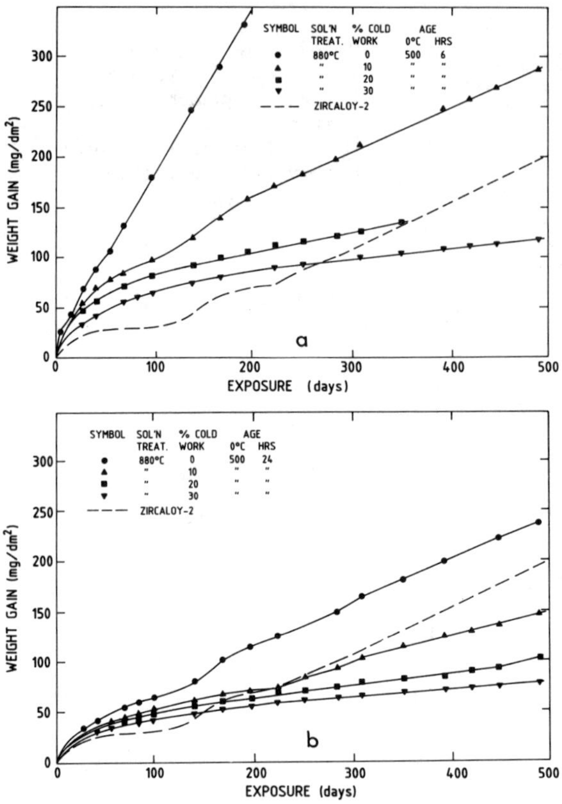

FIG. 9. Out-reactor corrosion of Zr-2.5Nb water quenched from 880°C, cold worked 0-30% and aged at 500°C for (a) 6 h and (b) 24 h in 360°C water.

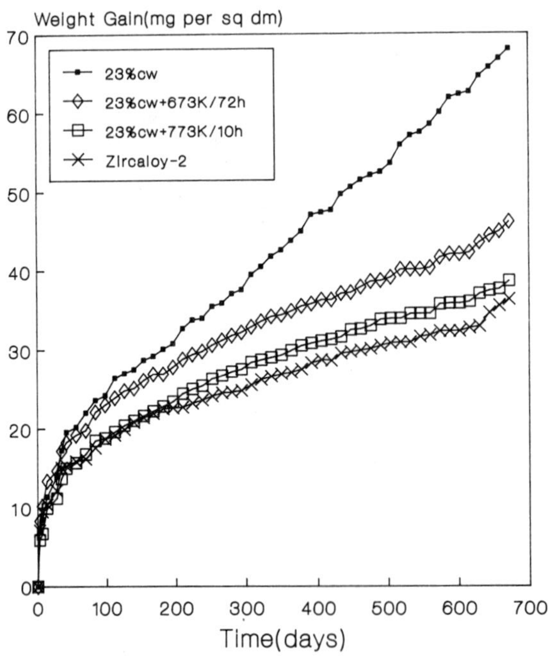

FIG. 10. Effect of thermal ageing on the corrosion of cold-worked Zr-2.5Nb pressure tubes in 316°C water.

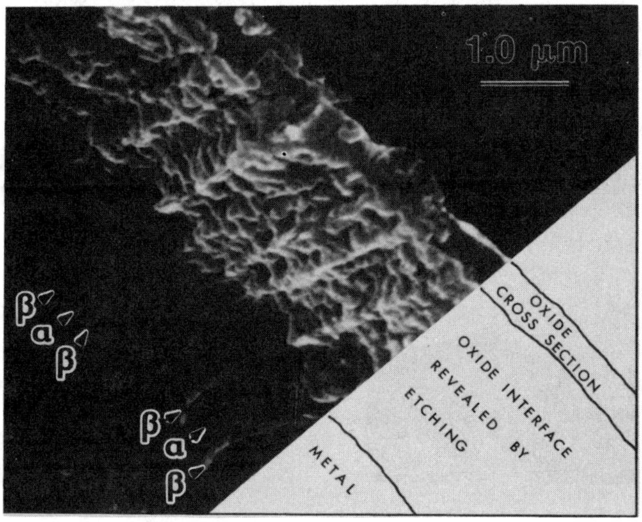

FIG. 11. Oxide interface of a 400°C steam-formed oxide on as-extruded Zr-2.5Nb revealed by etching away the underlying material.

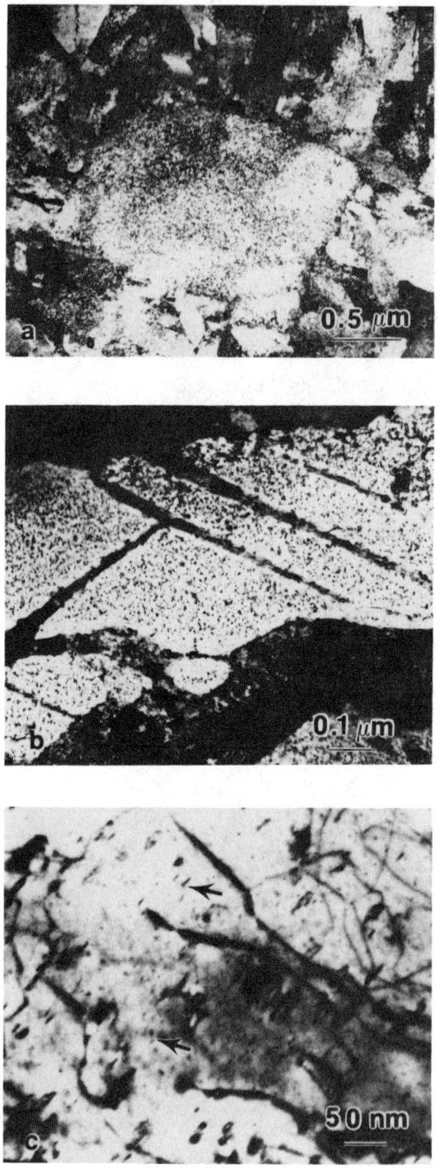

FIG. 12. Microstructure of heat treated Zr-2.5Nb pressure tube after irradiation showing (a) "black dot" irradiation damage, (b) Nb-rich precipitates in α' and (c) Nb-rich precipitates in equilibrium α-Zr.

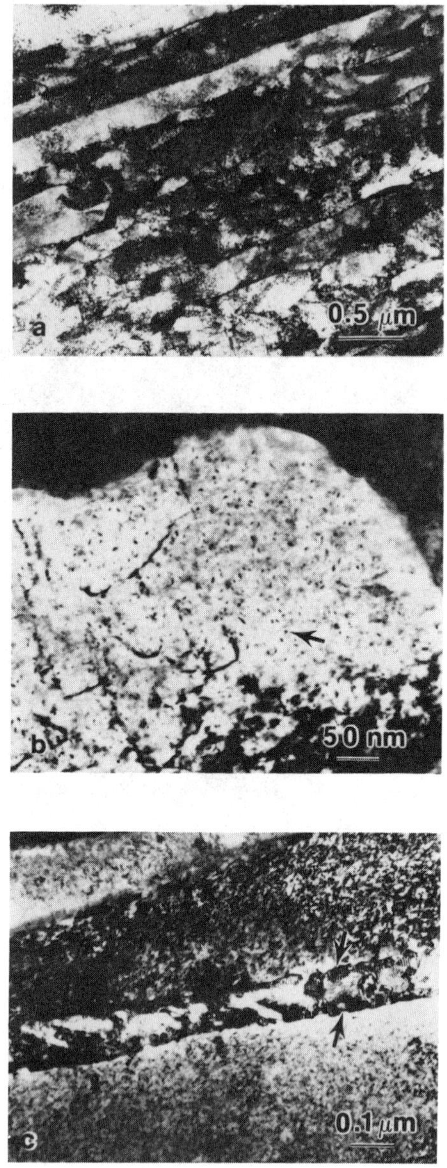

FIG. 13. Microstructure of extruded plus cold-worked Zr-2.5Nb pressure tube after irradiation showing (a) "black dot" irradiation damage, (b) Nb-rich precipitates in the α-Zr and (c) slight decomposition of a β-Zr filament.

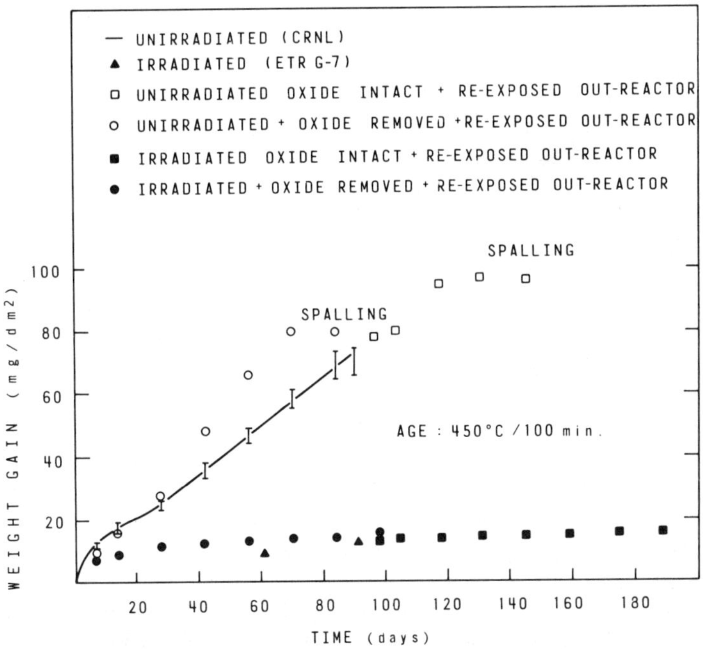

FIG. 14. Results of in-reactor, out-reactor and post-irradiation corrosion tests in 280°C water on Zr-2.5Nb alloy which has been water quenched from 870°C, cold worked 15% and aged at 400°C for 100 min.

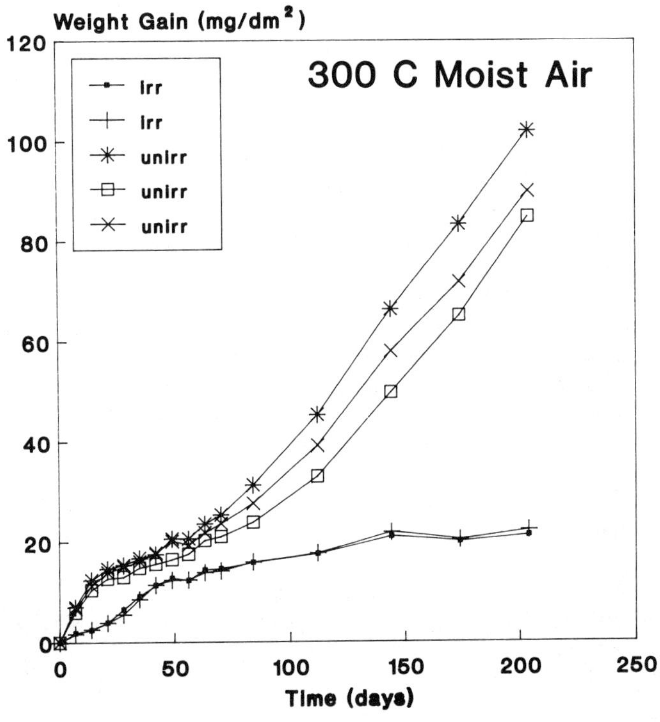

FIG. 15. Results of post-irradiation corrosion tests in 300°C moist air on cold-worked Zr-2.5Nb pressure tube material.

High-temperature oxidation behaviour of UO_2 in air and steam

D.S. Cox, C.E.L. Hunt, R.F. O'Connor, R.D. Barrand
Atomic Energy of Canada Limited, Research Company, Chalk River Nuclear Laboratories, Chalk River, Ontario, Canada, K0J 1J0

F.C. Iglesias
Ontario Hydro, Nuclear Safety Department, Toronto, Ontario, Canada, M5G 1X6

Abstract

The study of uranium dioxide (UO_2) oxidation behaviour in high temperature air and steam environments is part of an ongoing program in support of modelling fission product releases during postulated accidents in CANDU* nuclear reactors. This paper summarizes the results of oxidation and volatilization experiments which have been conducted in controlled atmospheres of steam/hydrogen at 500-800°C and in air at 1700-2100°C. The formation of U_3O_8 was observed in steam conditions below 800°C. The rate controlling process in low-temperature steam is discussed. The high-temperature volatilization rates in air are compared with other data from lower temperatures in steam, and a simple model is proposed to describe volatilization rates as a function of temperature and oxygen potential and flow conditions.

Introduction

One of the principal objectives of research relating to the safety of nuclear reactors is to predict the releases of radioactivity from the reactor core into containment for any postulated accident scenario. Since the radioactivity is primarily in the form of fission products which are produced inside the UO_2 fuel during irradiation, the first barrier to its release is the UO_2 oxide matrix.

During certain postulated accident sequences, the UO_2 can be exposed to an oxidizing environment such as air or steam. Under these conditions, the UO_2 can incorporate excess oxygen to form hyperstoichiometric UO_{2+X} or be converted to higher oxides such as U_4O_9, U_3O_7 or U_3O_8. Enhanced fission product releases from the fuel can result through oxidation, in addition to degradation of the fuel as a consequence of volume increases if U_3O_8 is formed. Thus, in order to develop adequate models for fission product release during reactor accident conditions, a model describing the composition and morphology of UO_2 oxidation products as a function of time, temperature and oxygen potential is required.

Volatilization of the UO_2 is a process by which the condensed phase is vapourized due to the formation of a volatile uranium bearing species. In the oxidizing conditions considered in this

* Canada Deuterium Uranium

paper, the process is incongruent vapourization with the formation of gaseous UO_3. The volatilization process can be rapid at high temperatures, and as such represents a possible mechanism for the release of low-volatility fission products by a matrix stripping process.

Uranium-Oxygen Phase Relations

The material used as CANDU nuclear reactor fuel is natural UO_2 which has been sintered to about 96-98% theoretical density. The phases which may form when this material is oxidized are defined by the equilibrium U-O phase diagram, a portion of which is shown in Figure 1. The UO_2 phase exists over a wide range of stoichiometry ($UO_{2\pm X}$) at high temperatures, and above 1677°C, UO_{2+X} is the highest solid oxide which is stable. Below this temperature, the highest oxide is U_8O_{21}/U_3O_8 (excluding the UO_3 phase which is regarded as metastable at atmospheric pressure). The U_8O_{21}-U_3O_8 region of the diagram has not been extensively studied and for the current discussion, this region will be subsequently referred to as U_3O_8. Below 1130°C, U_4O_9 is formed as an intermediate phase between UO_{2+X} and U_3O_8. Below about 510°C, U_3O_7 can be formed prior to U_3O_8.

Phase stability limits in the U-O system are also governed by oxygen potentials of the gas phase. Figure 2 (1-3) shows oxygen isobars superimposed on the phase diagram to illustrate the effect of oxygen partial pressures on phase stability limits. In atmospheric air (10^5 Pa), the oxidation limit is defined by U_3O_8 below about 1550°C. Above this temperature, UO_{2+X} is the highest oxide formed. The oxygen potential of steam is lower than air, and varies with temperature and total pressure. In 10^5 Pa steam, UO_{2+X} is the oxidation limit above about 1050°C, while U_4O_9 is formed at lower temperatures. U_3O_8 is predicted below about 900°C in steam conditions. With increased total steam pressures, the oxygen partial pressure increases and Olander (4) has shown that U_3O_8 could be formed at higher temperatures.

In this paper, some new experimental data is presented for steam oxidation of UO_2 at 500 to 800°C, and volatilization data is presented from 1700 to 2075°C in air and dilute O_2-inert gas mixtures. The purpose of the low-temperature steam oxidations was to assess the time required

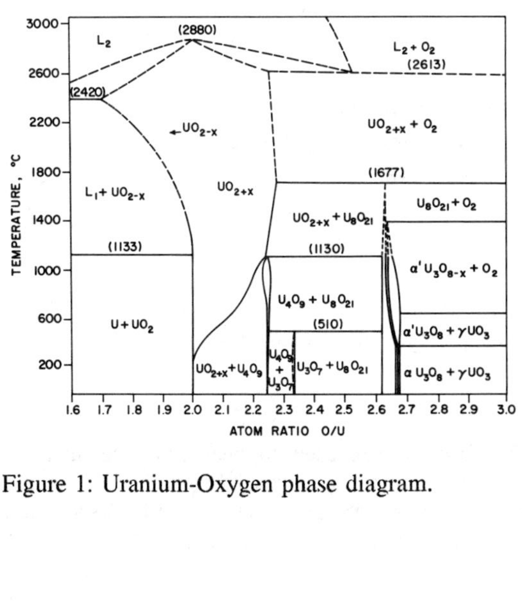

Figure 1: Uranium-Oxygen phase diagram.

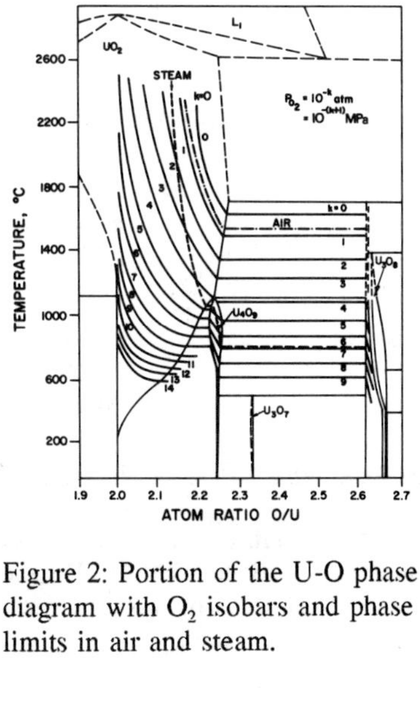

Figure 2: Portion of the U-O phase diagram with O_2 isobars and phase limits in air and steam.

to form U_3O_8, and thereby establish whether this reaction was relevant to the time scale of severe reactor accidents, fuel handling mishaps or long-term defected fuel operating conditions. This new data is interpreted with previously published data and simple models are proposed to describe the UO_2 behaviour.

Experimental Description

Low Temperature Steam Oxidation (500-800°C)

Four long-term steam oxidation tests were conducted with unirradiated fuel (designated as tests #1-4). Steam was generated by pumping deionized water at a controlled rate into a heated stainless steel pipe (steam generator). A carrier gas of blended argon/hydrogen was mixed with the steam, which flowed through a heated length of stainless steel tubing before entering the furnace. The UO_2 samples were held in a quartz tray with partitions located at measured positions in the temperature gradient of a horizontal tube furnace. The steam environment in the furnace was contained in a 25 mm diameter quartz tube with 'O'-ring connections to the steel piping. A partially-stabilized ZrO_2 (PSZ) oxygen sensor monitored the oxygen partial pressure of the gas after passing over the fuel. The sensor voltage was used to establish the mixing ratio of argon/H_2 in the carrier gas. The steam was condensed after exiting from the furnace, and the condensate was recycled by pumping to the steam generator. Steam flow rates were set by the pumping rate at 30 g/hr and the total flow rate of the argon/H_2 mixture was 100 STP ml/min. The carrier gas was vented from the condensate tank. Figure 3 shows the apparatus including the equipment for hot-cell testing.

The specimens were disks of UO_2, 12.15 mm in diameter and about 2 mm thick. Each sample was weighed before the test, and at intervals throughout the long-term steam exposures when the furnace was cooled briefly for this purpose.

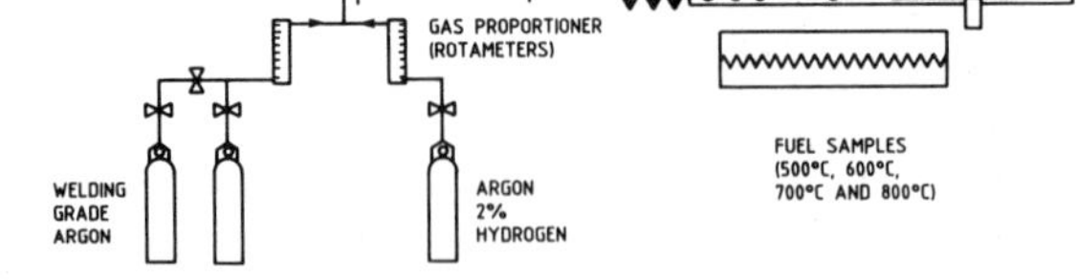

Figure 3: Apparatus for low-temperature steam oxidation tests.

Volatilization in Air (1700-2075°C)

A modified ARTCOR model 460-15 furnace was used to heat UO_2 specimens to temperatures above 2000°C. The thermal element was a resistance heated zirconia tube (16 mm outside diameter) which was preheated by a conventional nichrome heater. The UO_2 specimens were positioned on a thoria pedestal inside a vertical zirconia flow tube (6.4 mm outside diameter), concentric inside the heating element. The thoria boat was supported by a 3.8 mm diameter solid zirconia post. The preheater, main heater and zirconia flow tube were located in a canister which was flooded with welding-grade argon to limit the amount of oxygen in the hot region of the furnace and to provide some cooling of the preheater assembly. Gas-tight connections at both ends of the flow tube routed the gas flow down from the top of the furnace over the fuel sample and out the bottom of the tube which was connected to a PSZ oxygen sensor. The purpose of the oxygen sensor was to provide an accurate assessment of the Carrier gases of air and argon/2% hydrogen were supplied by pressurized gas cylinders connected to mass flow control valves. Figure 4 shows the experimental setup including the equipment for hot-cell tests.

A two-colour infrared pyrometer was used to measure the temperature of the external surface of the zirconia heating element at the elevation of the UO_2 sample. It was found during pre-test measurements, using representative gas flow rates, that close agreement was achieved between the pyrometer reading and the fuel temperature inside the reaction tube, as measured by Pt/Rh and W/Re thermocouples up to 1800°C. The pyrometer temperature was therefore regarded as a close approximation to the fuel temperature.

For all the tests, the UO_2 sample, the thoria boat, zirconia support rod and zirconia flow tube were loaded as a unit into the cold furnace. A new flow tube was used for each test. The specimens were heated in a carrier gas of argon/2% hydrogen flowing at 70 STP ml/min. The heating rate was 600°C/hour. In most of the tests, the carrier gas was switched to air (50 STP ml/min) after reaching the target temperature and allowing a stabilization period of approximately 5 minutes. The air flow was maintained for a period of about 15 minutes, after which it was replaced by argon/2%H_2 at 70 STP ml/min. Lowering of the temperature was started after a further five minutes. The cooling rate was about 10°C/min. After the furnace was cold, the reaction tube was removed and disassembled to recover the remaining UO_2 for weighing and ceramographic analysis.

Two types of samples were used in these tests. Unirradiated UO_2 parallelepipeds were cut from pellets using a oil-cooled diamond saw. These samples were about 2.5 mm by 2.5 mm by 6 mm in size. The second type of sample was from a previously irradiated fuel element and were irregular in shape. Magnified images of these samples were used to determine dimensions and to calculate surface areas.

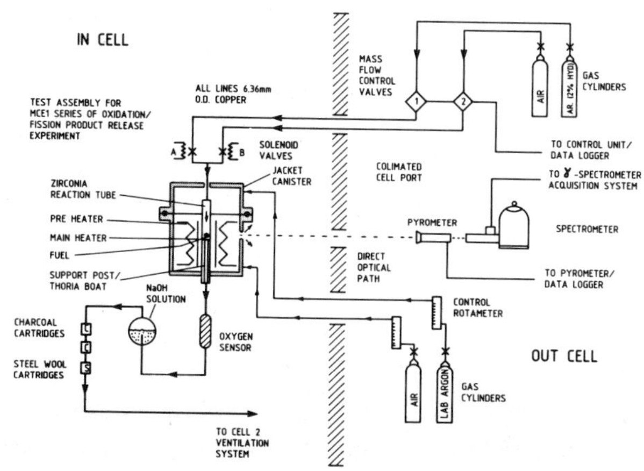

Figure 4: Apparatus for high-temperature volatilization tests.

Experimental Results

Low Temperature Steam Oxidation (500-800°C)

The results from all four unirradiated tests (tests #1, 2, 3 and 4) are summarized in Table 1. The first three tests showed that more degradation occurred at 700°C compared to 600 and 800°C. Test #1 was O_2-rich steam, and produced the most oxidation of any test in the series. Tests #2 and 3 were closer to a pure steam environment (although still slightly O_2-rich), and showed less oxidation. Table 1 indicates the predominant oxygen partial pressure for each test, and the ratio of oxygen content in the test to the oxygen content of pure stoichiometric steam. Test #4 was performed in slightly O_2-depleted conditions during the first 436 hours, and close to pure steam in the remaining period up to 901 hours. These conditions produced less oxidation than the previous tests which had excess O_2. The fastest oxidation in test #4 was observed at 600°C, as opposed to 700°C in the previous tests.

Figure 5 shows a summary of the oxidation kinetics for tests #1-4. The reaction rates decreased with time for all the tests. There was a six-fold increase in the oxidation rate between the kinetics at 700°C in test #1 (O_2-rich steam) compared to test #4 (close to pure steam). The same comparison showed only a 2.5 times increase in rate at 600°C.

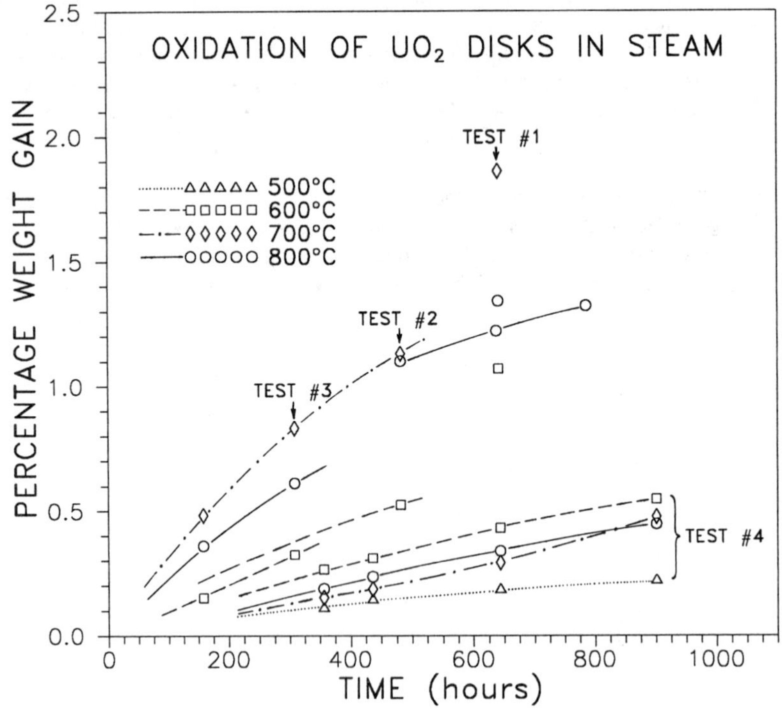

Figure 5: Oxidation kinetics for UO_2 disks in steam at 500, 600, 700 and 800°C. Disks were 12.15 mm diameter by about 2.0 mm thick.

TABLE 1: SUMMARY OF LOW TEMPERATURE STEAM OXIDATION EXPERIMENTS

TEST #	TOTAL TIME (hours)	TEMP. (°C)	INITIAL WEIGHT (g)	% WEIGHT CHANGE	DOMINANT O_2 PRESS. (Pa)	RATIO P_{O2}(test)/ P_{O2}(steam)	COMMENTS
1	641	600	2.42833	1.071	$3.0X10^{-1}$	476	oxidized
1	641	700	2.56393	1.864	$3.0X10^{-1}$	45	heavily oxidized
1	641	800	2.53012	1.337	$3.1X10^{-1}$	6.9	no visible oxid'n
2A	482	600	2.51090	0.523	$1.3X10^{-2}$	21	slightly oxidized
2A	482	700	2.50283	1.134	$1.7X10^{-2}$	2.5	oxidized
2A	482	800	2.42470	1.098	$5.6X10^{-2}$	1.2	no visible oxid'n
2B	640	600		N/A	$1.3X10^{-2}$	21	oxidized
2B	640	700		N/A	$1.7X10^{-2}$	2.5	oxidized
2B	640	800		1.22	$5.6X10^{-2}$	1.2	no visible oxid'n
2C	790	600		N/A	unknown	N/A	oxidized
2C	790	700		N/A	unknown	N/A	oxidized
2C	790	800		1.32	unknown	N/A	no visible oxid'n
3A	158	600	2.45768	0.151	$1.3X10^{-2}$	21	no visible oxid'n
3A	158	700	2.42769	0.483	$1.7X10^{-2}$	2.5	no visible oxid'n
3A	158	800	2.49746	0.360	$5.6X10^{-2}$	1.2	no visible oxid'n
3B	308	600		0.323	unknown	N/A	slightly oxidized
3B	308	700		0.831	unknown	N/A	oxidized
3B	308	800		0.611	unknown	N/A	no visible oxid'n
4A	356	500	3.09934	0.115	$4.3X10^{-9}$	$1.2X10^{-4}$	no visible oxid'n
4A	356	600	2.50424	0.263	$2.7X10^{-5}$	0.043	no visible oxid'n
4A	356	700	3.15004	0.152	$4.7X10^{-3}$	0.7	no visible oxid'n
4A	356	800	3.20114	0.187	$4.5X10^{-2}$	1.0	no visible oxid'n
4B	436	500		0.148	$4.3X10^{-9}$	$1.2X10^{-4}$	no visible oxid'n
4B	436	600		0.309	$2.7X10^{-5}$	0.043	no visible oxid'n
4B	436	700		0.186	$4.7X10^{-3}$	0.7	no visible oxid'n
4B	436	800		0.235	$4.5X10^{-2}$	1.0	no visible oxid'n
4C	644	500		0.189	$2.0X10^{-4}$	5.7	no visible oxid'n
4C	644	600		0.430	$7.8X10^{-4}$	1.2	slight oxidation
4C	644	700		0.291	$6.7X10^{-3}$	1.0	no visible oxid'n
4C	644	800		0.338	$4.5X10^{-2}$	1.0	no visible oxid'n
4D	901	500		0.224	$2.0X10^{-4}$	5.7	slight oxidation
4D	901	600		0.546	$7.8X10^{-4}$	1.2	slight oxidation
4D	901	700		0.476	$6.7X10^{-3}$	1.0	no visible oxid'n
4D	901	800		0.446	$4.5X10^{-2}$	1.0	no visible oxid'n

Cross-sections of all the oxidized samples were ceramographically examined at high magnification. A layer of U_3O_8 was identified on all of the samples which had appeared to be oxidized (ie flaky or fragmented). The nature of this U_3O_8 layer varied from a uniform, compact layer at 500 and 600°C in test #4 (Figures 6 and 7), to a loose, fragmented layer in O_2-rich conditions. Figures 8 and 9 show the appearance of the U_3O_8 layer and the underlying oxide after 790 hours at 700°C in test #2. The U_3O_8 layer formed initially along the grain boundaries of the lower oxide material (probably U_4O_9), and then slowly consumed the individual grains. No U_3O_8 layers were found on any sample at 800°C, nor on the 700°C sample in test #4.

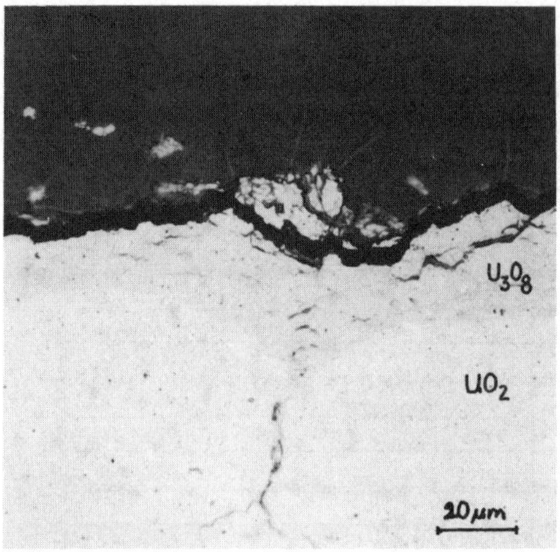

Figure 6: U_3O_8 layer on surface of UO_2 after 901 hours at 500°C (test #4).

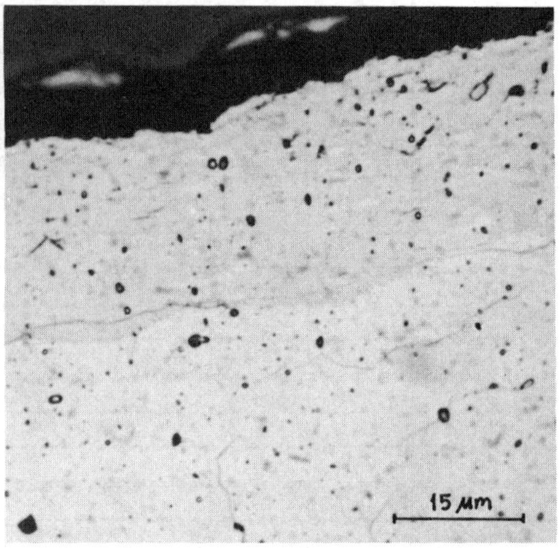

Figure 7: U_3O_8 layer on surface of UO_2 after 901 hours at 600°C (test #4).

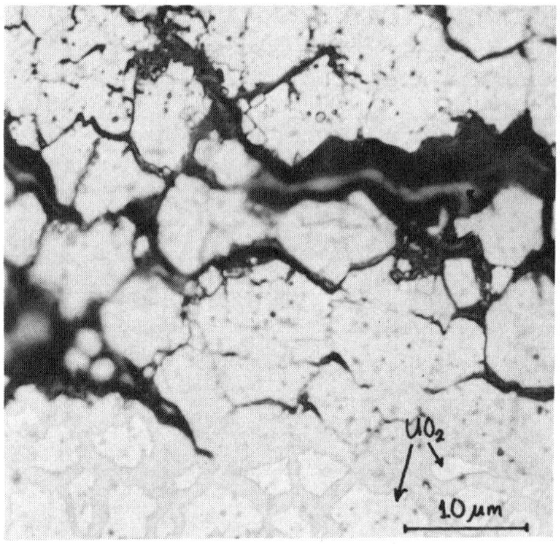

Figure 8: Fractured U_3O_8 layer and islands after 790 hours at 700°C (test #2).

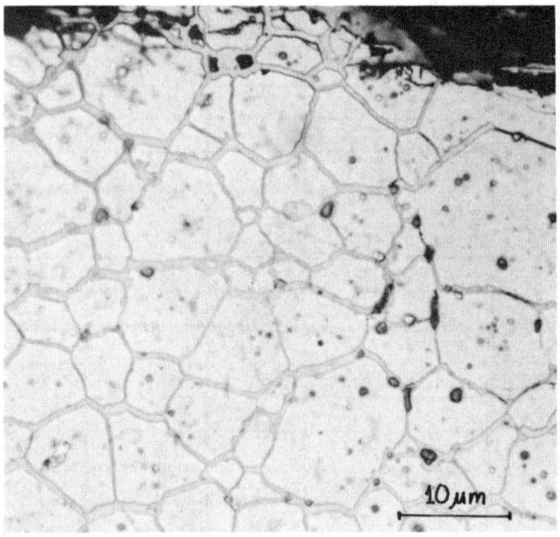

Figure 9: U_3O_8 formation along grain boundaries of UO_2 after 790 hours at 700°C (test #2)

Volatilization in Air (1700-2075°C)

A total of eleven tests were performed, nine in air and two in a dilute O_2/argon mixture. Five of these tests were performed on irradiated UO_2 as part of a fission product release experiment. A summary of the test results is given in Table 3. In all tests there was a net loss of UO_2 due to volatilization and the samples at 2005 and 2070°C were completely volatilized after 12 minutes in air. The experimental rates of volatilization reported in Table 3 were calculated assuming that the measured weight losses occurred over the total exposure time, and normalized to the initial geometric surface area of the specimen (as determined from pre-test dimensions). The rate of volatilization increased with temperature and P_{O2}.

TABLE 2: SUMMARY OF VOLATILIZATION TESTS

TEMP. (°C)	O_2 PARTIAL PRESSURE (Pa)	TIME (min)	% LOSS	EQUIL'M O/U RATIO	INITIAL WEIGHT (g)	VOLATILIZATION RATE ($g/cm^2/s$)		RATIO EXP/CALC
						EXPERIMENT	CALCULATION	
1705	$2.09X10^4$	15	19.5	2.227	0.256*	$1.14X10^{-4}$	$2.09X10^{-3}$	18.3
1719	$2.09X10^4$	20	15.2	2.225	0.241	$6.37X10^{-5}$	$2.11X10^{-3}$	33.1
1750	$2.09X10^4$	2	8.5	2.221	0.143	$2.58X10^{-4}$	$4.35X10^{-3}$	16.9
1804	$2.09X10^4$	15	47.3	2.215	0.261*	$2.79X10^{-4}$	$3.95X10^{-3}$	14.2
1813	$2.09X10^4$	20	30.0	2.213	0.276	$1.30X10^{-4}$	$3.46X10^{-3}$	26.6
1896	$2.09X10^4$	14	48.5	2.205	0.237	$2.74X10^{-4}$	$6.45X10^{-3}$	23.5
1911	$2.09X10^4$	2	34.2	2.203	0.188	$9.30X10^{-4}$	$8.08X10^{-3}$	8.69
2005	$2.09X10^4$	12	100	2.194	0.298*	$7.68x10^{-4}$	$1.11x10^{-2}$	14.5
2070	$2.09X10^4$	12	100	2.189	0.220*	$6.99x10^{-4}$ †	$2.09x10^{-2}$	29.9
1890	$6.91X10^{-4}$	15	18.3	2.080	0.380	$9.18X10^{-5}$ †	$1.99X10^{-4}$	2.17
2075	$1.05X10^{-3}$	15	90.6	2.063	0.191*	$7.64X10^{-4}$	$3.98X10^{-3}$	5.21

* Specimens from previously irradiated fuel. All other samples were unirradiated.
† Minimum rate, assumes complete exposure time required

Discussion of Results

Low Temperature Steam Oxidation (500-800°C)

Oxidation of UO_2 in steam above 1000°C has been studied previously by several authors. Carter and Lay (5) studied UO_2 in CO/CO_2 and proposed a rate model based on surface dissociation of CO_2. Their model was adapted by Cox et al (6) to oxidation in steam. According to this model, the rate of change of the deviation from stoichiometry (ie. x in UO_{2+X}) can be expressed as:

$$\frac{dx}{dt} = \frac{\alpha S}{V}(x^{eq} - x(t)) \qquad [1]$$

where α is the surface exchange coefficient, S is the surface area of UO_2, V is the UO_2 volume, and x^{eq} is the equilibrium deviation from stoichiometry. For a given temperature and mixture of gases, x^{eq} can be determined knowing the oxygen potential of the gas and partial molar thermodynamic properties of oxygen in UO_{2+X}.

By assuming that the UO_2 is stoichiometric at time, t, equal to zero, the solution to equation [1] becomes:

$$x(t) = x^{eq}(1 - \exp(-\frac{\alpha S t}{V}) \qquad [2]$$

Since no U_3O_8 was observed at 700 and 800°C after 901 hours in test #4, it is possible to calculate α for this data by assuming that x^{eq} is defined by the oxygen content of the UO_{2+X}

phase limit at these temperatures. Using values of x^{eq} equal to 0.170 and 0.180, and *S/V* equal to 11.1 and 10.9 cm^{-1}, α becomes $1.78X10^{-8}$ and $1.52X10^{-8}$ cm/s at 700 and 800°C respectively.

The surface exchange coefficient was found to vary with temperature, *T*>1000°C, according to the following Arrhenius relationship (6):

$$\alpha = 36.5\exp(-\frac{23500}{T(K)}) \qquad (cm \cdot s^{-1}) \qquad [3]$$

The expression in equation [3] was derived for CANDU UO_2 from data in the temperature range 1000 to 1600°C. This data is shown in Figure 10 compared to the values for α calculated from the current data at 700 and 800°C. The data point at 800°C is consistent with extrapolation of the high-temperature data, but the 700°C data is about one order of magnitude faster than the extrapolation. This may be due to the formation of higher oxides which were not detected in the ceramographic examination, but which contributed to a higher weight gain.

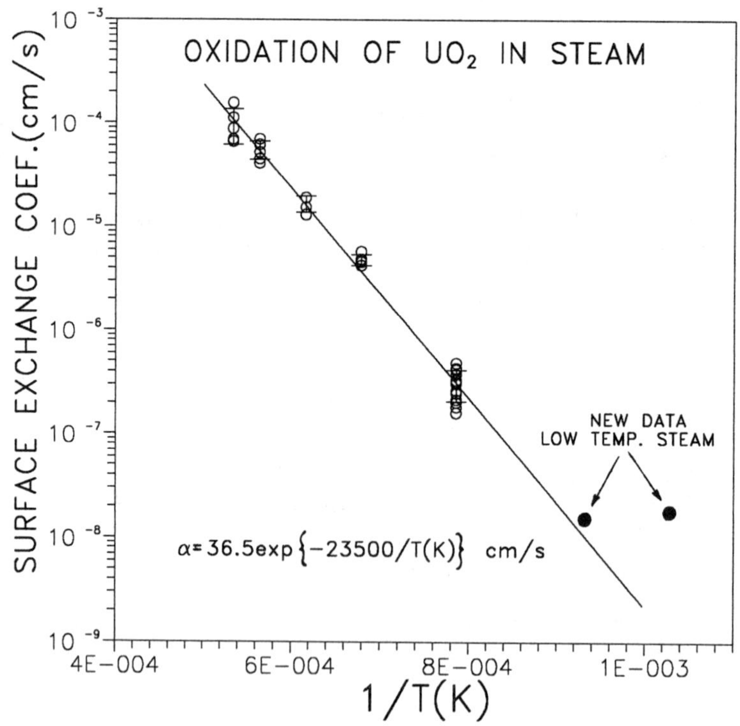

Figure 10: Arrhenius plot of surface exchange coefficient, α, according to (6), showing results of current study at 700 and 800°C.

It is currently possible to only qualitatively explain the effect of oxygen content in steam on the oxidation rate for cases were U_3O_8 was formed. This is accomplished using the available thermodynamic data for the substoichiometric U_3O_8 phase according to Ackermann et al (7,8). Figure 11 shows the relationship between oxygen partial pressure, degree of nonstoichiometry in U_3O_8 (ie oxygen content of U_3O_8), and temperature. Conditions for pure steam and rich or lean oxygen mixtures with steam are shown. Clearly, the composition of the U_3O_8 which forms at the surface of the fuel will vary significantly with the oxygen content of the steam. An obvious discrepancy between the current experimental data and this plot is the predicted formation of U_3O_8 at 800°C in pure steam (test #4). The information in Figure 11 must be combined with more accurate thermodynamic data and an oxygen diffusion model in order to predict the effect of temperature and oxygen concentration to rates of degradation. With the current information, it is not possible to predict the maximum oxidation rate at 700°C for O_2-rich steam, nor the maximum rate at 600°C for pure steam.

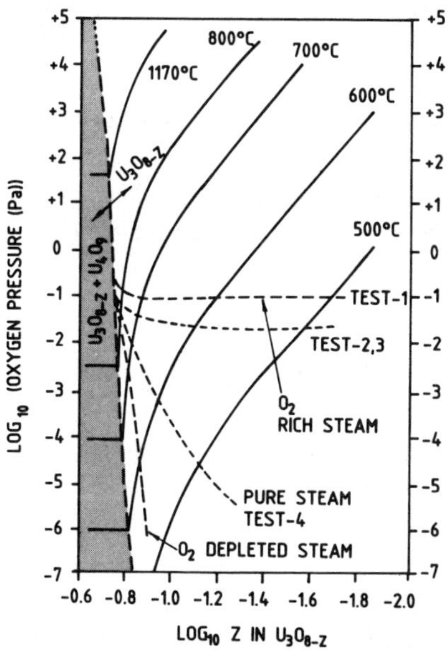

Figure 11: Phase stability limits for substoichiometric U_3O_8 in O_2-rich steam, pure steam and O_2-depleted steam. Calculated from data in references (7) and (8).

Volatilization in Air (1700-2075°C)

The calculations of volatilization rates shown in Table 2 were based on the assumption that UO_3(g) was responsible for the mass losses. The rate of mass loss (dm/dt) can be expressed as:

$$\frac{dm}{dt} = k_m (P^S_{UO_3} - P^{\infty}_{UO_3}) \frac{MW_{UO_3}}{RT} \qquad [4]$$

where k_m is the mass transfer coefficient, $P^S(UO_3)$ and $P^{\infty}(UO_3)$ are surface and distant vapour pressures of UO_3 and T is temperature (K). $MW(UO_3)$ is the molecular weight of UO_3.

The vapour pressure of UO_3 was calculated as outlined in (6), and from the data in (7), can be expressed as:

$$P_{UO_3} = P_{O_2}^{(\frac{1-x}{2})} \exp\left\{\frac{-19640}{T} + 7.21 + \frac{1}{2RT}\int_0^{x^{eq}} \Delta G_{O_2} dx\right\} \qquad [5]$$

Integration of the partial molar free energies of oxygen in UO_{2+X}, $\Delta G(UO_2)$, can be made from the data of Perron (9). P_{O2} is the oxygen partial pressure.

The forced convection mass transfer coefficients were evaluated according to Yovanovich (10).

$$k_m = \frac{D_{UO_3}}{\lambda A}\left\{2\sqrt{\pi} + (0.15\sqrt{\pi}\ \sqrt{Re} + 0.35\,Re^{0.566})\ Sc^{1/3}\right\} \qquad [6]$$

where λ is the characteristic length defined in (10) as the square root of specimen surface area. Re and Sc are the Reynolds and Schmidt numbers. $D(UO_3)$ is the diffusion coefficient of UO_3 in the gas phase, and is defined as (11):

$$D_{UO_3} \approx \frac{10^{-3}T^{1.75}}{P_{tot}(v^{1/3}_{UO_3} + v^{1/3}_{gas})^2}\sqrt{\frac{1}{MW_{UO_3}} + \frac{1}{MW_{gas}}} \qquad [7]$$

where v_i is the diffusion volume for molecule i, taken from (11).
The definition of the Reynolds number was taken as:

$$Re = \frac{U_{\infty}\lambda\rho}{\eta} \qquad [8]$$

where U_{∞} is the free stream gas velocity, ρ is gas density and η is gas viscosity. By using appropriate values for air or steam or argon, volatilization rates were calculated using equations [4-8].

A summary of the volatilization kinetics from the current tests, and from previous tests in steam and argon/O_2 (6) is shown in Figure 12. The calculations consistently over-predict experimental rates by a factor of 4 to 30, but show reasonable agreement for the three different gas mixtures (air, argon-O_2 and steam) over about 7 orders of magnitude.

The equilibrium oxygen contents during the high-temperature oxidation period for all of the tests were calculated from the measured oxygen partial pressures to be in the range of $UO_{2.063}$ to $UO_{2.227}$. The post-test oxygen contents of some of the samples were determined by oxidation to U_3O_8, and these results found oxygen levels less than $UO_{2.02}$. Thus, it is likely that the specimens were reduced at the end of the test during cooling in argon/2%H_2, and

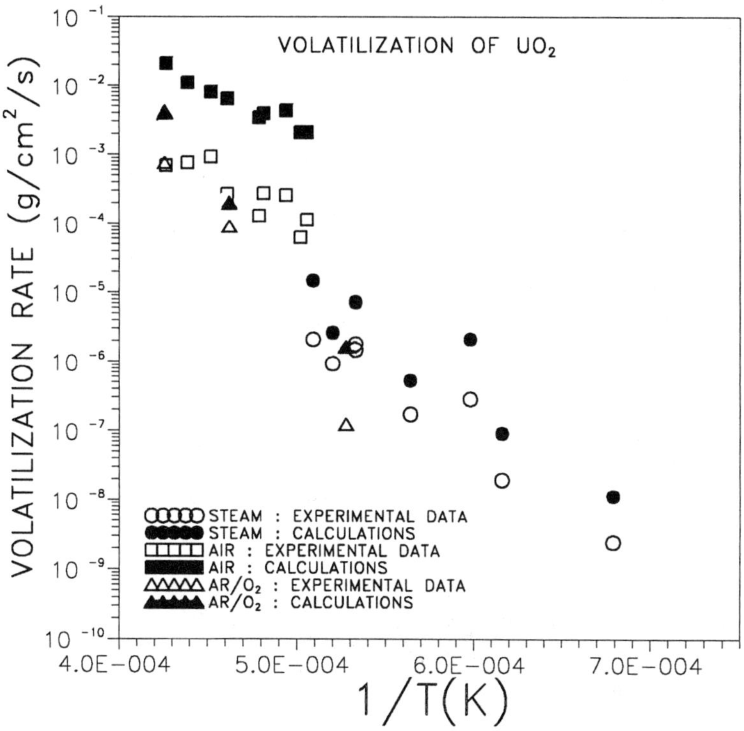

Figure 12: Arrhenius plot of volatilization rates of UO_2 in steam, air and argon/O_2 atmosheres, from 1000 to 2075°C. Calculations based on the proposed model are also shown.

therefore post-test O/U determinations could not be used to assess whether the equilibrium composition had been reached during the oxidation period. Strong evidence of reduction during cooling was noted in the data from the downstream oxygen sensor. Figure 13 shows the oxygen partial pressures and sample temperature measured during the air test at 1804°C. When the carrier gas was changed from argon/2%H_2 to air, the P_{O2} rapidly increased to a value representative of air. However, after 15 minutes, when air was replaced by argon/2%H_2 there was a delay of about 500 seconds during which the P_{O2} dropped slowly, after which there was a rapid drop in oxygen level representing the argon/2%H_2 mixture. Tests conducted with no specimen in the furnace showed only a minor delay of about 90 seconds. Thus it is possible that the delay resulted from oxygen which was evolved from the specimen (ie. reduction of the oxidized UO_{2+X}). Integration of the product of P_{O2} and flow rate over the 500 second delay period indicated that $1.4\pm.5\times10^{-4}$ moles of O_2 were evolved. This quantity of oxygen is about 1.5 times the amount required for oxidation to the equilibrium composition, and is therefore probably from reduction of the fuel upon changing atmospheres. The amount of oxygen evolved supports the conclusion that the UO_2 was completely oxidized during the 15 minute period.

Extrapolation of air oxidation rates from lower temperatures according to (6) predicts very rapid oxidation for these small samples. In fact, the extrapolated rates exceed the rate of O_2 supply from the flowing air, and thus oxygen starvation would be predicted. In such a case, the time to reach equilibrium O/U would be about 10-15 seconds for the sample sizes used in these tests. The occurrence of oxygen starvation could not be established from the oxygen sensor data during the change from argon/2%H_2 to air because of poor sensitivity under these conditions.

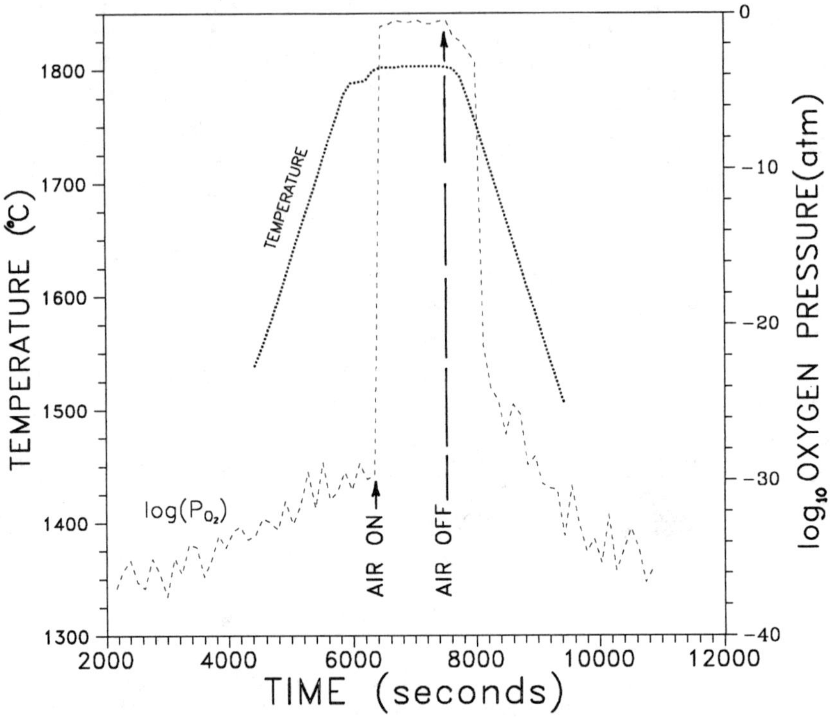

Figure 13: Oxygen partial pressures during air oxidation test at 1804°C, showing oxtgen evolution from UO_2 at end of test.

Summary and Conclusions

Significant degradation of unirradiated UO_2 can occur in steam at 500-800°C, as a result of higher oxide formation, with associated fragmentation and powdering. The most extensive degradation occurred at 700°C in slightly oxygen-rich steam. In pure steam, less oxidation was observed, with the maximum degradation at 600°C,

The rate of oxidation, and the reaction products which are formed, both depend strongly on the oxygen content of the steam. Higher oxygen levels produced more oxidation. In the range of oxygen partial pressures assessed in these experiments, a factor of 3 to 6 increase in oxidation rate was observed for oxygen-rich compared to oxygen-depleted steam.

Steam oxidation at 800°C was consistent with extrapolation of data from higher temperatures. A surface controlled reaction is implied.

More experimental data is needed to define the effect of temperature and oxygen pressure on the lower phase boundary limit of U_3O_8.

A simple model for UO_2 volatilization in steam and air was developed. The model over-predicts the volatilization rates, but is in general agreement over a wide range of conditions.

Acknowledgements

Financial support for the high temperature volatilization work was provided by the CANDU Owners Group under WPIR 492. Helpful discussions on mass transfer with J. DeVaal are acknowledged.

References

(1) K. Naito and K. Kamegashira, Adv. Nucl. Sci. Technol., 9 (1976) 99.

(2) A.M. Anthony, R. Kiyoura, T. Sata, J. Nucl. Mat., 10 (1963) 8.

(3) S. Aronson and J. Belle, J. Chem. Phys., 29 (1958) 151.

(4) D.R. Olander, Nucl. Technol., 74 (1986) 215.

(5) R.E. Carter and K.W. Lay, J. Nucl. Mat., 36 (1970) 77.

(6) D.S. Cox, F.C. Iglesias, C.E.L. Hunt, N.A. Keller, R.D. Barrand, J.R. Mitchell, R.F. O'Connor, "Oxidation of UO_2 in Air and Steam with Relevance to Fission Product Releases", presented at American Chemical Society Symposium on Chemical Phenomena Associated with Radioactivity Releases During Severe Nuclear Plant Accidents, Anahiem CA, 1986 Sept. 7-12.

(7) R.J. Ackermann and A.T. Chang, J. Chem. Phys., 5 (1973) 873.

(8) "Thermodynamics and Transport Properties of Uranium Dioxide and Related Phases", International Atomic Energy Agency, Vienna, Technical Report Series No. 39, 1965, p 36-37.

(9) P.O. Perron, "Thermodynamics of Non-Stoichiometric Uranium Dioxide", report AECL-3072, 1968 May.

(10) M.M. Yovanovich, "General Expression for Forced Convection Heat and Mass Transfer from Isopotential Spheroids", 26th AIAA Aerospace Sciences Meeting, Reno Nevada, 1988 Jan. 11-14, paper 88-0743.

(11) E.N. Fuller, P.D. Schettler, J.C. Giddings, Indust. and Eng. Chem., 58 (1966) 19.

Oxidation behaviour of iridium-aluminum and iridium-aluminum-silicon systems at very high temperatures (1500°C to 1800°C)

K.N. Lee and W.L. Worrell

Department of Materials Science and Engineering, University of Pennsylvania, Philadelphia, Pennsylvania 19104, U.S.A.

Abstract

Recent results for the oxidation behavior of iridium-aluminum and iridium-aluminum-silicon systems at temperatures between 1550 (2822) and 1800°C (3272°F) are summarized. Iridium-aluminum intermetallics with high aluminum concentrations (above 55 atomic percent) form protective alumina scales. The observed parabolic oxidation rates are controlled by grain-boundary transport of oxygen through the alumina scale. The addition of 10 atomic percent silicon decreases the aluminum concentration necessary for the formation of a protective alumina scale to 30 atomic percent. The ternary iridium-aluminum-silicon system initially exhibits parabolic growth of an alumina scale with subsequent decreasing rates, presumably due to the formation of an amorphous silica layer at the alumina/substrate interface. Thermal cycling does not cause spallation of the alumina scale formed on either the binary or ternary systems.

Introduction

Increasing demands for high-temperature materials have created much interest in the oxidation resistance of high-melting alloys and intermetallics. High-temperature alloys or intermetallics rely on the formation of oxide scales such as silica or alumina for oxidation resistance. (1,2). After an extensive literature survey, we have selected the iridium-aluminum system as having the highest potential for oxidation protection at temperatures above 1500°C. Iridium has a high melting point (2446°C), and it oxidizes by forming volatile iridium oxides at very high temperatures. However, an iridium-aluminum alloy or intermetallic compound could form a protective layer or alumina, which would prevent the formation of gaseous iridium oxides.

It is well know that incorporation of a third element with an oxygen affinity between that of solvent and solute element in a binary system promotes the selective oxidation of solute and thus the formation of a continuous scale at solute concentrations lower than otherwise predicted. For example, adding chromium significantly decreases the aluminum concentration necessary for the formation of a protective alumina scale in nickel-aluminum alloys (3-6). Wagner (7) has proposed an oxygen gettering mechanism by the third element in Cu-Zn-Al and Fe-Cr-Al alloys in which the rapid establishment of a surface oxide film prevents the diffusion of oxygen into the alloy allowing the outward migration of aluminum to form a continuous layer. In Ni-Cr-Al alloys, a rapid formation of $Ni(Cr,Al)_2O_4$ spinel and subsequent development of a discrete layer of Cr_2O_3 has been observed before a continuous protective alumina is formed (8). The oxygen affinity of silicon is between that of aluminum and iridium, and the growth rate of silica is slower than that of alumina at temperatures above ~ 1000°C (1). Thus, adding silicon in iridium-aluminum system could improve the oxidation resistance of the binary system.

This paper summarizes our recent investigations of the oxidation kinetics of iridium-aluminum and iridium-aluminum-silicon systems at temperatures between 1550 and 1800°C in 1 atm oxygen (9,10).

Experimental

Interrupted weight-change measurements (WGM) and thermo-gravimetric-analysis (TGA) experiments have been used to determine the oxidation behavior of the iridium-aluminum and iridium-aluminum-silicon systems (9,10). For the experiments below 1600°C, a vertical tube furnace with molydisilicide heating elements was used. For the experiments above 1600°C, a vertical tube furnace with molydisilicide and stabilized-zirconia heating elements was used. The oxidation temperature was monitored using Pt-6% Rh/Pt-30% Rh thermocouple.

Iridium-aluminum and iridium-aluminum-silicon compositions containing a wide range of aluminum and silicon have been prepared by arc melting pre-determined amounts of the pure metals. Specimens for the weight-change measurements were cut from the arc-melted coupons using a diamond saw. The specimens were then polished through 600 grit SiC paper, cleaned in an ultra-sonic cleaner, and rinsed in acetone. Selected iridium-aluminum specimens were sent to the Sandia Livermore Laboratory for quantitative chemical analysis using x-ray fluorescent analysis. The results of the chemical analysis at the Sandia Livermore Laboratory were within ±4 at % of the results determined by our EDAX analysis. The compositions of the various intermetallic compounds tabulated in this paper are those determined by EDAX analysis, and the overall compositions are determined from the weights of the constituents before arc melting.

Details of the TGA experiments are described elsewhere (9,10). After the oxidation experiments, selected specimens are sputter coated with gold, electroplated with copper, and pressure mounted with copper. The mounted specimens are polished through 0.25 micron diamond paste, coated with carbon, and then examined and characterized using scanning electron microscopy (SEM) and energy dispersive x-ray analysis (EDAX). X-ray diffraction is used to determine the oxide and intermetallic phases.

Results

Iridium-Aluminum System

The intermetallic phases observed in the iridium-aluminum system are summarized in Table 1. X-ray diffraction confirms that the single-phase region having 55-57 at % aluminum is the IrAl intermetallic. The 40, 60. 65. 70 and 74 at% aluminum compositions are two-phase mixtures. The diffraction pattern for the other single-phase region having 71-72 at aluminum is that of an unknown intermetallic, which is referred to as $IrAl_{2.5}$. Any further characterization of $IrAl_{2.5}$ is not possible at this time. It is interesting to note that the $IrAl_3$ diffraction patteren is observed only at the 74 at% aluminum composition.

Results for the oxidation of selected compositions in the iridium-aluminum system at 1650°C in 1 atm oxygen are shown in Fig.1. The results for the iridium-aluminum intermetallics containing 60 and 65 at %

TABLE 1
Observed Intermetallic Phases in the Iridium-Aluminum System

Ir/Al (at %)	60/40	45/55	40/60	35/65	30/70	28/72	26/74
Phases	Ir +IrAl	IrAl	IrAl $+IrAl_{2.5}$	IrAl $+IrAl_{2.5}$	IrAl $+IrAl_{2.5}$	$IrAl_{2.5}$	$IrAl_3$ $+IrAl_{2.5}$

aluminum have been obtained using thermogravimetric analysis, while those for pure iridium and for compositions between 5 and 55 at % aluminum are from individual weight-gain measurements. Alloys or intermetallics containing 55 or less at% aluminum do not form protective oxide scales, and lose weight by the formation of gaseous iridium oxides. However, the 60 and 65 at % aluminum compositions form a protective alumina scale, which prevents the formation of gaseous iridium oxides and is responsible for the weight gain (9). X-ray diffraction indicates that the protective oxide scale is α-Al_2O_3 Since 55 at% aluminum is the aluminum-rich boundary of the IrAl intermetallic phase (see Table 1), these results indicate that $IrAl_{2.5}$ is necessary for the formation of a protective alumina scale.

To examine its cyclic oxidation behavior, an iridium-60 aluminum specimen has been removed rapidly (about 5 seconds) from the furnace every 25 hrs, cooled down to room temperature, and inserted back into the furncace. Such thermal cycling does not result in scale spallation. The coefficient of thermal-expansion (CTE) for alumina (9×10^{-6} in./in.°C, up to 1000°C) (11) is in good agreement with that measured for the iridium-60 aluminum composition (1.0×10^{-7}) at temperatures above 500°C (12). The close match of the thermal-expansion coefficients explains the excellent adherence of alumina on the iridium-60 aluminum composition.

Iridium-Aluminum-Silicon System

The addition of 10 at% silicon in iridium-aluminum system reduces the aluminum concentration necessary for the formation of a protective alumina scale from above 55 to 30 at %. (10) Iridium-aluminum -silicon compositions containing 30 to 50 at% aluminum and 8 to 10 at% silicon consist of two phases. Our X-ray diffraction results indicate that one phase is the IrAl intermetallic, which may contain up to about 5 at % silicon. Silicon concentrations below 5 at% in the IrAl intermetallic cannot be accurately determined, because the silicon peak is located near the strong iridium peak. Although the second phase has not been identified, its composition is tabulated in Table 2.

TABLE 2
Composition of the Second Phase for Selected Starting Compositions (atomic %)

Starting Composition			Second Phase Composition (± 4%)		
Ir	Al	Si	Ir	Al	Si
42	50	8	33	44.5	22.5
50	40	10	51.7	18.7	29.6
52	40	8	55.3	22.7	22
60	30	10	63.3	12.3	24.4

Figure 2 shows the TGA results for the oxidation of some compositions in the iridium-aluminum-silicon system at 1600°C in 1 atm oxygen. The iridium-50 aluminum-8 silicon composition exhibits parabolic oxidation for 350 hr, at which time the experiment is terminated. The iridium-40 aluminum-8 silicon composition exhibits parabolic oxidation for ~ 250 hr. The ternary iridium-aluminum-silicon system containing 10 at% silicon initially exhibits parabolic oxidation, with the rates decreasing after ~ 100 hr. As shown in Fig.2, the parabolic-oxidation period increases as the aluminum concentration increases and/or the silicon concentration decreases. The parabolic-rate constants of the ternary iridium-aluminum-silicon system are in excellent agreement with those of the binary iridium-aluminum system. X-ray diffraction confirms that the oxide scale formed on the ternary system during the parabolic-oxidation period is also α-Al_2O_3.

EDAX analysis shows that the oxide scale formed on the iridium-50 aluminum-8 silicon composition after 350 hrs of oxidation at 1600°C in 1 atm oxygen is pure alumina, which is consistent with the observed parabolic-oxidation rate. However, a thin and discontinuous silica layer is observed at the oxide/substrate interface of the iridium-40 aluminum-8 silicon composition.

A cross-section of the oxide scale formed on the iridium-30 aluminum-10 silicon composition after 360 hrs of oxidation at 1600°C in 1 atm oxygen is shown in Fig.3.a. Figures 3.b and 3.c are EDAX maps of aluminum and silicon for the cross-section shown in Fig.3.a. A continuous silica layer containing ~ 5 mole% alumina is observed at the oxide/substrate interface, and pure α-Al_2O_3 is present at the gas/scale interface. The variation of the alumina composition (mol%) across the oxide scale, which has been determined by EDAX analysis at locations 1 to 7 in Fig.3.a, is tabulated in Table 3. The ~ 5 mol% alumina concentration at the oxide/substrate interface is the eutectic composition (mp = 1587 ± 10°C) of the silica-alumina system (13). As one moves away from the substrate interface, the scale becomes richer in alumina and eventually becomes pure alumina at ~ 10 μm (Fig.3.c).

TABLE 3
The Alumina Concentration (mol%) across the Oxide Scale at the Points Shown in Fig.3.a

Points (Figs.3.a)	1	2	3	4	5	6	7
mol.% (Alumina)	5.5	5.5	10	64	83	90	100

The morphology and composition of the oxide scale formed on the iridium-40 aluminum-10 silicon composition is similar to that of the iridium-30 aluminum-10 silicon composition, as is expected from their similar weight-gain behavior.

Isothermal and cyclic oxidation results for the iridium-50 aluminum-8 silicon composition are in excellent agreement, indicating that thermal cycling does not lead to scale spallation. The coefficient of thermal-expansion for the iridium-50 aluminum-8 silicon ternary is presumably very similar to that of the iridium-60 aluminum binary, which would explain the excellent adherence of the alumina scale on both the ternary and binary systems.

Discussion

The activation energy for the parablolic oxidation of the iridium-60 aluminum composition has been determined in the temperature range between 1550°C and 1800°C. Oxide-scale thicknesses have been measured using SEM micrographs, and rate constants have been calculated assuming 100% dense alumina (9). The results are shown in Fig.4. Parabolic-rate constants on nickel-25 aluminum (14) at temperatures 1000-1300°C are included for comparison. The iridium-aluminum system has the same activation energy (76 ± 5 kcal/mol) as that for the nickel-25 aluminum. This agreement of the activation energy indicates that the same transport controls the growth of alumina on both the iridium-aluminum system and nickel-25 aluminum.

Aluminum oxide is a p-type conductor at high oxygen pressure ($> 10^{-4}$ atm), and an n-type conductor at low oxygen pressure ($< 10^{-6}$ atm)(15). Inert marker experiments at oxygen pressure above 0.1 atm have indicated that alumina scales grow by oxygen diffusion (16,17). Since oxygen diffusion in polycrystalline alumina occurs primarily by grain-boundary transport (18,19), it is generally accepted that oxygen transport in alumina scales also occurs primarily via grain boundaries. Thus, the parabolic-growth rates of the alumina scales on iridium-aluminum and iridium-aluminum-silicon systems are controlled by the grain-boundary diffusion of oxygen.

The chemical stability of silica and alumina at the substrate/oxide interface can be calculated using Eq.(1), where $\underline{Al}$ and $\underline{Si}$ are the activities of aluminum and silicon in the ternary system.

$$3\ \underline{Si} + 2\ Al_2O_3 = 3\ SiO_2 + 4\ \underline{Al} \qquad (1)$$

The standard free energy for reaction (1) is 110,495 cal, and the equilibrium constant is 1.3×10^{-13} at 1600°C (20). Thus, the initial oxide scale formed on the ternary system is expected to be alumina. As the oxidation continues, the aluminum activity decreases, and the silicon activity increases. Eventually the aluminum and silicon activities at the oxide/substrate interface become such that reaction (1) can proceed, and silica can form at the oxide/substrate interface. The discontinuous silica layer observed for the iridium-40 aluminum-8 silicon composition indicates that more than 8 at% silicon is necessary for the formation of a continuous silica layer underneath the outer alumina scale.

The oxidation rates of the ternary system containing 10 at% silicon decrease after ~ 100 hr. For the iridium-40 aluminum-10 silicon composition, the oxidation rate between 100 and 300 hr is about one order of magnitude higher than the growth rate for amorphous silica, while the rate from 300 to 800 hr agrees with that for oxygen transport through amorphous silica (10). The decrease in the growth rate between 100 and 300 hr is presumably due to the formation of a discontinuous silica layer below the outer alumina layer. After ~ 300 hr, the silica layer is continuous, and the oxidation rate is controlled by transport through the amorphous silica.

Summary

The oxidation behavior for the iridium-aluminum and iridium-aluminum-silicon system has been measured at temperatures between 1550 and 1800°C. In the iridium-aluminum binary system, the aluminum concentration necessary for the formation of a protective alumina scale is above 55 at%. The activiation energy for the growth of alumina is 76 ± 5 kcal/mol°K, which is in good agreement with the activation energy for nickel-25 aluminum at temperatures between 1000 and 1300°C. Grain-boundary transport of oxygen through the alumina scale is the rate-controlling step for the oxidation of both the iridium-aluminum and nickel-aluminum intermetallics at temperatures between 1000 and 1800°C.

The addition of 10 at% silicon decreases the aluminum concentration necessary for the formation of a protective alumina scale from above 55 to 30 at%. Scale lifetimes are also increased significantly. For example, at 1600°C in 1 atm oxygen, the lifetime of the iridium-50 aluminum-8 silicon ternary system is at least ten times longer than that of the iridium-65 aluminum binary system. The ternary iridium-aluminum-silicon system initally exhibits a parabolic growth of an alumina scale, with subsequent decreasing rates due to the formation of an amorphous silica layer at the alumina/substrate interface.

The alumina scale formed on either the binary or ternary system does not spall during thermal cycling, presumably due to the good agreement between the CTE values for alumina and the iridium-aluminum intermetallics.

References

1. P. Kofstad, High Temperature Corrosion, (Elsevier Applied Science, New York, 1988).
2. H. Hindam and D.P. Whittle, Oxid. Met. 18, 245 (1982).
3. C.S. Giggins and F.S. Pettit, J. Electrochem. Soc. 118, 1782 (1971).
4. F.H. Stott and G.C. Wood, Corros. Sci. 11, 799 (1971).
5. I.A. Kvernes and P. Kofstad, Met. Trans.3, 1511 (1972).
6. A. Kumar, M. Nasrallah and D.L. Douglass, Oxid. Met. 8, 227 (1974).
7. C. Wagner, J. Electrochem. Soc. 99, 369 (1952).
8. B.H. Kear, F.S. Pettit, D.E. Fornwalt and L.P.Lemaire, Oxid. Met. 3, 557 (1971).
9. K.N. Lee and W.L. Worrell, Oxid. Met. 32, 357 (1989).
10. K.N. Lee and W.L. Worrell, (in preparation).
11. W.D. Kingery, H.K. Bowen and D.R. Uhlmann, Introduction to Ceramics, Ch.IV (John Wiley & Sons, New York, 1976).
12. S. Szaruga, Wright Patterson Air Force Base, Private Communication.
13. I.A. Aksay and J.A. Pask, J. Am. Ceram. Soc. 58, 507 (1975).
14. F.S. Pettit, Trans. Metall. Soc. AIME 239, 1296 (1967).
15. J. Pappis and W.D. Kingery, J. Am. Ceram. Soc. 44, 459, (1961).
16. J.K. Tien and F.S. Pettit, Met. Trans. 3, 1587 (1972).
17. H.M. Hindam and W.W. Smeltzer, Oxid. Met. 14, 337 (1980).
18. Y. Oishi and W.D. Kingery, J. Chem. Phys. 33, 480 (1960).
19. K.P.R. Reddy, J.L. Smialek and A.R. Cooper, Oxid. Met. 17, 429 (1982).
20. JANAF Thermochemical Tables, D.R. Lide, Jr., ed., (American Institute of Physics, New York, 1985).

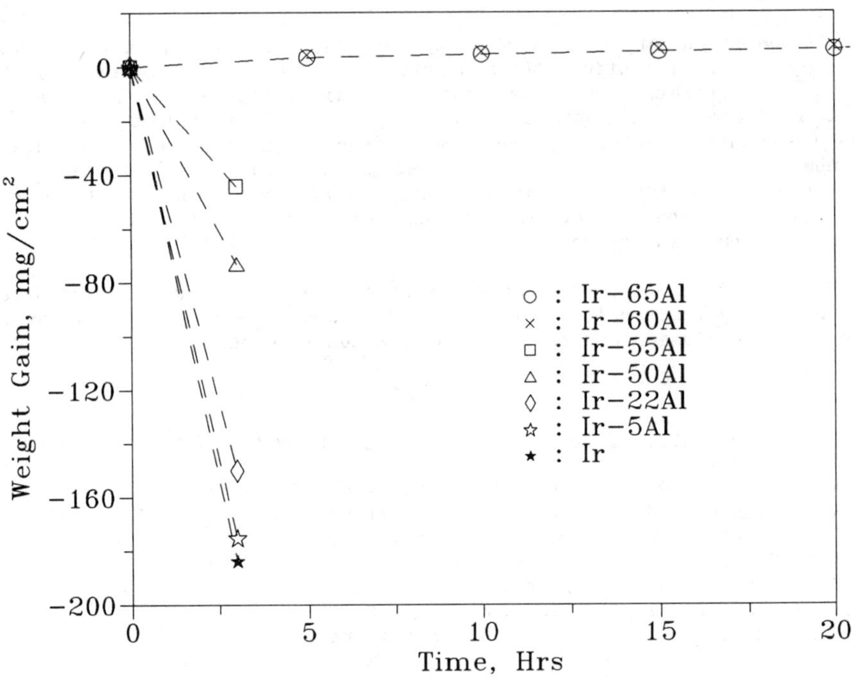

Fig. 1. Oxidation of Iridium-Aluminum System at 1650°C in 1 atm Oxygen (TGA: Ir-60 and 65 Al Composition, WGM: Ir, Ir-5, 22, 50, and 55 Al Composition)

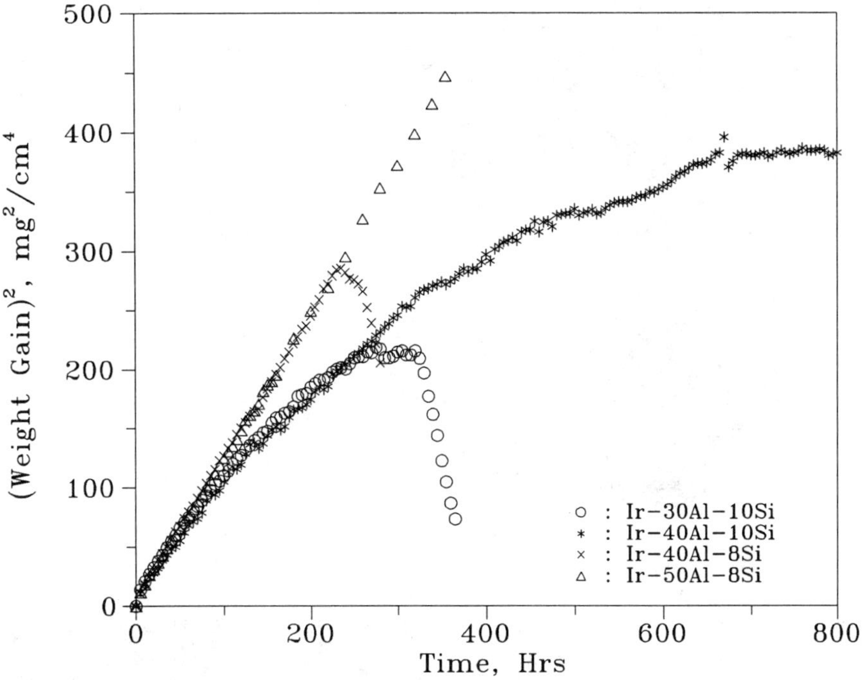

Fig. 2. Oxidation of Ir–Al–Si Ternary System at 1600°C in 1 atm Oxygen. Specimen thicknesses are 300 um.

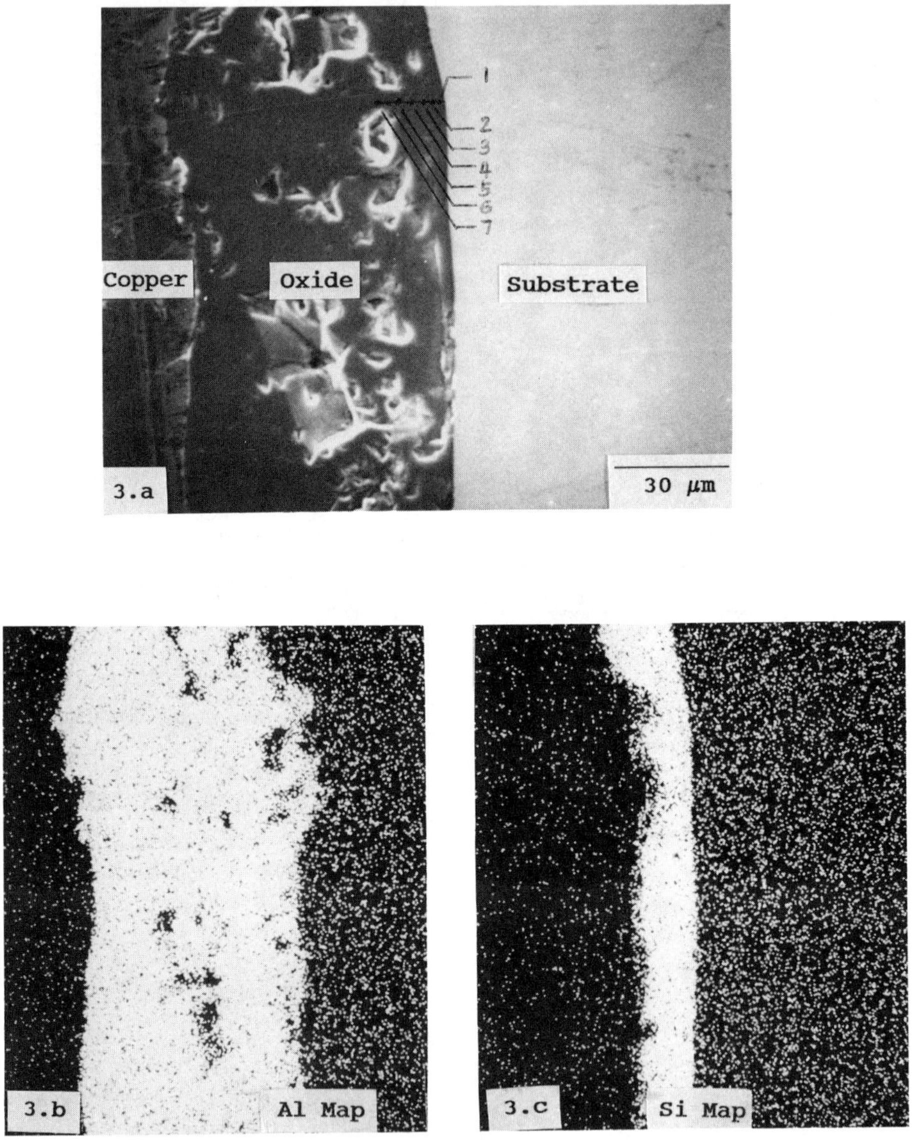

Fig. 3. Oxide Scale Formed on the Iridium–30 Aluminum–10 Silicon Composition at 1600°C in 1 atm Oxygen after 360 hr. (3.a: Cross–Section, 3.b: Al Map, 3.c: Si Map)

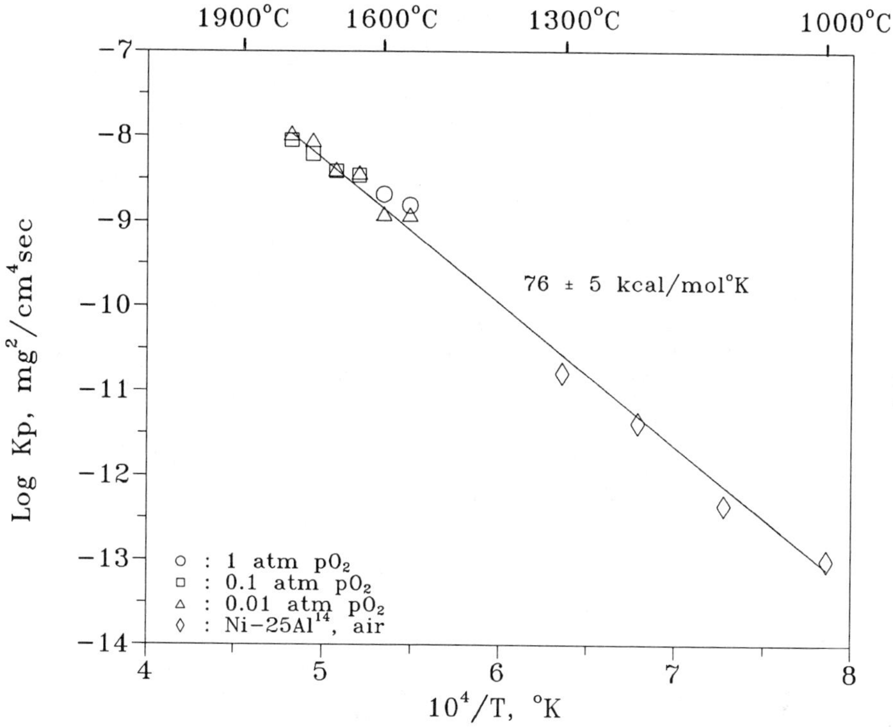

Fig. 4. The Variation of Log (Parabolic-Rate Constant) with Temperature for the iridium-60 aluminum Intermetallic.

Oxidation of nickel-aluminum alloys

P. Byrne and C.A. Pickles
Department of Metallurgical Engineering, Queen's University, Kingston, Ontario, Canada, K7L 3N6

ABSTRACT

Nickel aluminides have considerable potential as high temperature engineering materials because of their elevated temperature strength and excellent oxidation resistance. In this work the oxidation behaviour of two nickel-aluminum alloys was studied by thermogravimetry. Oxidation testing of a 50/50 mass percent nickel-aluminum alloy and a 50/50 atomic percent nickel-aluminum alloy showed that the rates of oxidation increased slowly with temperature up to about 1050 to 1100°C. Above this temperature the oxidation rate increased rapidly. The effect of impurity oxide additions was also studied. Vanadium oxide resulted in catastrophic oxidation while yttrium oxide reduced the rate of oxidation.

INTRODUCTION

The study of intermetallic compounds began in the early nineteenth century. In the course of the first systematic studies of these alloy systems, researchers observed that at certain compositions the behavior of these alloys was similar to that of ordinary chemical compounds. The first intermetallic was discovered when a discontinuity was noticed in the action of acids on alloys of copper and zinc at the equiatomic composition. It is now known that a compound does exist at this composition (beta brass, CuZn). Other intermetallic compounds were postulated on the basis of similar discontinuities in electrical, mechanical, and magnetic, as well as chemical properties [1].

Intermetallics are generally characterized by their brittleness and high hardness relative to the pure components. These properties are usually associated with harmful second-phase embrittlement in most alloy systems. In recent years, however, there has been an increased interest in the use of intermetallics in bulk form as result of the demand for improved high-temperature engineering materials [1]. Thus, the high-temperature oxidation resistance of these intermetallic compounds is of considerable interest.

The Nickel-Aluminum System

The Ni-Al phase diagram is shown in Figure 1. Four intermetallic compounds are present: $NiAl_3$, Ni_2Al_3, NiAl, and Ni_3Al. All of the intermetallic compounds have substantially higher melting points than pure aluminum. Ni_2Al_3, NiAl, and Ni_3Al occur over a range of compositions (as opposed to the "line" or stoichiometric compound $NiAl_3$). These compounds illustrate the difficulty in defining a "pure" intermetallic compound. $NiAl_3$, Ni_2Al_3, and Ni_3Al are incongruently melting compounds.

Ni_3Al has a melting point of 1385°C, a high yield point and a short-period creep limit at temperatures up to 1100°C. NiAl has a melting point of 1650°C, and is deformable at temperatures above 1300°C. Because of

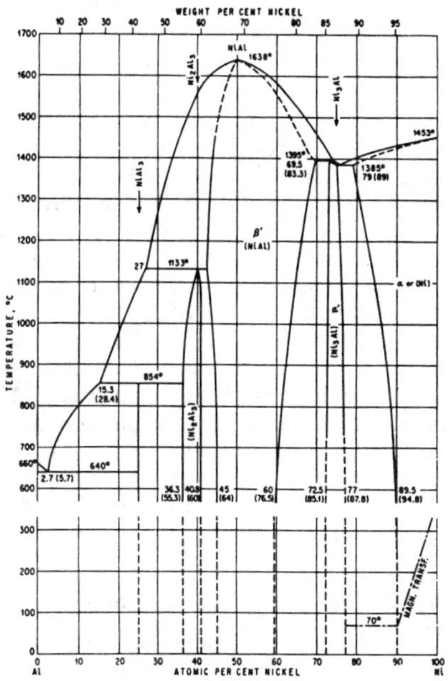

Figure 1: Nickel-Aluminum Phase Diagram (from Constitution of Binary Alloys by M. Hansen, McGraw-Hill, 1958, p. 119).

their high temperature oxidation resistance both of these materials can be employed at high temperatures in an oxidizing environment [2].

Preparation of Intermetallics

The most common method of forming intermetallics involves the direct union of the elements at temperatures as high as 3000°C. Most metals are highly reactive at high temperatures and hence possible reactions involving the container materials and the atmosphere present in the vessel must be considered. Thus, crucibles should be of a material that does not react with any of the reactants or products at the temperature in question. In addition, the crucible should have a high resistance to thermal shock and a softening point in excess of the reaction temperature. To avoid reactions with the atmosphere, the reaction should be carried out either under vacuum or under an inert atmosphere.

In actual practice, nickel aluminides have been produced using several different techniques. Hindam and Smeltzer [3] prepared samples of 32.2 weight percent (w/o) aluminum by electric arc melting nickel (99.99 w/o pure) and aluminum (99.95 w/o pure) in an argon atmosphere gettered of oxygen by titanium chips. The samples were homogenized at 1473 K for 24 hours.

Taniguchi and Shibata [4] prepared their specimen alloy (Ni_3Al-0.1B) by melting high-purity metals and a Ni-15.5% B alloy in a vacuum-induction furnace, followed by casting into a steel die. The ingot was homogenized at 1500 K for 48 hours in an argon atmosphere.

Kuenzly and Douglass [5] produced 100 gram buttons of their sample (Ni_3Al with 0.5 wt.% yttrium) by arc-melting high purity nickel (99.99 w/o pure) and aluminum (99.99 w/o pure) along with 0.5 w/o yttrium (99.99 w/o pure) on a water cooled hearth in an inert argon atmosphere. The buttons were turned and remelted six times. Gettering of residual oxygen and nitrogen was accomplished by premelting a titanium button. The samples were homogenized at 1200°C for 24 hours.

Oxidation Theory

The classical Wagner treatment of oxidation rates is based on the diffusion of ions and electrons through a growing oxide layer. Once an oxide layer is established on the surface of a metal (or alloy), one or both of the reactants must penetrate the scale for the reaction to proceed further. That is, either metal must reach the oxide-gas interface or oxygen must reach the metal-oxide interface [6].

The first step in the oxidation process is the adsorption of oxygen and the formation of the initial oxide. The kinetic gas theory predicts that a monolayer of adsorbed gas will form in about 2 seconds on a clean metal surface at a pressure of 10^{-6} torr at room temperature [7]. In most metals, adsorbed oxygen will quickly react to form an oxide. Diffusion through this thin oxide layer will be rapid, virtually establishing equilibrium with the metal at the oxide-gas interface. Thus, the metal activity is maintained at a high value (close to unity) at the interface and the rate controlling step is how fast the metal and oxygen react to form the oxide. This is a constant, and hence the rate of oxidation can be represented by:

$$X = k_1 t \tag{1}$$

where x is the scale thickness, t is time, and k_1 is the linear rate constant. The units of the constant depend on the method used to measure the rate of the reaction. If the oxide thickness is measured, then the units are cm/s; if the weight gain is measured, the units are $g/(cm^2\text{-}s)$. Equation (1) is known as the linear rate law.

As the oxide scale thickens, the flux of ions through the scale must be equivalent to the surface reaction rate. In order for this flux of ions to remain constant, the activity of the metal at the oxide-gas interface must decrease as the scale thickens. Eventually, the activity reaches the value in equilibrium with the atmosphere, and further increases in scale thickness result in a reduction in the metal activity gradient across the scale. As a result, the ionic flux and reaction rate decrease. At this point, the transport of ions across the oxide scale becomes the rate controlling process. Assuming thermodynamic equilibrium at each interface, the process can be analysed as follows. It is assumed that the scale is growing at the scale-gas interface due to cation transport through the oxide. The outward cation flux, j_m^{2+}, must be equal and opposite to the inward flux of cation defects (assumed to be cation vacancies). Thus

$$j_m^{2+} = -jv_m = Dv_m \left(\frac{Cv_m'' - Cv_m'}{x}\right) \tag{2}$$

where x is the oxide thickness, Dv_m is the diffusion coefficient for cation vacancies, and Cv_m'' and Cv_m' are the vacancy concentrations at the scale-gas and scale-metal interface respectively.

Since thermodynamic equilibrium is maintained at each interface, the value $(Cv_m'' - Cv_m')$ is constant and thus

$$jv_m = (\text{const.}) \frac{dx}{dt} = Dv_m \left(\frac{Cv_m'' - Cv_m'}{x}\right) \tag{3}$$

$$\text{i.e. } \frac{dx}{dt} = \frac{k'}{x} \quad \text{where } k' = Dv_m(Cv_m'' - Cv_m')/(\text{const.})$$

Intergrating with the condition that x = 0 at t = 0, yields

$$x^2 = k_p t + C_p \tag{4}$$

where C_p is the intergration constant and k_p is known as the parabolic rate constant. If the reaction is monitored by the weight gain of the sample, the parabolic rate constant has units of $g^2\ cm^{-4}\ s^{-1}$. Equation (4) is known as the parabolic rate law.

When studying the oxidation behaviour of an alloy, the situation is more complex, due to factors such as [6]:

1) the metals in the alloy will have different affinities for oxygen (reflected by the different free energies of formation of the oxides);
2) ternary and higher oxides may be formed;
3) there may be some solid solubility between the oxides;
4) the different metal ions will have different mobilities in the oxide phase(s);
5) the different metals will have different diffusivities in the alloy; and
6) the dissolution of oxygen into the alloy may result in subsurface precipitation of oxides of one or more alloying elements (internal oxidation).

In intermetallic compounds, the situation is further complicated by factors such as [8]:

1) generally lower diffusivities in the alloy;
2) lower interstitial solubilities;
3) smaller differences in the stabilities of the oxides of the components;
4) tight stoichiometry of many compounds; and
5) transient oxides which grow with linear or rapid parabolic kinetics.

Oxidation Behaviour of Intermetallics

Aluminide intermetallic compounds offer promise for use at high temperatures, but they have rarely been employed in load-carrying applications due to their brittleness. Instead, most of these compounds have been used as coatings over a ductile base metal. Recently, there has been an increased amount of interest in the oxidation behaviour of these intermetallic compounds.

Theories for the oxidation of metals are also pertinent to the oxidation of intermetallic compounds. The process of oxidation involves the absorption, dissociation, and ionization of oxygen molecules. The ions are then incorporated to form the oxide layer as a thin film, then as a thick film or scale. Later stages of oxidation can lead to changes such as breakaway, spalling, or cracking. Oxidation should proceed spontaneously on nearly all metals when they are exposed to oxygen, since the free energy change involved is negative for most metals. Beyond the monolayer stage, an activation barrier is encountered, and any further growth is kinetically controlled. The controlling processes can be diffusion of cations, anions, vacancies, interstitial atoms, electrons, or some combination of the above.

Mozhukhin, Pivovarov, and Umanskiy [9] studied the effect of composition on the oxidation behaviour of NiAl. Aluminum-rich NiAl oxidized to a lesser extent at 1150°C than electrolytic nickel at 1000°C. Nickel-rich NiAl oxidized at a rate comparable to that of pure nickel.

Young, Riviere, and Welch [10] studied the effect of the heating procedure (i.e. pre-oxidized samples versus direct insertion into a preheated furnace versus preheating in vacuum) on the oxidation behaviour of NiAl. They found that there are no kinetic barriers to the formation of alumina scales on clean NiAl. However, if the initial oxide contains appreciable Ni (probably as $NiAl_2O_4$), then the subsequent formation of alumina is hindered. Thus, if slow growing scales of Al_2O_3 are desired, then a procedure which prevents the incorporation of Ni in the early stages of oxidation should be followed.

Horton, Cathcart, and Liu [11] studied the effect of chromium on the early stages of oxidation of Ni_3Al. The rate of oxidation of the alloy with chromium decreased rapidly with time and was very small, while the alloy without chromium continued to show a significant weight gain increase after 30 minutes. These results indicate that, as expected, the oxide layer is thinner and more protective on the chromium-modified alloy than on the unmodified alloy.

Allaverdova et. al. [12] investigated the effect of 3d transition elements (titanium, vanadium, chromium, manganese, iron, cobalt, and copper) on the high temperature oxidation behaviour of NiAl alloys. They found that additions of titanium, chromium, iron, and cobalt (1 wt.%) stabilized the oxidation resistance of both stoichiometric NiAl and of solid solutions of nickel in NiAl. Increasing the amount of the additive (especially if it replaced aluminum in the alloy) impaired the heat resistance of the alloys. Alloys with vanadium additions oxidized catastrophically.

Kuenzly and Douglas [5] studied the oxidation behaviour of Ni_3Al containing yttrium. Their results indicated that the addition of yttrium to Ni_3Al did not significantly alter the oxidation kinetics. The steady-state scaling rates of pure Ni_3Al and Ni_3Al containing 0.5 wt.% Y were identical. The presence of yttrium, however, allowed a disproportionate amount of NiO to form during the early stages of oxidation. The subsequent solid-state reaction of NiO with Al_2O_3 resulted in a composite scale consisting of $NiAl_2O_4$ and Al_2O_3.

Taniguchi and Shibata studied the effect of reactive element additions on the oxidation behaviour of Ni_3Al [13]. It was found that zirconium and chromium increased the oxidation rate while hafnium decreased the oxidation rate. Similarly, Doychak et al. found that zirconium reduced the oxidation rate of nickel-rich Ni-Al intermetallics [14]. Mrowec and Jedlinski observed a decreased rate of oxidation with yttrium additions [15].

EXPERIMENTAL

In this study the oxidation behaviour of two nickel-aluminum powders was investigated. The characteristics of the powders are summarized in Table I.

TYPE	COMPOSITION	SIZE	PURITY
Cerac Nickel Aluminide (NiAl)	50(a/o)Ni 50(a/o)Al	-150 + 325 mesh	99.5%
BDH Nickel-Aluminum Alloy Powder (Ni/Al)	50(m/o)Ni 50(m/o)Al	-200 mesh	99.5%

TABLE I: Nickel-Aluminum Alloy Powder Characteristics

The briquettes were prepared by compressing 3 grams of the powder in a one-half inch hardened alloy steel die to a pressure of 24,000 psi. When reagent grade oxide additions were made the powdered mixtures were tumbled in a mill for at least three hours.

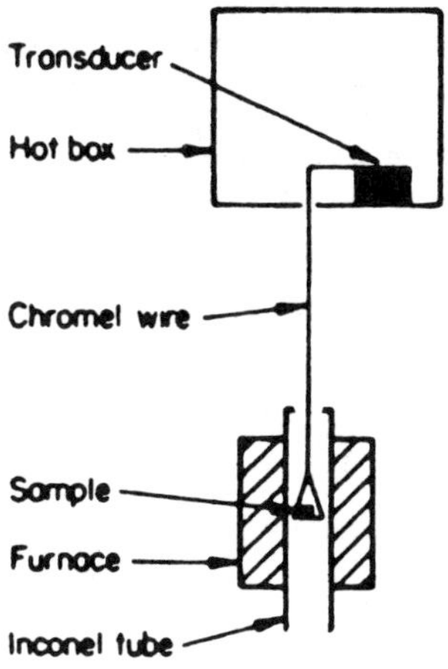

Figure 2: Schematic Diagram of the Continuous Weight Change Apparatus.

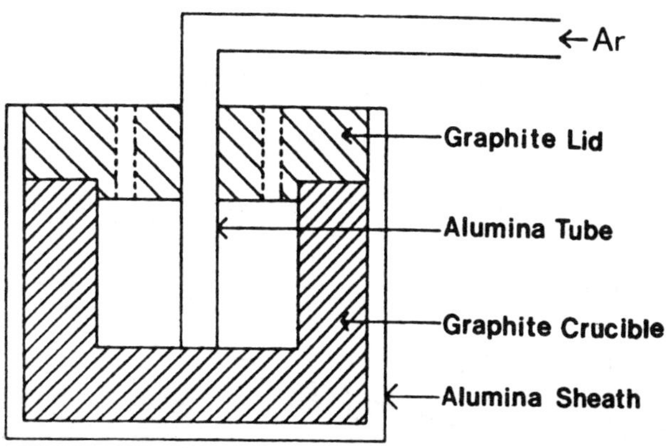

Figure 3: Schematic Diagram of the High Temperature Apparatus.

In the continuous weight change experiments the briquettes were suspended in air in a resistance furnace. The weight change was monitored with a transducer which was connected to a chart recorder. A schematic diagram of this apparatus is shown in Figure 2. In another series of experiments the briquettes were placed in a high temperature furnace under an argon atmosphere in the crucible assembly shown in Figure 3. In this case the experimental temperatures were over 1200°C and commercial purity argon gas (<1ppmH_2O, <0.5ppmO_2) was passed over the samples. In all the tests the surface area of the briquettes was taken to be the nominal surface area of the sample.

The oxidized samples were examined by optical and scanning electron microscopy. Heavily oxidized samples were examined by x-ray powder diffraction.

RESULTS AND DISCUSSION

Figure 4 shows a typical plot of the specific weight gain (weight gain/nominal surface area) squared versus time for the 50/50 mass percent Ni/Al sample which was oxidized at 1097°C. There is an initial large increase in weight due to the rapid growth of a thin oxide film on the unoxidized specimen. This initial stage is followed by parabolic growth. In this region the growth rate is linear, Figure 5. The slope of the line gives the parabolic rate constant. The parabolic rate constants are plotted against temperature in Figure 6. The oxidation rate increases slowly until about 1060°C. Above this temperature the oxidation rate increases rapidly.

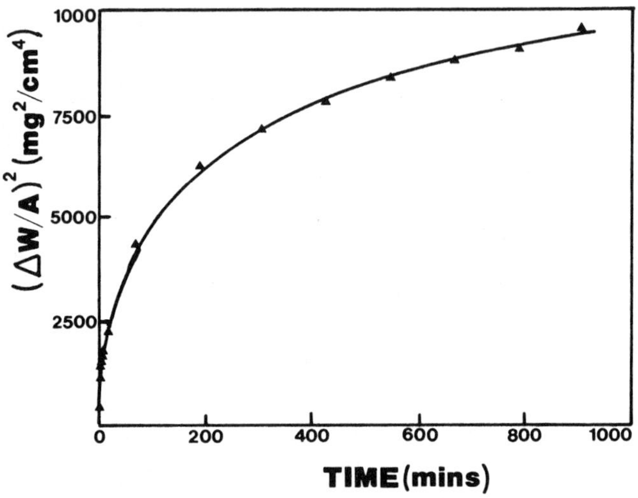

Figure 4: Typical $(\Delta W/A)^2$ Plot for a Briquette of Ni/Al Oxidized at 1097°C in Air.

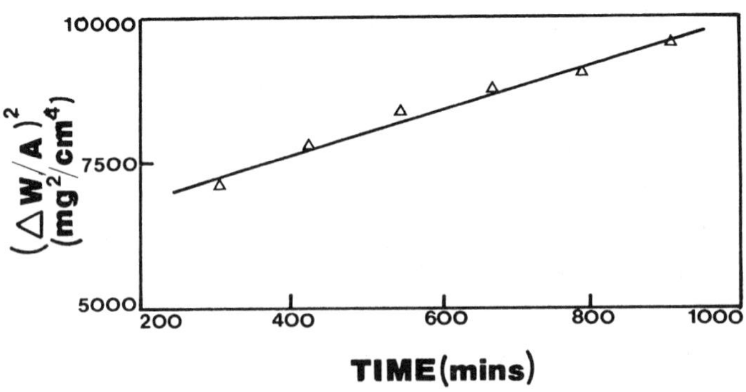

Figure 5: Straight Line Plotted Through the Linear Region of Figure 4.

Figure 7 shows some SEM micrographs of the surfaces of the briquettes which were oxidized at 836 and 1108°C. It is clear from these micrographs that the oxide which formed at the higher temperature was considerably more porous. X-ray diffraction showed that alumina was the major phase in the oxide with some $NiAl_2O_4$. Figure 8 shows the parabolic rate constants for the

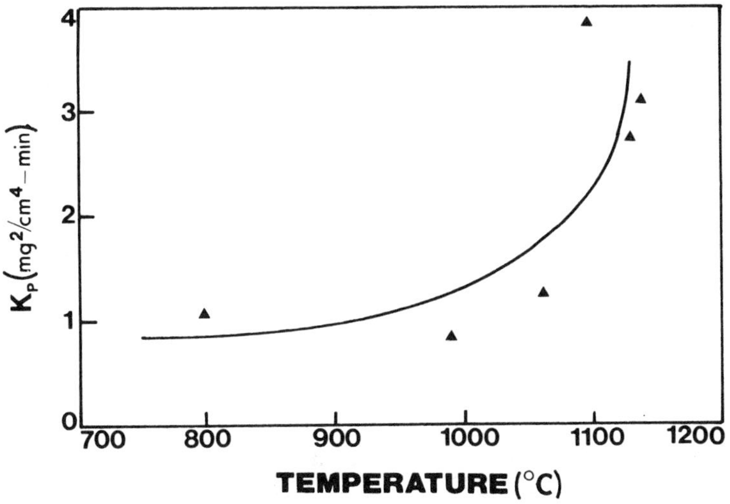

Figure 6: Parabolic Rate Constants Versus Temperature for Ni/Al in Air.

Figure 7: SEM Micrographs of Briquette Surfaces of Ni/Al Oxidized at 836°C and 1108°C in Air.

NiAl alloy. As was observed for the alloy discussed above there is only a relatively small increase in the oxidation rate below 1075°C. Above this temperature the oxidation rate increases rapidly with temperature. Scanning electron micrographs of the sample surfaces are shown in Figure 9. Some fine whiskers could be observed at the higher temperatures. Again the porosity of the surface oxide increased with temperature. X-ray diffraction studies showed that alumina was the major oxide present with some $NiAl_2O_4$.

The effects of impurity oxide additions on the oxidation behaviour of the two alloys are shown in Figures 10 to 15. Zirconium oxide did not have a significant effect on the oxidation rate. On the other hand, both chromium oxide and vanadium oxide additions resulted in increased rates of oxidation of the NiAl. However, chromium oxide reduced the initial rate of oxidation of the Ni/Al alloy. The composition of the oxide produced in the tests with vanadium and chromium oxides are summarized in Table II. It can be seen that chromium appears to concentrate in the oxide. The results for vanadium oxide are similar to those observed by Allaverdora et al. [12]. They postulated that vanadium pentoxide can form and this oxide travels

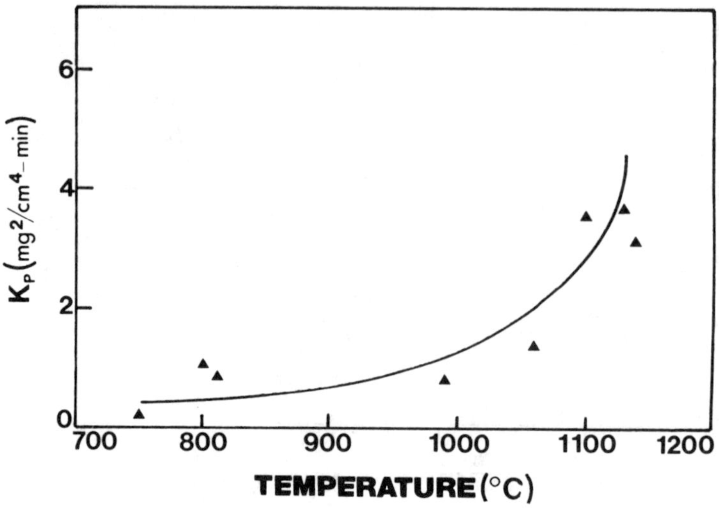

Figure 8: Parabolic Rate Constants Versus Temperature for NiAl in Air.

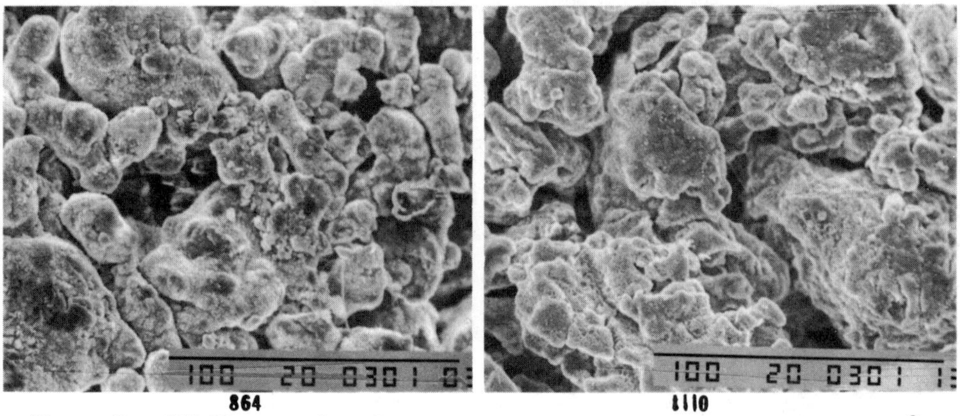

Figure 9: SEM Micrographs of Briquette Surfaces of NiAl Oxidized at 864°C and 1110°C in Air.

Alloy	%Ni	%Al	%Cr	%V
Ni/Al	39.71	60.11	0.00	0.00
Ni/Al+10%Cr_2O_3	29.53	24.19	46.09	0.00
Ni/Al+10%V_2O_5	83.13	10.73	0.00	6.14
NiAl	56.10	43.74	0.00	0.00
NiAl+10%Cr_2O_3	33.80	47.22	18.98	0.00
NiAl+10%V_2O_5	43.30	42.85	0.28	13.57

Table II: Composition of the Oxide in Briquettes Both with and without Vanadium and Chromium Oxide Additions (from Figure 13).

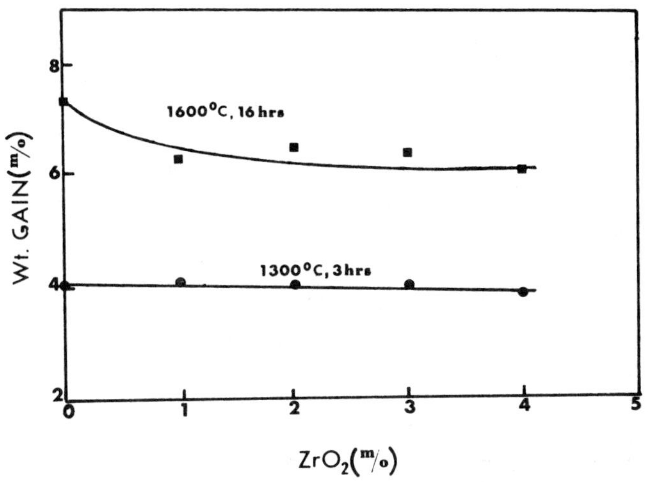

Figure 10: Weight Gain of NiAl Briquettes with Zirconium Oxide Additions in Argon.

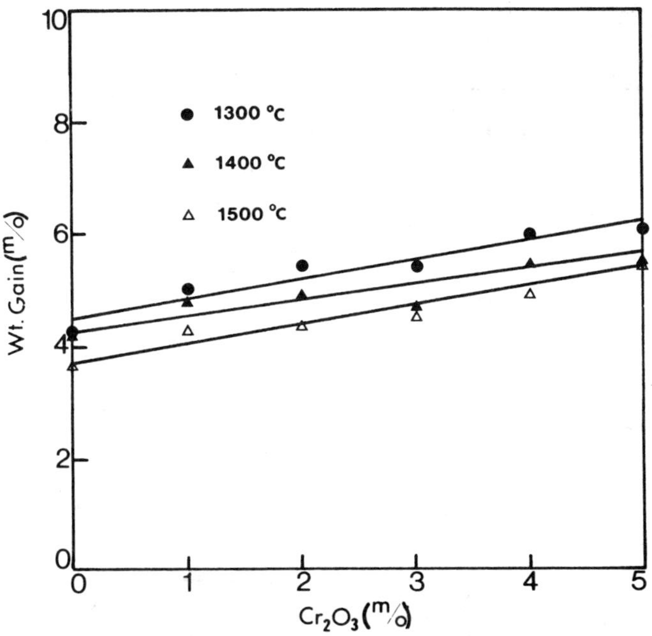

Figure 11: Weight Gain of NiAl Briquettes with Chromium Oxide Additions after Three Hours in Argon.

along the grain boundaries of the oxide film to the surface. This results in a porous scale with poor oxidation resistance. However, in the present work, since briquettes were employed, the samples were quite porous and thus it appears that vanadium oxide accelerates the rate of diffusion of reactants through the oxide layer. Yttrium oxide additions resulted in decreased oxidation rates at all the temperatures studied.

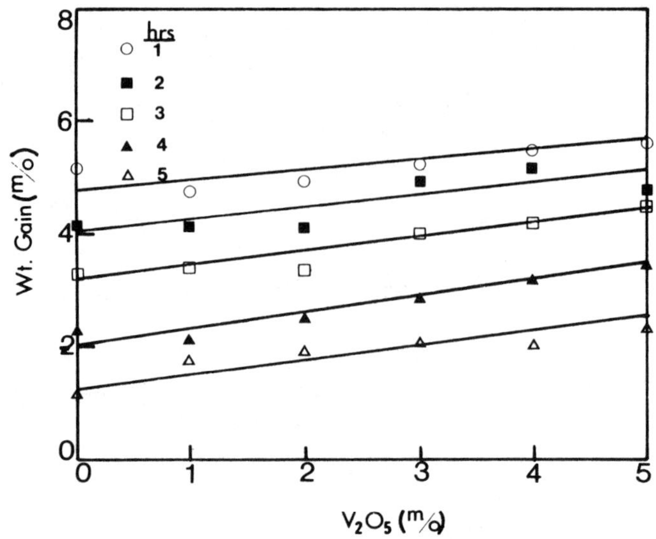

Figure 12: Weight Gain of NiAl Briquettes with Vanadium Pentoxide Additions at 1300°C in Argon.

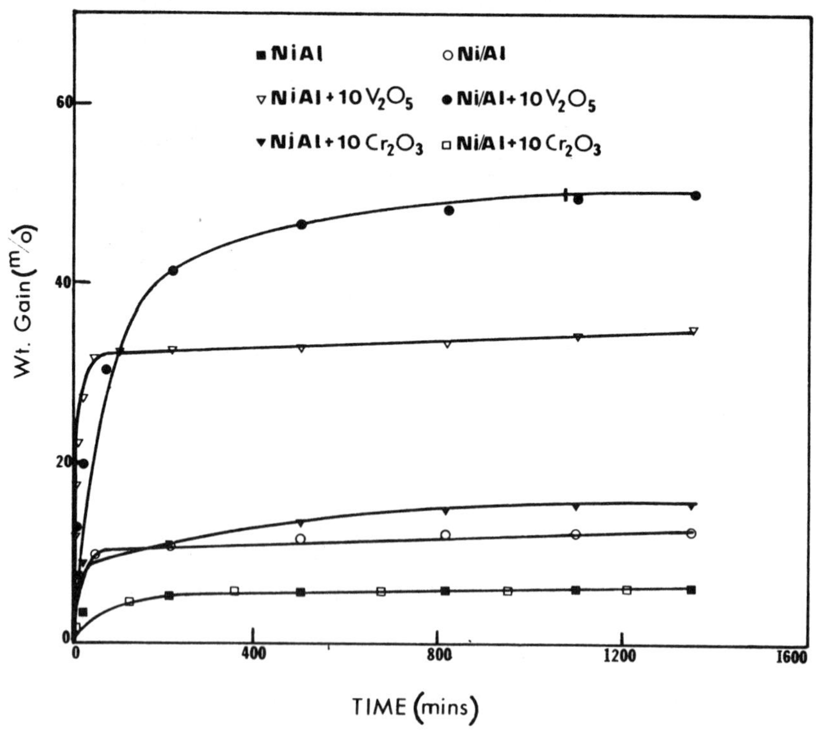

Figure 13: Weight Gain of NiAl and Ni/Al Briquettes Containing Vanadium and Chromium Oxides at 982°C in Air.

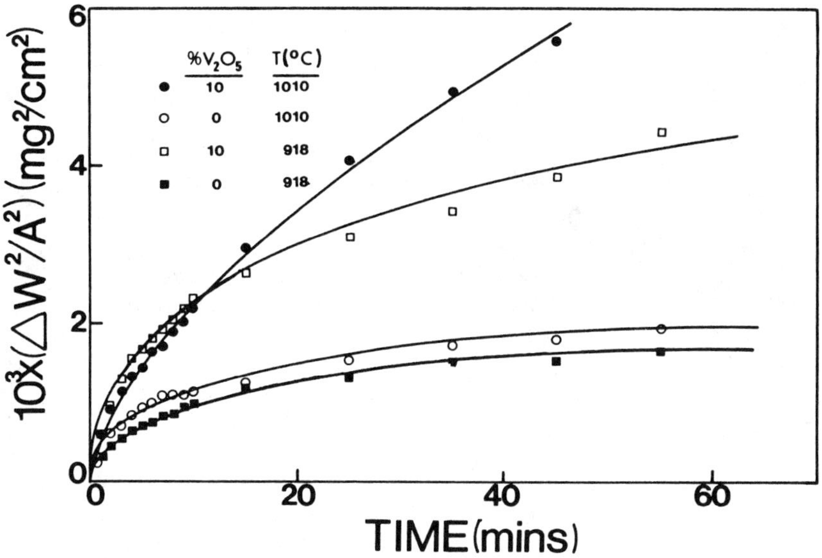

Figure 14: Effect of Vanadium Pentoxide in the Oxidation Rate of Ni/Al in Air.

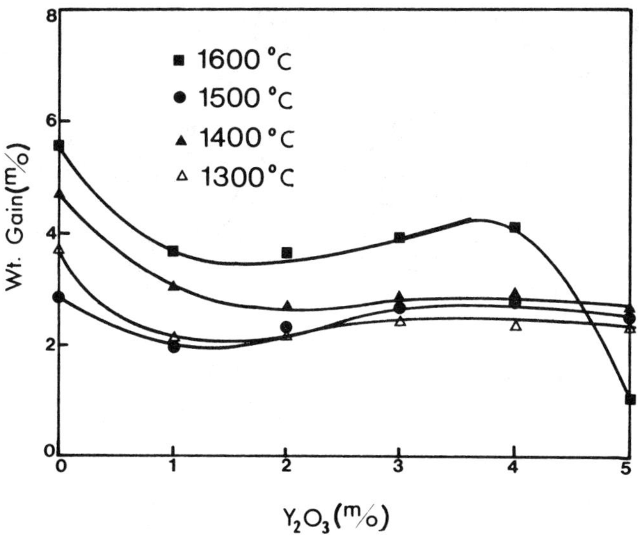

Figure 15: Weight Gain of NiAl Briquettes with Yttrium Oxide Additions after Three Hours in Argon.

Scanning electron micrographs of the oxidation products are shown in Figure 16.

At high temperatures, without any oxide additions, alumina whiskers were observed on the NiAl briquettes. A number of these whiskers had secondary arms. The addition of chromium oxide resulted in the growth of ribbon-like structures. With vanadium oxide there was evidence of the formation of a liquid oxide phase.

10% Cr_2O_3 10% V_2O_5

Figure 16: SEM Micrographs of the Oxidation Products in Argon.

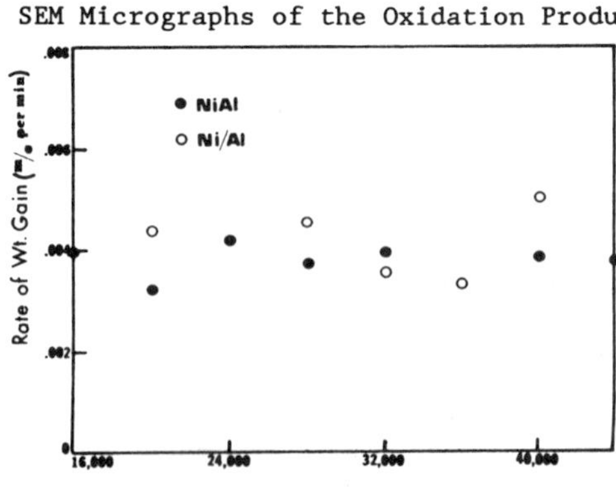

Figure 17: Effect of Briquetting Pressure on the Rate of Weight Gain at $1077^{\circ}C$ in Air.

The effect of briquetting pressure is shown in Figure 17 for the two nickel-aluminum powders. In both cases, there was no significant effect of briquetting pressure on the amount of oxide formed.

CONCLUSIONS

(1) The oxidation rates of a 50/50 mass percent nickel-aluminum alloy (Ni/Al) and a 50/50 atom percent nickel-aluminum (NiAl) were studied. It was found that the oxidation rate increased rapidly above about 1050°C due to the increased porosity of the oxide layer.

(2) At high temperatures, alumina whisker growth was observed on the surface of the NiAl samples.

(3) Vanadium oxide additions resulted in catastrophic oxidation of the sample. Chromium oxide increased the oxidation rate of the NiAl alloy but reduced the initial rate of oxidation of the Ni/Al alloy. Yttrium oxide additions reduced the oxidation rate of the NiAl alloy while zirconium oxide did not appear to have a significant effect.

ACKNOWLEDGEMENTS

The authors would like to thank the Natural Sciences and Engineering Research Council of Canada (NSERC) for support of this work.

REFERENCES

1. Westbrook, J.H., editor, Intermetallic Compounds, John Wiley and Sons, Inc., New York, 1967.

2. Kornilov, Ivan Ivanovich, The Chemistry of Metallides, Consultants Bureau, Plenum Publishing Corporation, New York, 1966.

3. Hindam, H.M., and Smeltzer, W.W., "Growth and Microstructure of Alpha Al_2O_3 on Ni-Al Alloys: Transition to External Scale," J. Electrochem. Soc., Vol. 127, No. 7, 1980, pp. 1622-1630.

4. Taniguchi, S., and Shibata, T.,"Oxidation Behaviour of Ni_3Al Containing Yttrium", Oxidation of Metals, Vol. 28, Nos. 3/4, 1987, pp. 155-163.

5. Kuenzly, J.D., and Douglass, D.L., "The Oxidation Mechanism of Ni_3Al Containing Yttrium", Oxidation of Metals, Vol. 8, No. 3, 1974, pp. 139-177.

6. Birks, N., and Meier, G.H., Introduction to High Temperature Oxidation of Metals, Edward Arnold, London, 1983, p. 31.

7. Kofstad, P., High-Temperature Oxidation of Metals, John Wiley and Sons, New York, 1966, p. 21.

8. Meier, G.H., "Fundamentals of the Oxidation of High-Temperature Intermetallics", Oxidation of High-Temperature Intermetallics, Grobstein, Toni, and Doychak, Joseph, editors, The Minerals, Metals, and Materials Society, Warrendale, 1988, pp 1-16.

9. Mozhukhin, E.I., Pivovarov, L. Kh., and Umanskiy, Ya. S., "Oxidation of Alloys Containing the Compound NiAl",J. Appl. Chem., (USSR), 30, 1957, pp 1593-1599.

10. Young, E. W.A., Riviere, J.C., and Welch, L.S., "The Influence of the Heating Procedure on the Oxidation Behaviour of NiAl Studied by XPS", Applied Surface Science, 31, 1988 pp 370-374.

11. Horton, J.A., Cathcart, J.V., and Liu, C.T., "Effect of Chromium on Early Stages of Oxidation of Ni_3Al Alloys at 600^oC", Oxidation of Metals, Vol. 29, Nos. 3/4, 1988, pp. 347-365.

12. Allaverdova, N.V., Kucherenko, L.A., Portnoy, V.K., and Sokolovskaya, Ye. M., "High Temperature Oxidation of NiAl Alloys with the Addition of 3d Transition Elements", Journal of the Less-Common Metals, 114, 1985, 299-304.

13. S. Taniguchi and T. Shibata, "Influence of Reactive Element Additions on the Oxidation Behaviour of Ni_3Al Base Alloys" Workshop on the Oxidation of High Temperature Intermetallics, TMS, 1988, pp. 31-40.

14. J. Doychak, J.L. Smialek, and C.A. Barrett, "The Oxidation of Ni-rich Ni-Al Intermetallics", Workshop on the Oxidation of High Temperature Intermetallics, TMS, 1988, pp. 41-55.

15. S. Mrowec and J. Jedlinski, "The Influence of Implanted Yttrium and Lanthanum on the Kinetics and Mechanism of High Temperature Oxidation of β-NiAl Intermetallic Compounds", Workshop on the Oxidation of High Temperature Intermetallics, TMS, 1988, pp. 57-66.

High-temperature oxidation studies of titanium disilicide

D. Mallette and C.A. Pickles
Department of Metallurgical Engineering, Queen's University, Kingston, Ontario, Canada, K7L 3N6

Abstract

The effect of impurities on the oxidation resistance of $TiSi_2$ was studied. Iron additions caused a slight increase in the oxidation rate, while samples containing aluminum or copper showed no useful resistance to oxidation. The oxidized samples were examined by optical and SE microscopy, EDS, and thin film diffractometry.

Introduction

Because of their excellent oxidation resistance and ability to withstand very high temperatures, metal silicides have potential use as a conductor in microelectronics or as a coating or bulk refractory material. Extensive research has been made to investigate the oxidation behaviour of refractory silicides such as those of molybdenum, columbium, tantalum, and titanium. However, very little is known about the oxidation of such silicides with metallic impurities.

Several studies have been made on the oxidation behaviour of pure titanium silicide (1-9). Various mechanisms have been proposed to describe this behaviour, all of which are based on a qualitative analysis of the oxidized silicide. Conflict in models arises from the complexity of behaviour encountered over a temperature range, as considerations must be made for the thermodynamic, kinetic, physical, and transport properties of the system. It is, however, agreed in general that the resistance to oxidation is due to the formation of a protective film of oxides on the surface of the silicide that is effective in reducing further oxidation of the substrate material.

One proposed method of producing titanium disilicide ($TiSi_2$) is by the reduction of ilmenite ($FeO \bullet TiO_2$) in the presence of silica (SiO_2) and a reducing agent such as carbon, aluminum, magnesium, or calcium. Titanium silicide can also be precipitated from a copper melt (10). However, the effect of an inherent impurity in the silicide on its resistance to oxidation at high temperatures must be known before this method of production can be considered feasible. The present study was undertaken in order to determine the effect of iron, copper, and aluminum impurities on the oxidation resistance of titanium disilicide.

The time and temperature dependence of the oxidation of bulk $TiSi_2$ has been studied by Schwettman et al (1) in order to provide data and a mechanism for oxidation. The proposed typical oxidation reaction,

$$TiSi_2 + O_2 \Longrightarrow Ti_xO_y + SiO_2 + Ti_5Si_3 \qquad \text{(R1)}$$

was shown to be significant only above 300°C, with the amount of Ti_5Si_3 too small to be positively identified.

The controlling mechanism of oxidation in the range 300°C to 700°C is a combination of the oxidation reaction (R1) at the solid-gas interface and diffusion of oxygen through the oxide film. In the initial phase of oxidation, diffusion through micropores in the forming oxide film is rapid, and thus reaction at the surface is rate-controlling, as depicted in Figure 1.

FIG. 1. Oxidation of $TiSi_2$ in the Initial Phase

Because the free energy of formation of titanium oxides is very close to that of silica, selective oxidation of titanium or silicon is unlikely to occur. Instead, an amorphous titania-silica glass will form on the surface of the silicide. As oxidation proceeds and this oxide film grows in thickness, the micropores become blocked and diffusion of oxygen through the film becomes rate-controlling. Since diffusion through a silica film at low temperatures is slow, the rate of oxidation gradually decreases with time, as shown in Figure 2 (1).

It has been found that, at temperatures below 600°C, somesilicides exhibit an increase rather than a decrease in oxidation rate. Molybdenum disilicide, although highly oxidation-resistant at temperatures above 600°C (2), oxidizes quite rapidly at temperatures under 600°C (3). This "pest corrosion" is due to the presence of molybdenum oxide, MoO_3, which is incompatible with SiO_2 and, upon formation, causes severe volume expansion in any surface crack in which it may form. The subsequent collection of oxides causes crack propogation, which promotes deterioration of the silicide. Above 600°C, MoO_3 becomes volatile, allowing for the formation of a film of pure SiO_2 that tends to bridge across any surface cracks in the silicide, forming a very protective film. However, pest corrosion has not been observed during the oxidation of titanium silicide at any temperature. Since TiO_2 is not volatile in the range of temperatures of interest, and initial surface cracks are inevitable, it must be that the TiO_2 is compatible enough with the SiO_2 to form the observed bridge-like film over the surface cracks, thus eliminating pesting at low temperatures. Schwettman et al (1) have suggested that this is partly due to the ability

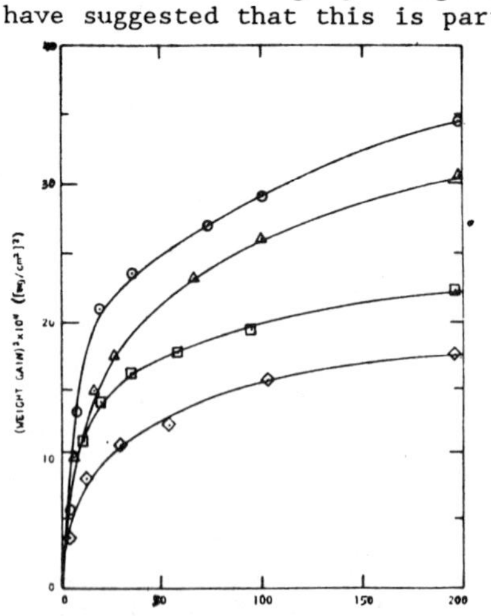

FIG. 2. Rate Data for $TiSi_2$ Oxidized at 600°C

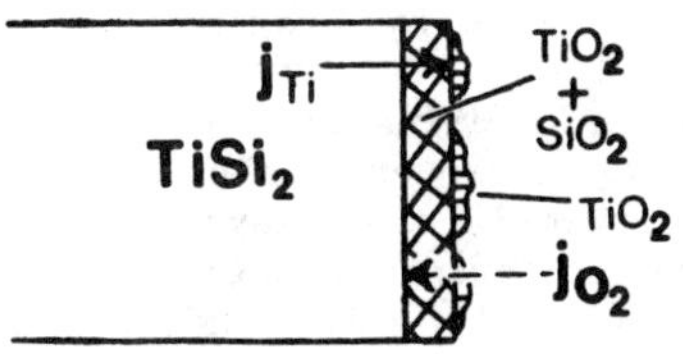

FIG. 3. Continued Oxidation of $TiSi_2$ at high temperatures

of titanium to assume fourfold coordination and replace silicon in the growing silica film without disrupting the tetrahedral network.

At temperatures of 1000°C and above, the initial stage of oxidation is unchanged (i.e. reaction at the surface is rate-controlling). However, as the oxide film continues to grow in thickness, counter-diffusion of Ti to the gas-oxide interface occurs as the solubility limit of titanium in silica is surpassed. Crystalline clumps of TiO_2 are formed at that interface (1). As before, a virtually inpenetrable layer of amorphous SiO_2 forms, slowing further diffusion and hence the oxidation rate. This phenomenon is represented schematically in Figure 3.

However, as diffusion increases with temperature, it can be expected that the amount of material gained through oxidation will continue to grow as temperature increases. Schwettman et al (1) observed parabolic kinetics at temperatures of 1000°C and above, as shown in Figure 4 (1).

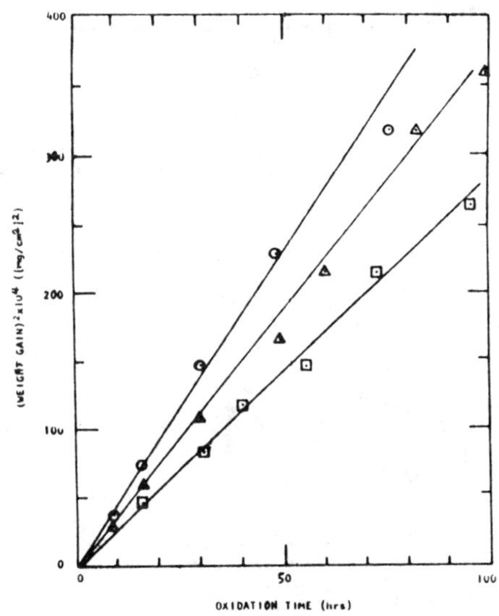

FIG. 4. Rate Data for $TiSi_2$ Oxidized at 1000°C

Experimental

Iron powder (J.T. Baker Chemical Co., 97.9%), aluminum powder (BDH Chemicals, 99.9%), and copper powder (BDH Chemicals, 99.9%) were used as the added impurities to $TiSi_2$ (Alfa Products, 95%). These elements were added separately in concentrations varying from 0 wt.% to 10 wt.% and mixed in a rolling mill. Binder (H_2O) was added and 1.25 in. diameter wafers were pressed in a hardened steel die at 4-6 tsi. Sample thickness was typically

0.5 in. The binder was removed by heating samples for several hours at 100°C.

Wafers were placed on an alumina dish in a graphite crucible containing zirconium chips as an oxygen getter during sintering, carried out in an argon atmosphere. The extremely thin layer of oxide which formed on the faces of the wafers during sintering was removed by grinding successively on carbide paper of 120, 240, 320, 400, and 600 grit coarseness. Final preparation of the wafers involved polishing with 5 μm diamond paste.

Oxidation of the wafers was carried out in a high-temperature furnace with a PID controller. The oxidizing atmosphere for all oxidations was air ($p_{O_2} \cong 0.21$ atm.). Samples were placed on a porous alumina brick in the middle of the furnace.

Oxidation data were obtained by measuring the gain of mass of the samples on a Mettler AE50 analytical balance. Some samples were analyzed by thin film diffractometry using a high-intensity x-ray diffractometer to determine the nature of the oxide product. Oxidized samples were gold-coated using a Varian VE-10 vacuum evaporator in preparation for qualitative analysis by scanning electron microscopy (SEM) and energy dispersive spectroscopy (EDS). A JEOL JSM-840 was used for the SEM work. EDS was performed with a Tracor Northern X-ray tube (TN5550-10) in conjunction with a Tracor Northern X-ray analyzer (TN5500) and Flextran software package.

Results and Discussion

Sintering of Powder Compacts

The theoretical density of $TiSi_2$ as determined by x-ray diffraction is 4.13 g/cm^3 (11). Powder compacts containing pure $TiSi_2$ could be densified to 93-97 % of the theoretical value by heating at 1420°C for 20 minutes. The surface of a polished and lightly etched sintered sample is shown in the optical micrograph of Figure 5. It can be seen that pores are for the most part concentrated at the triple points of the grains. A second phase occurring in small amounts, again mostly along grain boundaries, is most likely free silicon. Average grain size is approximately 20 μm.

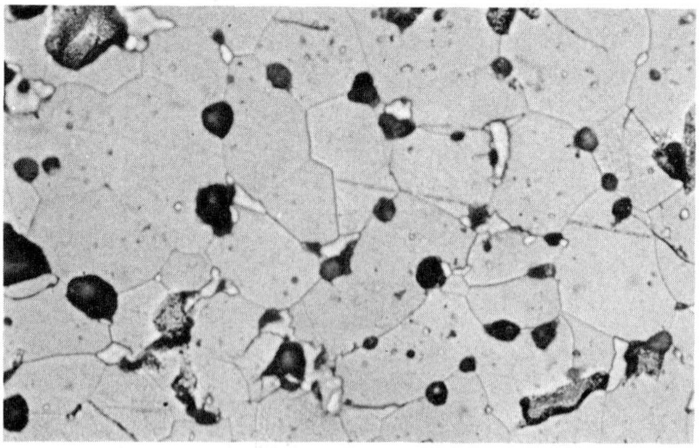

FIG. 5. Optical Micrograph of Surface of Sintered $TiSi_2$ (500x)

Powder compacts of $TiSi_2$ containing varying amounts of iron could be densified at 1420°C for 20 minutes to provide samples containing surface porosity of at most 10 percent. There was little variation in porosity among samples with different levels of iron as an impurity. However, the pores that formed in samples containing iron were several times larger than those found in a sample of pure $TiSi_2$, as shown in the optical micrograph of Figure 6. It is apparent that at least one other phase forms and diffuses along the grain boundaries, separating grains of pure $TiSi_2$, and that the majority of pores are formed in this second phase. When this sample was investigated under SEM, it was found that there were in fact three phases formed (see Figure 7). The chemical analysis of these phases as determined by EDS can be found in Table 1. It is at present unknown whether all phases exist at high temperatures. Average size of $TiSi_2$ grains in a sample containing 10 wt.% Fe is about 40 μm.

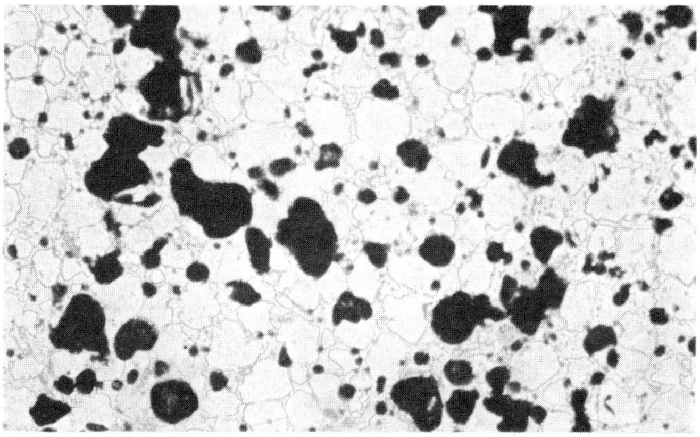

FIG. 6. Optical Micrograph of Sintered $TiSi_2$ with 10 wt.% Fe (100x)

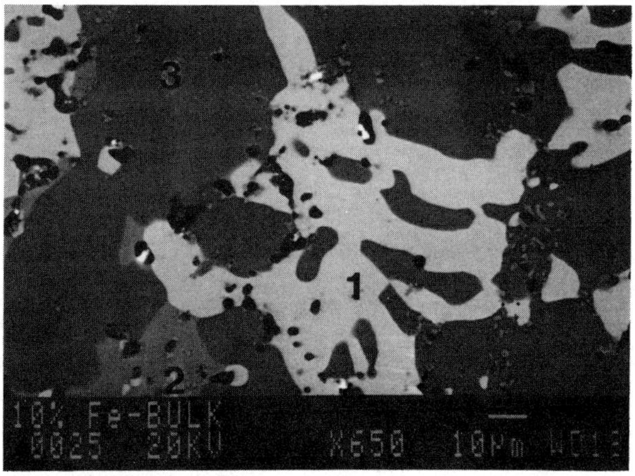

FIG. 7. Backscattered Image of Sintered $TiSi_2$ with 10 wt.% Fe

Due to the low melting point of aluminum, sintering of powder compacts of $TiSi_2$ containing aluminum could not be achieved above 1300°C without deformation of the samples. Longer sintering times of up to 4 hours were required to obtain a densified product. These samples showed a slight increase in porosity with the addition of Al, up to a maximum of about 16 percent. As was observed in the case of iron, pores in these samples were larger than in pure samples. It is again apparent, as seen in the scanning

TABLE 1
Chemical analysis of phases present in sample of Figure 7

Phase no.	EDS analysis	standard deviation
1	28.29 at.% Ti	3.27
	41.27 at.% Si	2.71
	30.44 at.% Fe	0.99
2	42.75 at.% Ti	2.96
	50.94 at.% Si	1.84
	6.31 at.% Fe	0.52
3	38.57 at.% Ti	5.07
	61.30 at.% Si	4.94
	0.13 at.% Fe	0.11

electron micrograph of Figure 8, that a separate phase forms and diffuses along the grain boundaries to eventually separate the grains of $TiSi_2$, and that most pores are formed in this second phase. Although it could be determined that this phase was high in aluminum content (see Table 2), it was not possible to get a consistent analysis by EDS. It was observed, however, that this phase was much softer than that containing little or no aluminum, resulting in an uneven surface due to accelerated wear of the phase during polishing. Average size of $TiSi_2$ grains in a sample containing 10 wt.% Al is about 40 μm.

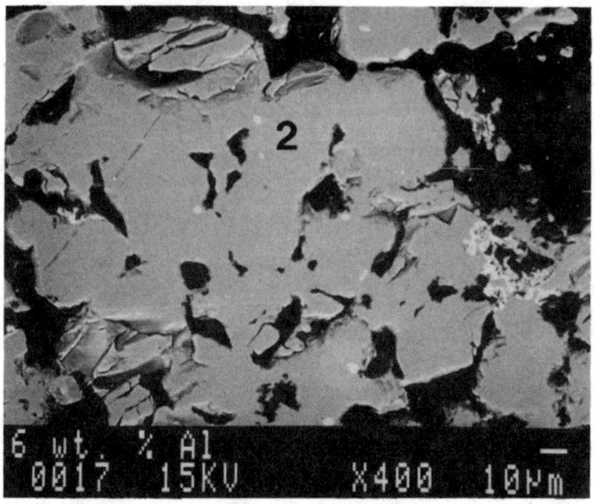

FIG. 8. Backscattered Image of Sintered $TiSi_2$ with 6 wt.% Al

TABLE 2
Chemical analysis of phases present in sample of Figure 8

Phase no.	EDS analysis	standard deviation
1	20.12 at.% Ti	10.22
	40.25 at.% Si	11.78
	39.63 at.% Al	11.54
2	45.84 at.% Ti	2.75
	52.82 at.% Si	1.98
	1.34 at.% Al	0.43

Powder compacts of $TiSi_2$ containing varying amounts of copper could be densified at 1350°C for 60 minutes to provide samples with no deformation and surface porosity of at most 8 percent. Pores in these samples were again larger than in pure samples, as seen in the scanning electron micrograph of Figure 9. Some samples high in copper content contained a few larger, spherical pores. The bright phase of the micrograph of Figure 9 has formed and diffused along grain boundaries of $TiSi_2$ and contains most of the pores, suggesting that this phase may have been in a liquid state at the sintering temperature. The chemical analysis of these phases as determined by EDS can be found in Table 3. It is apparent that the bright phase is composed almost entirely of copper, with some dissolved silicon. The presence of dissolved silicon in the phase suggests that some $TiSi_2$ dissociates during sintering to a lower silicide and free silicon. This is supported by the chemical analysis of the dark phase, which indicates less silicon than that which would correspond stoichiometrically to pure $TiSi_2$. Average size of $TiSi_2$ grains in a sample containing 10 wt.% Cu is approximately 50 μm.

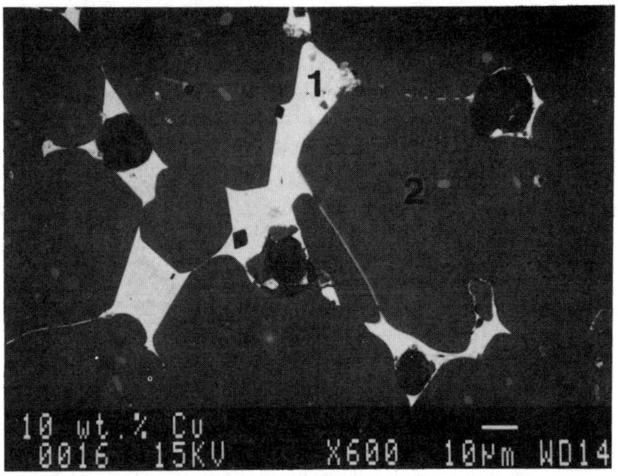

FIG. 9. Backscattered Image of Sintered $TiSi_2$ with 10 wt.% Cu

TABLE 3
Chemical analysis of phases present in sample of Figure 9

Phase no.	EDS analysis	standard deviation
1	0.99 at.% Ti	0.83
	17.09 at.% Si	3.19
	81.92 at.% Cu	3.46
2	45.37 at.% Ti	2.59
	54.63 at.% Si	2.11
	0.12 at.% Cu	0.38

Oxidation of Pure $TiSi_2$

A scanning electron micrograph of the surface of a wafer of pure $TiSi_2$ oxidized for 260 hours at 1200°C is shown in Figure 10. It can be seen that a thin but uniform protective oxide layer forms on the surface of the

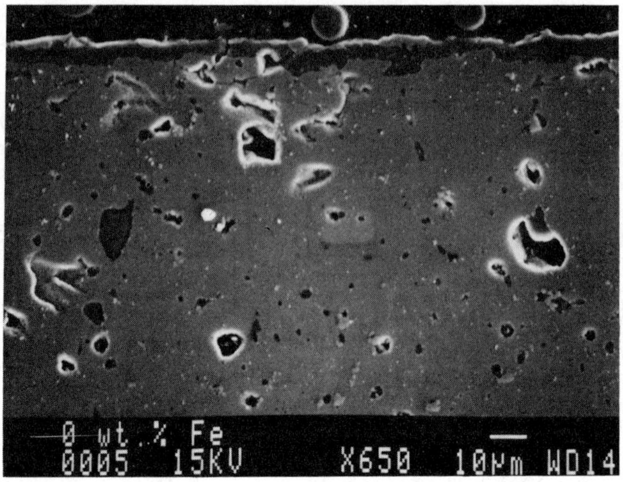

FIG. 10. SEM Micrograph of $TiSi_2$ Oxidized for 260 h at 1200°C

sample. EDS analysis of this film indicated that this oxide contains up to 95 at.% silicon, most likely in the form of silica, with a small amount of titanium in the form of titania or elemental Ti. If a stoichiometric reaction occured during oxidation, the oxide should contain twice as many Si atoms as Ti atoms. The phase diagram for the SiO_2-TiO_2 system (12) indicates a solid solubility of about 10 wt.% TiO_2 in SiO_2 between the temperatures of 300 and 1500°C. It is possible that at oxidation temperatures, the oxide film contains the solid solubility limit of TiO_2 in SiO_2, with the remaining titanium having reacted to form a lower silicide:

$$6\ TiSi_2 + 10\ O_2 \Longrightarrow TiO_2 + 9\ SiO_2 + Ti_5Si_3 \qquad \text{(R2)}$$

$$2\ TiSi_2 + 4\ O_2 \Longrightarrow TiO_2 + 3\ SiO_2 + TiSi \qquad \text{(R3)}$$

It is also possible that some Ti atoms counter-diffuse through the oxide to the gas-oxide interface as the solubility limit in silica is surpassed. This is supported by the presence of crystals on the surface of the sample, analyzed by EDS to contain about 98 at.% Ti, undoubtebly in the form of titania. The presence of crystalline titania on the surface of these samples was also confirmed by x-ray diffraction. While the kinetics of the oxidation of pure $TiSi_2$ will be discussed in the next section, it is very important to mention here one very important feature of these oxidized samples. As seen in Figure 11, the presence of surface pits does not prohibit the formation of a uniform glassy phase. Thus, initial weight gains of a porous sample may be high due to a relatively high surface area in comparison to a pore-free sample. These pits may initially act as a nucleation site resulting in rapid localized diffusion. However, once the oxide film grows to a point where the rate becomes diffusion controlled, the effect of these surface pits will be minimized.

Oxidation of $TiSi_2$ with an Iron Impurity

Typical weight gain data for the oxidation at 1200°C of pure $TiSi_2$ and with varying iron additions are shown in Figures 12 and 13. When the square of weight gain is plotted against time, straight lines are obtained, as shown in Figures 14 and 15. Some general comments can be made at this point. While some samples may have experienced very large weight gains in

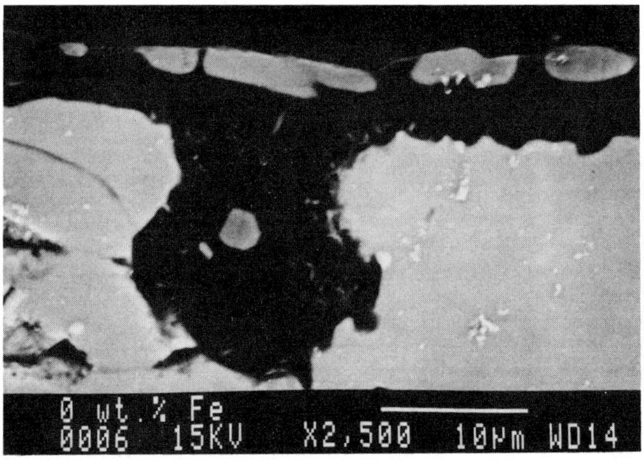

FIG. 11. Formation of Protective Oxide Layer Over Surface Pit of $TiSi_2$ Oxidized at 1200°C

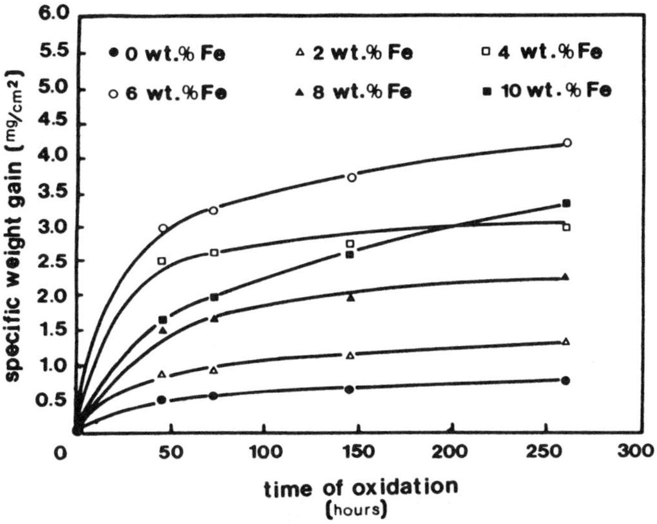

FIG. 12. Effect of Time on Oxidation Resistance of $TiSi_2$ with an Iron Impurity (1200°C)

the first 50 hours, subsequent gains are often relatively low, such that parabolic rate constants (the slope of the lines in Figures 14 and 15) have a low value compared to some samples experiencing lower initial weight gains. This suggests that, like samples of pure $TiSi_2$, surface porosity can lead to high initial rates of oxidation, but this does not hinder the formation of a uniform barrier of silica that minimizes the effect of initial surface defects and leads to slower rates of oxidation. This is the point at which the rate in these localized areas changes from surface-controlled to diffusion-controlled. To determine the final rate at which samples can be expected to oxidize, parabolic rate constants were determined by least squares regression, and the effect of iron additions on the rate constant at 1200°C is shown in Figure 16.

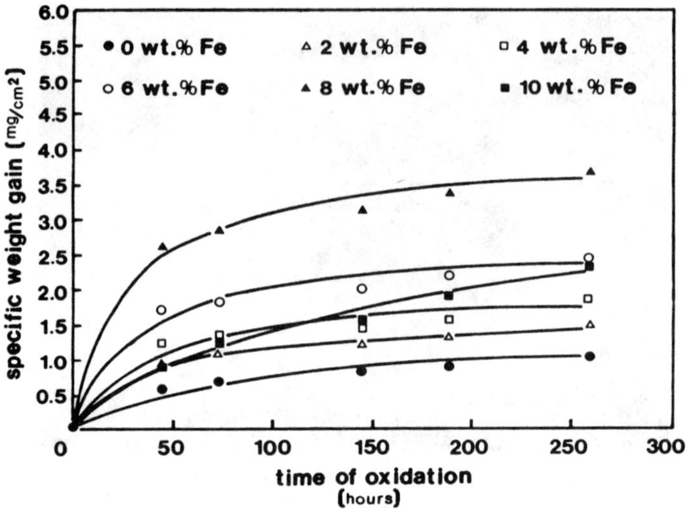

FIG. 13. Effect of Time on Oxidation Resistance of $TiSi_2$ with an Iron Impurity (1200°C)

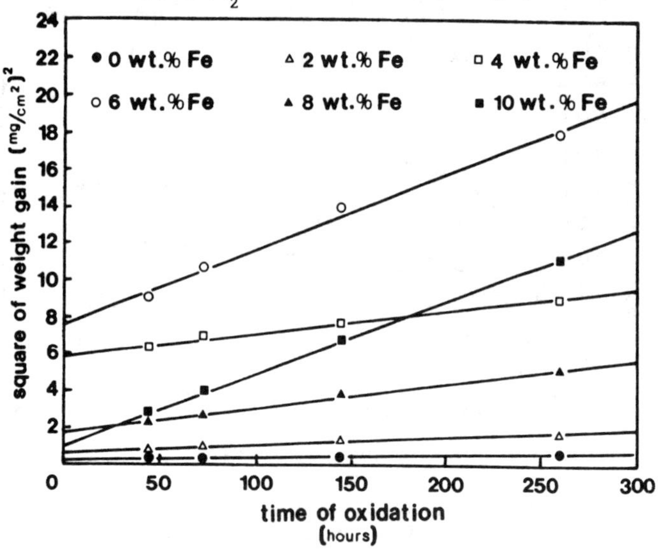

FIG. 14. Effect of Time on Square of Weight Gain Using Data of Fig. 12

Weight gain data for oxidation at 1300°C, shown in Figure 17, was very similar to that obtained at 1200°C except that parabolic rate constants, again determined from the plot of the square of weight gain versus time (see Figure 18), were slightly higher at 1300°C. This can be attributed to the higher rates of oxygen diffusion through the silica layer that can be expected at higher temperatures. The effect of iron additions on the rate constant at 1300°C is shown in Figure 19.

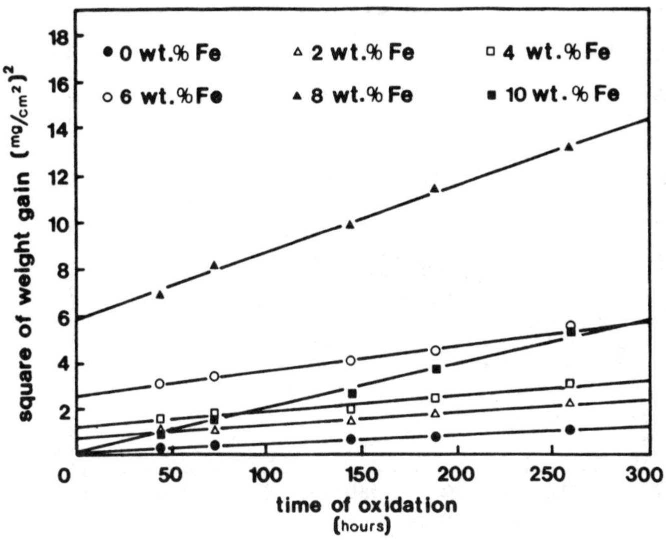

FIG. 15. Effect of Time on Square of Weight Gain Using Data of Fig. 13

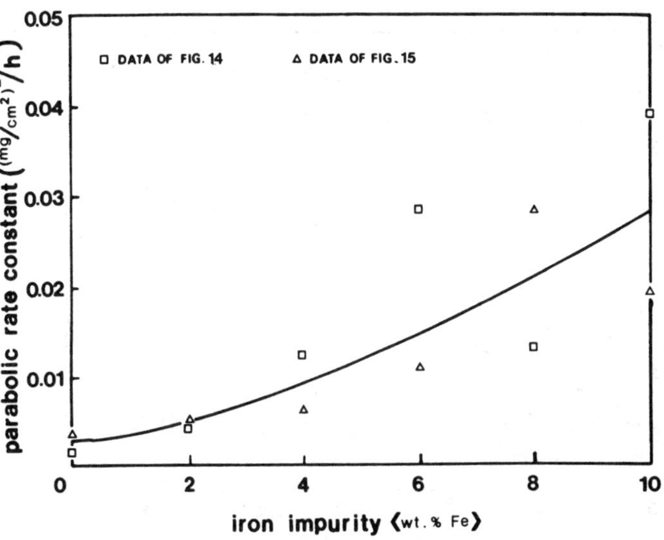

FIG. 16. Effect of Iron Additions on Parabolic Rate Constant (1200°C)

A scanning electron micrograph of a cross-section of a wafer of $TiSi_2$ containing 10 wt.% Fe oxidized for 260 hours at 1200°C is shown in Figure 20. It can immediately be seen that the oxide layer formed is similar to that formed on a sample of pure $TiSi_2$ except in its uniformity. It is apparent that localized oxidation occurs at a higher rate, such that the oxide penetrates the sample surface more in some areas than others. The boxed area of Figure 20 is shown magnified several times in Figure 21.

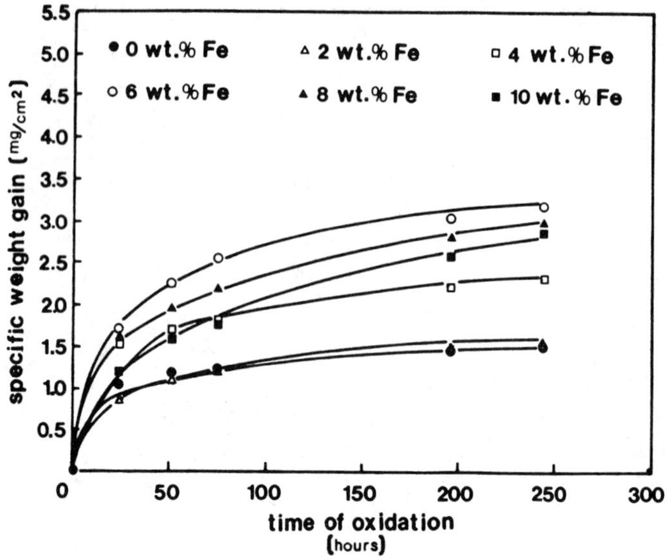

FIG. 17. Effect of Time on Oxidation Resistance of $TiSi_2$ with an Iron Impurity (1300°C)

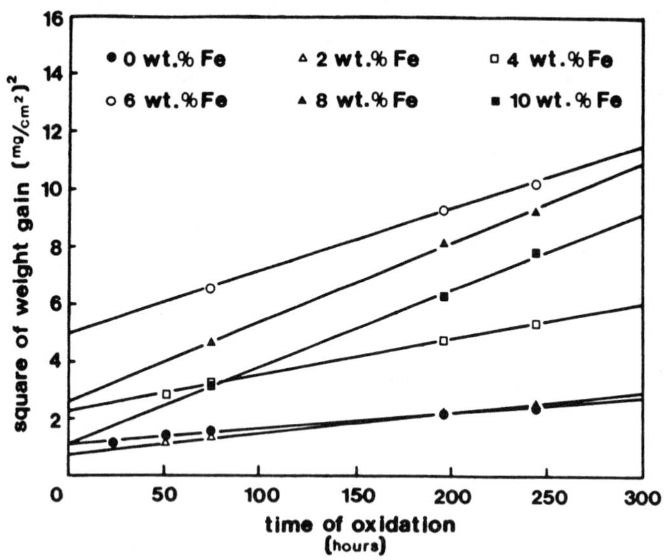

FIG. 18. Effect of Time on Square of Weight Gain Using Data of Fig. 17

Preferrential oxidation of the lighter phase can be explained in terms of its chemical composition as tabulated earlier. Searcy (13) recognized that in order to have a consistent, protective silica coating, it is necessary that metal silicides have a silicon-to-metal atomic ratio of 3:2 or higher. In the preferrentially oxidized phase here, the silicon-to-metal ratio is only 2:3. If this phase cannot in fact provide sufficient silica upon oxidation to form a protective layer, it might be expected that more

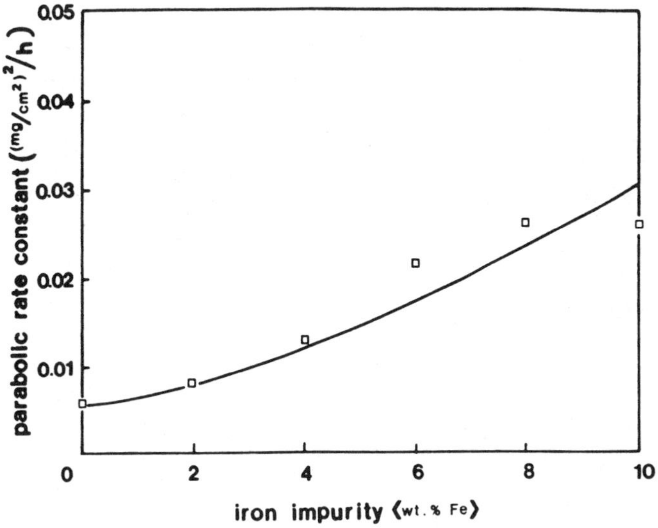

FIG. 19. Effect of Iron Additions on Parabolic Rate Constant (1300°C)

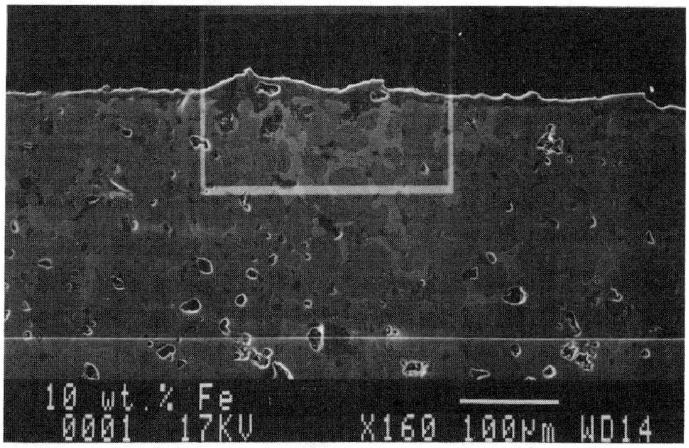

FIG. 20. SEM Micrograph of $TiSi_2$ with 10 wt.% Fe Oxidized for 260 hours at 1200°C

titanium oxides might be produced, since they are more stable than iron oxides, than in a sample containing no iron. The large pieces of material concentrated on the surface of the oxide layer, immediately above the pits formed in the phase deficient in silicon, have been found by EDS to contain 98-100% Ti atoms, most likely in the form of titania crystals. The surface of such oxidized samples visibly contain an increasing amount of titania crystals with increasing iron content. An SEM micrograph of such surface crystals can be found in Figure 22.

Further evidence of this phenomenon was found through an investigative chemical analysis of the bulk oxide. The oxide formed above the dark phase containing high amounts of Si was found to contain only 6 at.% Ti atoms, the

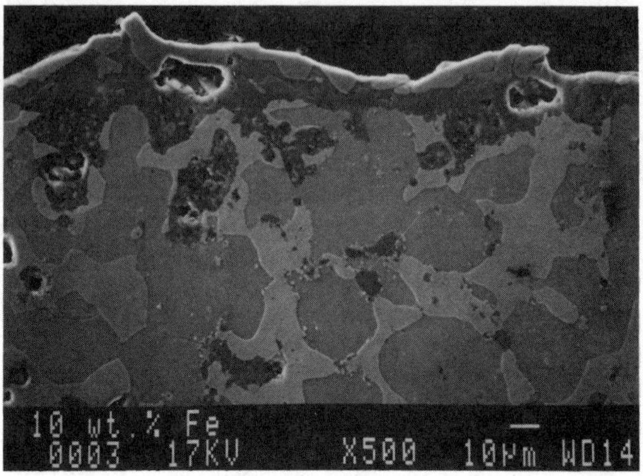

FIG. 21. SEM Micrograph Showing Magnification of Boxed Area of Figure 20

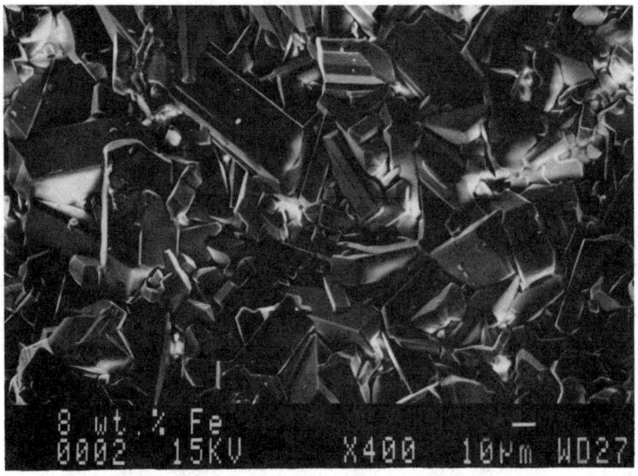

FIG. 22. SEM Micrograph of Surface Crystals Grown on Surface of Oxidized $TiSi_2$ Containing 8 wt.% Fe

remainder being Si atoms. However, the oxide formed above the phase low in Si and high in Fe and Ti atoms was found to contain 22 at.% Ti atoms, the remainder again being Si atoms. Furthermore, the oxide formed within the pits of this same phase contained about 40 at.% Ti and 60 at.% Si atoms. This represents the same atomic ratio of Si:Ti as the bulk phase from which the oxide formed. This oxide phase thus provides little protection to the bulk phase, so that all the Ti atoms are oxidized along with those of Si resulting in rapid localized oxidation.

It can be seen in the micrograph of Figure 23 that the oxide formed upon preferrential attack of this phase also contains a large number of crystals that seem to be migrating toward the surface of the sample. These were found to be very high in titanium. It is possible that these have precipitated as Ti crystals once the saturation limit in silica is

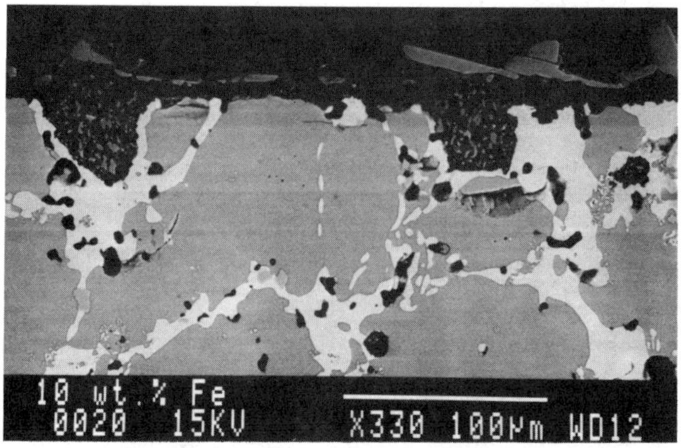

FIG. 23. Backscattered Image Showing Preferrential Oxidation of One Phase in $TiSi_2$ Containing 10 wt.% Fe

surpassed, or that they are crystals of titania that precipitate once the saturation limit of TiO_2 in SiO_2 is surpassed. A higher rate of oxidation in these localized areas might be expected due to faster oxygen diffusion along the surface of these crystals, as proposed in an earlier paper by Schlichting (14), or by some modification of the protective properties of the silica layer by the high level of impurity.

An x-ray map for iron in an oxidized sample containing 10 wt.% Fe is shown, along with the corresponding micrograph, in Figure 24. The presence of iron in surface crystals indicates that counter-diffusion of iron through the oxide layer must occur during oxidation at high temperatures. Chemical analysis showed an absence of Si atoms and a Ti:Fe atomic ratio of approximately 2:1, indicating that pseudobrookite or crichtonite may form.

After an extended period of oxidation, preferrential attack of the phase low in silicon eventually results in the depletion of Si and Ti, leaving grains high in iron, which could oxidize. This is accompanied by the separation of many $TiSi_2$ grains initially connected by the preferentially oxidized phase. While this indicates a loss of the material's strength, it also aids in holding the oxide layer together so that spalling, the flaking of the oxide resulting in rapid transport of oxygen, does not occur during thermal cycling.

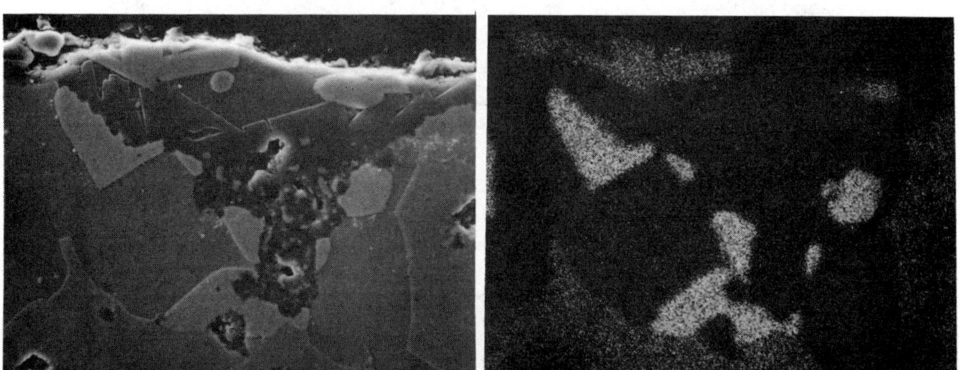

FIG. 24. SEM X-ray Map for Iron at Surface of Oxidized $TiSi_2$ Containing 10 wt.% Fe (850 x)

Oxidation of $TiSi_2$ with an Aluminum Impurity

Typical weight gain data for the oxidation at 1200°C of pure $TiSi_2$ and varying aluminum additions are shown in Figure 25. When the square of weight gain is plotted against time, straight lines are again obtained, as shown in Figure 26. Once more, surface porosity is likely responsible for high initial rates of oxidation in localized areas. Oxidation of samples containing aluminum required approximately twice as long as samples containing iron to exhibit parabolic kinetics.

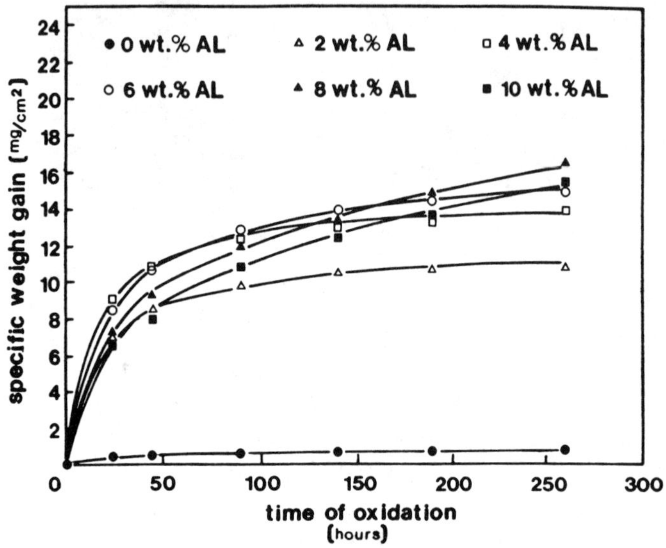

FIG. 25. Effect of Time on Oxidation Resistance of $TiSi_2$ with an Aluminum Impurity (1200°C)

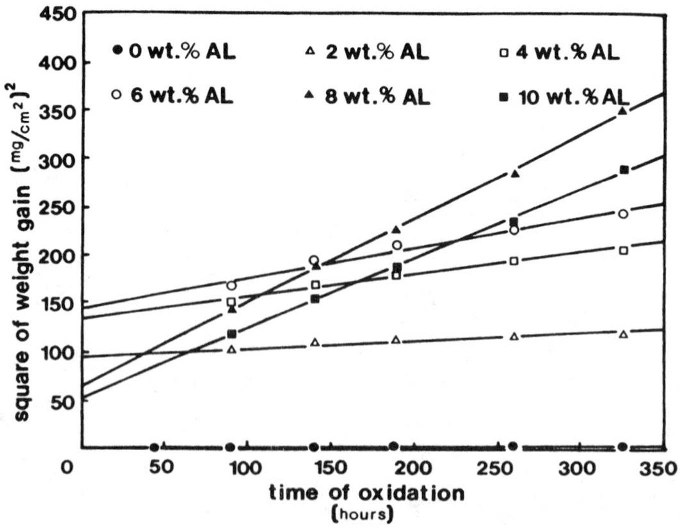

FIG. 26. Effect of Time on Square of Weight Gain Using Data of Fig. 25

Parabolic rate constants are plotted as a function of aluminum content of the samples in Figure 27. It is apparent that even small additions of Al result in a drastic reduction in oxidation resistance. Samples high in aluminum content showed no useful resistance to oxidation at this temperature, and oxidation of such samples resulted in a complete loss of strength after only two hundred hours.

A scanning electron micrograph of a cross-section of a wafer of $TiSi_2$ containing 10 wt.% Al oxidized for over 300 hours at 1200°C is shown in Figure 28. The oxide layer formed on samples containing aluminum was not uniform or thick enough to provide much protection of the substrate. In contrast, the phase high in aluminum appeared to have been almost complete-

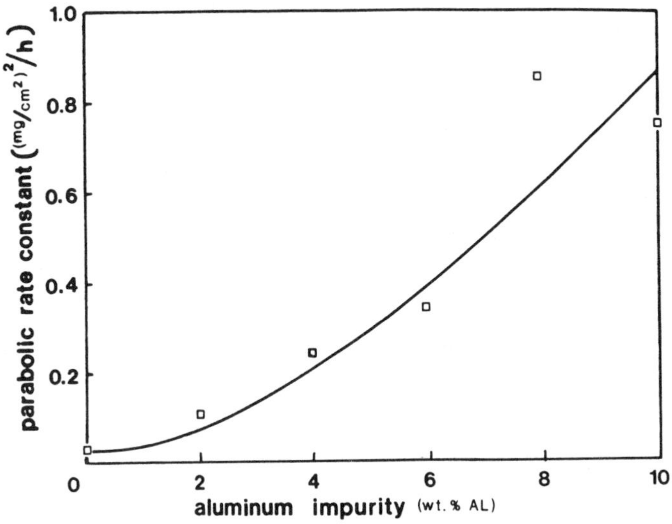

FIG. 27. Effect of Aluminum Additions on Parabolic Rate Constant (1200°C)

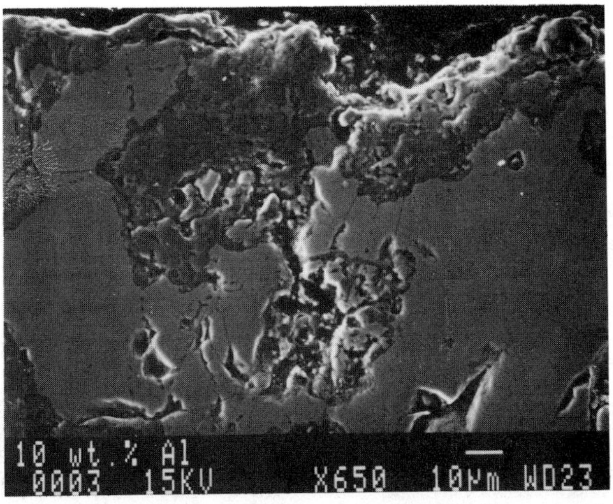

FIG. 28. SEM Micrograph of $TiSi_2$ with 10 wt.% Al Oxidized for 300 hours at 1200°C

ly oxidized preferentially, accounting for the material's drastic reduction in strength. This preferential oxidation can be attributed to two factors. First, the silicon-to-metal atomic ratio is only 2:3 and, as discussed earlier, such a phase cannot provide sufficient silica upon oxidation to form a protective layer. It is even possible that this phase is in a liquid state, which would oxidize faster than the solid, at the oxidation temperature. Second, because aluminum oxide is thermodynamically slightly more stable than silicon or titanium oxides, it could be expected that the aluminum in the phase will oxidize and join any silica-titania glass, possibly forming aluminum silicates that may result in a marked reduction in viscosity. The chemical composition of the oxidized phase was very close to that of the bulk phase deficient in Si, again suggesting that this phase is completely oxidized preferentially to the phase deficient in aluminum. Eventual separation of grains of pure $TiSi_2$ due to intergranular oxidation can be expected.

Oxidation of $TiSi_2$ with a Copper Impurity

The high-temperature oxidation of $TiSi_2$ with a copper addition proceeds without resistance due to the formation of a low-viscosity liquid phase, resulting in extremely high weight gains even after only 24 hours. This has been observed at temperatures as low as 1000°C. At 1200°C, this liquid oxide or copper silicate has a low enough viscosity that it can flow off the sample.

A scanning electron micrograph of a cross-section of a wafer of $TiSi_2$ containing 10 wt.% Cu oxidized for 24 hours at 1200°C can be found in Figure 29. It shows the collection of the oxide in a liquid droplet on the surface of the sample, and the absence of the oxide on the underside from which it was able to flow off during oxidation. A magnified portion of the "droplet" of Figure 29 can be found in the micrograph of Figure 30. It can be seen that there are several different products formed, some of which appear to have precipitated during sample cooling. The dark area directly above the substrate surface was analyzed by EDS to contain about 95 at.% Si, 2 at.% Ti, and 3 at.% Cu. These species are all likely in oxide form. The large grey crystals were analyzed by EDS to contain 99 at.% Ti. It is reasonable to expect that these are crystals of titania.

In addition to these features similar to those observed in pure samples and those containing iron, there are the large white crystals and smaller, whisker-like features assumed to have precipitated during sample cooling. EDS analysis of the large white crystals showed a composition of 98 at.% Cu

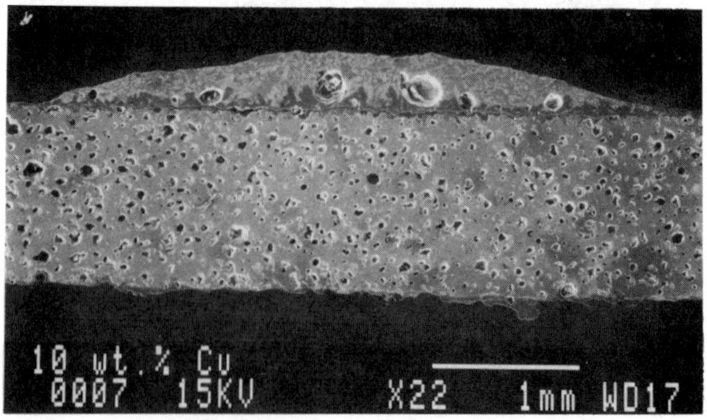

FIG. 29. SEM Micrograph of $TiSi_2$ with 10 wt.% Cu Oxidized for 24 hours at 1200°C

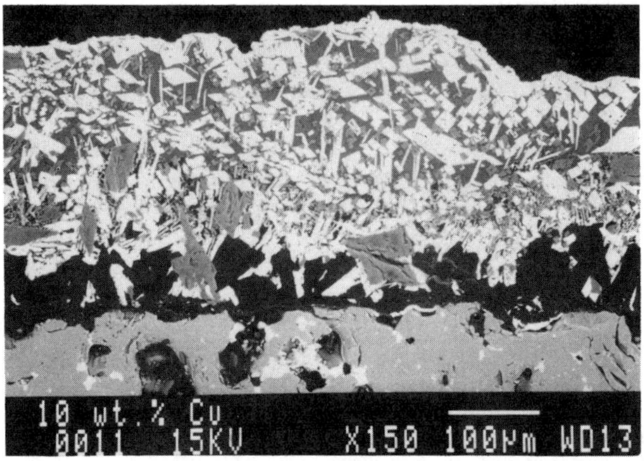

FIG. 30. Backscattered Image Showing Magnification of Oxide Features of Figure 29

and 2 at.% Si. Since the surface of the oxidized sample was black in colour, it can be expected that these crystals are CuO, which presumably grew from those of Cu_2O during cooling. The whisker-like features contain 78 at.% Cu and 22 at.% Ti.

It is apparent that, since copper oxides thermodynamically should not form before silicon or titanium oxides, the bulk phase that is rich in copper (over 80 at.%) cannot be protected from oxidation at high temperatures. Copper is oxidized in localized areas and joins the silica-titania glass, forming a low-viscosity modified silica network or possibly copper silicates of low viscosity. The phase diagram for Cu_2O-SiO_2 (15) indicates an 8% Cu_2O eutectic at 1060°C. In either case, the layer formed at 1000°C and above is not protective in nature, and diffusion of oxygen through the layer proceeds at a very high rate.

Conclusions

The oxidation resistance of pure $TiSi_2$ at high temperatures can be attributed to the formation of a uniform, protective silica-titania glass that tends to minimize the effect of initial surface defects. Parabolic rate constants are typically 2.4 x 10^{-3} $(mg/cm^2)^2/h$ at 1200°C and 4 x 10^{-3} $(mg/cm^2)^2/h$ at 1300°C.

Titanium disilicide containing up to 10 wt.% iron can be sintered at 1420°C for 20 minutes to provide samples with maximum surface porosity of 10 %. Such samples show a slight decrease in oxidation resistance, with parabolic rate constants increasing with iron content to a value of about 3 x 10^{-2} $(mg/cm^2)^2/h$ for a sample containing 10 wt.% Fe oxidized at 1200°C. Iron additions have the identical effect on oxidation resistance at 1300°C, with all samples experiencing an increase in the parabolic rate constant of about 2 x 10^{-3} $(mg/cm^2)^2/h$ over corresponding samples oxidized at 1200°C. The decrease in oxidation resistance with the addition of iron is attributed to the presence of a phase low in silicon content which cannot provide sufficient silica upon oxidation to provide a protective film. This results in more rapid localized oxidation, eventually leading to the separation of

$TiSi_2$ grains.

Titanium disilicide containing up to 10 wt.% aluminum can be sintered at a maximum temperature of 1300°C for 4 hours to provide samples with maximum surface porosity of 16 %. Such samples show a dramatic decrease in oxidation resistance, with parabolic rate constants increasing with aluminum content to a value of about 8×10^{-1} $(mg/cm^2)^2/h$ for a sample containing 10 wt.% Al oxidized at 1200°C. The decrease in oxidation resistance is again attributed in part to the presence of a phase of low Si:Me atomic ratio. In addition, the aluminum oxide is believed to act as a network modifier in the silica glass, reducing its protective properties considerably. This results in the complete preferrential oxidation of the phase high in Al and low in Si that separates grains of $TiSi_2$. This intergranular oxidation leads to a drastic reduction in the strength of the material.

Titanium disilicide containing up to 10 wt.% copper can be sintered at a maximum temperature of 1350°C for 60 minutes to provide samples with maximum surface porosity of 8 %. Such samples show no useful resistance to oxidation at temperatures as low as 1000°C due to the formation of a liquid oxidation product upon localized oxidation of one phase very high in copper content but very low in silicon content.

References

1. Schwettman, F.N., R.A. Graff, and M. Kolodney, *Journal of the Electrochemical Society*, **118** [12], 1973 (1971).
2. Wirkus, C.D., and D.R. Wilder, *Journal of The American Ceramic Society*, **49** [4], 173 (1966).
3. Berkowitz-Mattuck, J.B., P.E. Blackburn, and E.J. Felten, *Transactions of the Metallurgical Society of AIME*, **233**, 1093 (1965).
4. Valyi, G., L. Zhaokang, A. Steffen, M. Maier, and P. Balk, *Thin Solid Films*, **116** [4], 383 (1984).
5. Yang, W.Y., H. Iwakuro, H. Yagi, T. Kuroda, and S. Nakamura, *Japanese Journal of Applied Physics*, **23** [12], 1560 (1984).
6. Lavendel, H.W., and A.G. Elliot, *Transactions of the Metallurgical Society of AIME*, **239**, 143 (1967).
7. d'Heurle, F., E.A. Irene, and C.Y. Ting, *Applied Physics, Letters*, **42** [4], 361 (1983).
8. Stone, H.E.N., *Journal of Materials Science, Letters*, **2**, 609 (1980).
9. Chen, J.R., M.P. Houng, S.K. Hsiung, and Y.C. Liu, *Applied Physics, Letters*, **37** [9], 824 (1980).
10. Baas, H., C.A. Pickles, and J.M. Toguri, "Recovery of Titanium Silicide Crystals from a Copper Melt", to be published.
11. Samsanov, G.V., "Silicides and Their Uses in Engineering", Translation from Russian FTD-TT-62 430 1&2 (1962).
12. Ricker, W., and F.A. Hummel, *Journal of The American Ceramic Society*, 34 [9], 271 (1951).
13. Searcy, A.W., *Journal of The American Ceramic Society*, **40** [12], 431 (1957).
14. Schlichting, J., *Ceramurgia International*, **4** [4], 162 (1978).
15. Berezhnoi, A.S., L.I. Karyakin, and I.F. Dudavskii, *Doklady Akad. Nauk S.S.S.R.*, **83** [3], 401 (1952).

Acknowledgements

This work was funded by the Department of Energy, Mines and Resources (Canada) through DSS Contract No. 55SS.23440-7-9161.

Mixed oxidant corrosion in coal combustion and conversion systems: manifestations and mechanisms

J. Stringer
Electric Power Research Institute, Palo Alto, California, U.S.A.

Abstract

High temperature corrosion in coal combustion and conversion systems frequently involves more than one oxidant. Typically, oxygen, sulfur, and carbon are involved, but other species such as chlorine may also play a role. Alloys for high-temperature applications generally depend for their resistance to degradation by reaction with the environment on the formation of a protective oxide layer on their surface. To be protective, the oxide must be slow growing, have slow transport rates for various species through it, and be continuous and adherent to the surface. It must also be able to retain these qualities under thermal cycling conditions. The presence of other oxidizing species may have adverse effects on the protectiveness of the oxide. This paper discusses some of these effects.

Introduction

Metals and alloys for high temperature applications are generally exposed to environments in which they are not stable. In the majority of cases, the stable species is the oxide, and this is the only situation which will be considered in this paper. Reaction with the environment can degrade the properties of the alloy in several ways. Clearly, as metal is consumed by the oxide-forming reaction, there will be a loss of section. In some cases, the oxidation does not occur uniformly, but oxide penetrations may develop into the alloy: these produce a disproportionate loss of useful section, and also may produce stress concentrations. In addition, when this type of intrusive oxide growth is combined with thermal cycling, the differences in specific volume of the oxide and metal and the differences in thermal expansion may result in a progressive 'growth' of the alloy. One component of the alloy may be selectively removed by the oxidation process, and the surface depletion may also result in a disproportionate degradation of mechanical properties. Migration of components of the environment into the alloy grain boundaries can result in severe embrittlement. This list is not intended to be exhaustive, but it illustrates that the possible adverse effects of reaction with the environment may be quite complex.

To resist these effects, the alloy must be capable of forming a protective oxide layer. To be protective, the oxide layer must be continuous, and it must have good mechanical strength and be adherent to the surface of the metal. Since in use it is expected that the component will be heated from room temperature to its use temperature and cooled down again many times, the protective oxide layer must be resistant to the stresses induced by the differences in thermal expansion coefficient between the oxide and the metal substrate. There are circumstances when it must also be able to tolerate some strain: metallic components frequently undergo elastic strains in normal use, and the oxide layer should have a fracture strain high enough to tolerate these. Abrasion or erosion resistance may also be required. The protective oxide must also be resistant to the other environments it may experience: for example, while the component is cool it may be exposed to a moist atmosphere, or even to water: it should not be hygroscopic, or water-soluble. There are other possibilities. The protective oxide must be stable throughout the temperature range expected during operation, since phase changes or decompositions will probably be associated with volume changes which will lead to fracture.

The protective oxide must be slow growing. This is not a well-defined statement, but generally if the growth is uniform, as required by the previous conditions, then the section loss expected during service as a result of the formation (and possible reformation) of the oxide must be tolerable. The actual acceptable rate will thus depend on the temperature, the tolerable section loss, and the required component lifetime. It should be noted that nothing is to be gained by having an oxide that grows slower than is required by these conditions. `Slow growing' implies that the transport rates of both oxygen and the metal component of the oxide are slow.

In addition, transport of other species through the protective oxide may be required to be slow: this does not necessarily follow from the previous condition. For example, it has been shown (1) that when a rutile TiO_2 scale grows on a Ti-Mo alloy, the transport of both the oxygen and the titanium through the oxide is relatively slow, and the oxide grows protectively. However, the molybdenum appears to be capable of moving very rapidly through the oxide. Similarly, Cr_2O_3 grows slowly, and both Cr and O are transported relatively slowly through the oxide. It appears, however, that nitrogen can move relatively quickly through the oxide, and chromium oxidizing in air can suffer severe embrittlement as a result of nitride formation beneath the oxide. The word `transport' is used here rather than diffusion. Bulk diffusion is only one way that species may be transported through an oxide, and work by Atkinson (2) and others suggests that it is the exception rather than the rule. Other transport paths include grain boundary transport (which may or may not resemble a diffusion process); surface transport along the walls of cracks or pores; gaseous transport down cracks or pores; and transport through second phase paths or along phase boundaries.

If a protective oxide satisfies all these conditions, it will generally grow at a rate which diminishes with time. If the appropriate transport process through the oxide is rate controlling, and it does not vary with the oxide thickness, the rate will vary with the reciprocal of the thickness: the growth rate will follow a parabolic rate law. This is true whatever the transport process is, and the appearance of a parabolic rate law cannot be used as evidence that bulk diffusion is involved.

For practical purposes, for the highest temperature applications, the only acceptable protective oxides are chromia, Cr_2O_3, and alumina, Al_2O_3. SiO_2 may also be an acceptable protective oxide, but examples of its use are few, and it will not be expressly considered here. Compound oxides, such as $FeCr_2O_4$, cannot normally be regarded as potential protective oxides for reasons which are outside the scope of this paper. The remainder of this paper will discuss mainly the problems that second oxidants present in the formation and maintenance of protective chromia and alumina scales; although one or two other specific cases will be mentioned.

The Formation of a Protective Oxide on an Alloy

To form one of these protective oxides, an alloy must contain sufficient of the oxide-forming element, Cr or Al. For a variety of reasons the alloy may have limited ability to tolerate these elements. Thus, Al embrittles steel at relatively low concentrations; Cr cannot exceed certain values in nickel-base superalloys because of a need to maximize the solubility of Ti and Al to produce as high a volume fraction as possible of the precipitate $Ni_3(Ti,Al)$. In some cases, the corrosion resistance of the material selected may be inherently inadequate, and in this event coatings or claddings may be used; for these, the need to limit the content of the stable-oxide-forming element is less of a restriction, although even in these cases the choice of composition is not free. For example, it may be necessary to have a sufficiently low ductile-to-brittle transition temperature to avoid extensive coating cracking in service, and this will result in limits on the acceptable composition range.

The amount of the stable-oxide-forming element that is sufficient depends on several criteria, which are generally outside the scope of this paper. However, the crucial criteria are not generally thermodynamic, and thus it is not the _activity_ of the Al or Cr that is important (within certain broad ranges) but the concentration. This distinction is important in understanding the role of second oxidants, so it is worth explaining what is meant in a little more detail. Generally, a Co-Cr alloy must contain 35at%Cr if it is to form a protective Cr_2O_3 scale at (say) 1000^oC. If an alloy of Co-Cr-C is prepared, the carbon reacts with the chromium to form a relatively stable carbide. Once the carbon content reaches 2.8 wt% or so, all the chromium is present as carbide in a cobalt matrix that is essentially chromium-free. _Provided_ the distribution of the carbide is fine enough, the alloy will still form a Cr_2O_3 scale, at least in the short term (3). However, if an alloy is prepared with a dispersion of Cr_2O_3 in a cobalt matrix instead, although the total amount of Cr is still equivalent to 35at%, the alloy will _not_ be capable of forming an external chromia scale, at least in the short term. For an alloy to form a protective oxide, when the protective-oxide-forming element is concentrated in in a compound, the compound must be less stable than the oxide, and it must be finely dispersed. This principle is the basis of a method of designing oxidation-resistant alloys called the `dispersed reservoir' method (4,5).

Many commercial alloys consist of more than one phase, and the oxide-forming element is, in general, not uniformly distributed amongst the phases (the _activity_ will be uniform, but remember that this is not the criterion of importance). As a result, the protective oxide will be capable of forming on some of the phases, and not on others (taken in isolation). The fact that protective oxides can nevertheless form on alloys of this type appears to be because the oxide can spread laterally over the surface beyond the extent of the phase on which it is growing (6). It is important for protection to be achieved that the phases over which the oxide will not spontaneously form should be fine. This situation arose some years ago with a superalloy which suffered severe hot corrosion attack. This was traced to the existence of coarse refractory metal carbides. Changing the heat treatment to refine the size of these carbides eliminated the hot corrosion problem (7).

The problem of adhesion of these protective oxides under thermal cycling and cyclic straining conditions has been an issue for many years. It is well-known that minor amounts of reactive elements such as yttrium, cerium, or hafnium, or dispersions of their oxides, can enormously improve the ability of the oxide scale to maintain its integrity under these conditions. As a result of the work of Smeggil and co-workers (8,9,10) and Smialek (11) amongst others, it seems probable that a major factor in determining the adhesion of the scale is the segregation of impurities, and in particular sulfur, to the metal/oxide interface. They present convincing evidence to show that the oxide/metal bond is intrinsically strong, and that the primary role of the reactive elements is to getter the sulfur (and perhaps other harmful species) to the interface. In their studies, the sulfur segregated to the surface from the bulk of the alloy, but it is also possible to introduce harmful contaminants from the environment: recently, Hou has shown that chlorine, present as HCl, can have the same effect on scale adhesion (12), and essentially the same point was made several years ago by Hancock (13).

The Role of Second Oxidants: Mechanisms

The important second oxidants in practice in coal combustion and conversion systems are sulfur, carbon, nitrogen, and chlorine. In particular cases other oxidants may be involved, but the general principles are more or less the same for all. As was stated at the beginning, this paper is only concerned where the most stable product is the oxide. There are some special circumstances where a sulfide or a carbide may be the most stable species (sulfur boilers and pyrolysis plants, for example), but these situations will not be explicitly considered here; the general principles are, however, the same.

The Phase Stability Diagram and the Reaction Path.

Figure 1 shows a thermodynamic phase stability diagram for the Fe-O-S system at 621°C (14). As presented here, and as normally drawn, the activities of all the condensed phases are unity. The diagram can be drawn for cases where the activity of one or more of the phases is different to unity, but it is difficult to present the situation where the activity varies. By extending it to three dimensions, it is possible to show some of the relationships for binary alloys (15). The axes in Figure 1 represent the activities of oxygen and sulfur, expressed as partial pressures, on a logarithmic scale; lines of constant SO_2 partial pressure are also drawn, and it is apparent that it would be simple to draw the diagram in terms of axes representing oxygen and SO_2 activities, for example, if this was more convenient from the point of view of the practical system to be analysed.

This type of diagram, which appears to have been first developed for use in extraction metallurgy, has been widely used for discussing mixed oxidant reactions in recent years. Because of the limitations indicated above, it must be used with considerable caution, but some important points can be qualitatively illustrated with its aid.

A second useful concept is that of the `reaction path' (16,17). Suppose that an iron specimen is exposed to an environment containing sulfur and oxygen, whose activities can then be represented by a point such as A in the simplified diagram drawn in Figure 2. Furthermore, suppose that it is possible to determine the activity of sulfur and oxygen at every point in the system. For simplicity, assume as usual that the activities vary only with the distance normal to the surface. The iron as received will have an activity of sulfur and oxygen determined by the method of manufacture, and in particular the source of the materials and the type and quantity of desulfurizer (eg. manganese) and deoxidant (eg. aluminum) used. These activities will be represented by the point B, which obviously will lie somewhere in the Fe stability field. Note that if no desulfurizer, or an inadequate amount had been use so that the as-received iron contained a dispersion of iron sulfide, however little there was of it, the point B would lie on the Fe/FeS boundary. If the iron also contained a dispersion of iron oxide, the point B would be located at the Fe/FeS/FeO corner. As will be seen, this would have very important consequences for the oxidation reactions.

Assuming for the moment that this is not the case, then as oxygen and sulfur are dissolved in the iron, the activities of these two species will increase at the surface. Starting at point B, there will be a gradual movement as the analysis moves out towards the surface. At the surface, a second phase will be present. If that phase is an oxide, and assuming local equilibrium, the surface layer of the metal must have an activity of O and S that lies somewhere along the Fe/FeO equilibrium line, and the activities on the innermost surface of the product layer must be the same. Equally, the activities on the outermost surface of the product layer must be identical to those in the atmosphere (this also assumes no gas boundary layer: this issue will be addressed in a moment) and thus also represented by the point A. As one goes out through the product layer, the activities will change, tracing out a path on the diagram which is continuous everywhere. This is the reaction path. It has certain similarities to a diffusion path, but a diffusion path is traced out in a phase diagram, and is usually drawn in terms of compositions. The reaction path is drawn on the stability diagram, and is drawn in terms of activities. The portion of the reaction path that lies within the Fe-stable field, for example, would be essentially invisible on a diffusion path, at least as conventionally drawn. Figure 3 shows different possible paths from the same values of A and B (17): path 1 is the case where only an oxide layer forms, and there is no sulfide formation, although the sulfur activity in the metal may increase significantly. Path 2 represents the case where a sulfide layer forms under the oxide layer. Path 3 is more interesting, and illustrates how a two-phase oxide and sulfide layer may form. If the path enters the Fe field at the corner then no single phase layer will form adjacent to the metal. Generally, absent major interactions, the activities of both the sulfur and the oxygen must decrease continuously as the path goes from A to B. The consequence of this is that no sulfide should form anywhere if A lies below the horizontal line which is the extrapolation of the Fe/FeS stability line. None of this says anything particularly new: these points have been made earlier on purely thermodynamic grounds, as is discussed in (17). However, the use of the reaction path makes it easier to see some of these points. Note that if the iron does contain oxide and sulfide, so that point B lies at the corner, it becomes impossible to form a single phase sulfide layer. This point does not appear to have been made before, and it should be relatively easy to test experimentally.

The restrictions on sulfide formation listed above depend on the assumptions made. In particular, the existence of a gaseous boundary layer may allow the gas composition at the oxide surface to be different from that in the bulk. An example that is easy to understand derives from the experience with the oxidation of iron in CO_2. The situation is sketched in Figure 4. There are apparently pores in the oxide through which gaseous CO_2 diffuses inwards. The oxidation reaction releases CO, which then migrates outwards. The CO/CO_2 ratio thus increases towards the interior of the oxide, which is consistent with a decline in the oxygen activity. However, the carbon activity is given by

$$a_C = k_C \, p_{CO_2} / (p_{CO})^2$$

and the carbon activity can increase towards the metal surface producing carbon deposition in the case of iron and internal carburization in the case of Fe-Cr steels (18). Similar processes are possible in the sulfur system, and the `safe' region is then that between the Fe/FeO vertical and the SO_2 line which passes through the corner point. This region is shaded in Figure 2. Other related processes that may be important include chemisorption: the chemisorption of sulfur at grain boundaries in the oxide or at the metal/oxide interface may produce apparent enhancements in activity. Equilibrium may not be established at the interfaces. Thus, if SO_2 adsorbs on the oxide surface, and loses oxygen to form oxide, resulting in a chemisorbed S, the removal of the S by reaction with oxygen from the atmosphere to form SO_2 or the desorption as S_2 may be slow, and the effect will be to produce an increase in sulfur activity at the surface driving sulfur inward.

The Effect of the Second Oxidant on the Oxidation Process.

The second oxidant can have several possible effects.

(1) The second oxidant may enter the lattice of the oxide, modifying its transport properties, in a manner essentially identical to the well-known `Wagner-Hauffe doping'.

(2) The second oxidant may segregate in the oxide to the grain boundaries, modifying the grain boundary transport properties. It is less easy to predict, even qualitatively, how the transport properties might be affected.

(3) The second oxidant may form filaments or sheets of its own product within the oxide. These may then act as short-circuit paths for cation transport: only a very small volume fraction may be required to produce a major effect on the overall growth rate.

(4) As described above, the second oxidant may accumulate at the scale-metal interface, and impair the adhesion of the scale.

(5) Also as discussed above, the second oxidant may establish a continuous layer of its product phase (or, less probably, a compound product phase containing oxygen as well) between the outer oxide and the metal. This may or may not be a bad thing: often, the product phase may oxidize to form the protective oxide anyway. Thus, chromium sulfide or chromium carbide will oxidize to form a continuous layer of chromium oxide. The presence of the internal layer may not therefore automatically preclude the formation of a protective oxide, and it may even reduce the oxidation growth stresses.

(6) The second oxidant may enter the alloy beneath the interface and form an internal precipitate of its product phase. As is clear from the previous discussion, this does not necessarily preclude the formation of an external protective oxide either: to be harmful, one of two things must happen: either the second phase must be coarse, or the second phase must oxidize in situ, producing an internal dispersion of the oxide of the element required to form the external protective oxide.

In practical terms, most important effect is probably the last; but the formation of internal second-phase short-circuit paths is also probably important.

Breakaway Corrosion.

Perhaps the most important aspect of the role of second oxidants is that they can promote "breakaway" oxidation. This is the situation where the initial oxidation appears to be protective: an apparently continuous scale of the desired composition forms, the same as in oxidation in the absence of the second oxidant. However, eventually the protectiveness breaks down and the oxidation rate accelerates, with the formation of a much thicker oxide, usually of the base metal. Often, this breakdown is localized, either starting at edges and corners, or appearing as blister-like growths on the flat surfaces. The danger of this type of attack is that short-term testing may give entirely over-optimistic lifetime estimates for the components. Furthermore, it may be difficult to detect by accelerated testing: for example, raising the temperature may accelerate the initial protective rate, but may suppress entirely the breakaway process. This is a matter of importance in coal combustion and conversion equipment, where lives of at least six years and preferably twenty years are required: materials selection must frequently be based on relatively short-term testing.

Choi (19,20) studied the effect of sulfur on the oxidation of Fe-Cr alloys. In oxygen atmospheres, a Fe-20%Cr alloy will normally oxidize protectively at 1223 K, forming a Cr_2O_3 scale which grows parabolically. Choi showed that the presence of sulfur could lead to an eventual breakdown in the protectiveness of the chromia scale, after which rapid growth of iron oxides took place. Figure 5 shows an example of this in environments intended to simulate those in an atmospheric pressure fluidized bed combustor (AFBC) burning coal. For the lower sulfur environment, there was no formation of internal sulfides. However, with the higher sulfur partial pressure, sulfides were formed inside the alloy from the start of the oxidation. However, the depth of internal sulfidation was not in itself a factor in the onset of breakaway. Figure 6 shows the microstructural changes in the alloy. Up to breakaway the sulfides are relatively fine, and there is little evidence of internal oxidation. However, with time the sulfides appear to coarsen, and at 100 h (Figure 6(c)) fingers of oxide can be seen extending into the alloy. It appears that these have resulted from the oxidation of the coarser chromium-rich sulfides. The breakaway occurs in the form of large oxide blisters on the specimen surface, which gradually spread to cover the whole surface. Figures 6 (d) and (e) are from regions of the specimen after 150 h that correspond to non-blistered and blistered regions respectively. The structure shown in Figure 6 (d) can therefore be regarded as the immediate precursor to the catastrophic breakaway, and the extensive internal oxidation is apparent. Choi proposes that the onset of breakaway attack requires the development of a `critical microstructure' with coarse second phase particles involving the second oxidant intersecting the metal/oxide interface.

Figure 7 shows one of the oxide fingers extending in from the oxide/metal interface; the metal has been dissolved away with bromine-methanol. Energy-dispersive X-ray analysis of the region A reveals the presence of sulfur at the tip of the intrusion: there is no iron present in the intrusion at all, and it seems probable that the sulfur is present as chromium sulfide, CrS. This is a general feature of these intrusions. Hitherto, there was a problem in understanding how the chromium oxide intrusions could grow so quickly. Choi suggests that the oxide formed as a result of the oxidation of a sulfide will be porous, since the Pilling-Bedworth ratio is larger for CrS than for Cr_2O_3. Choi further suggests that this may explain why apparently sulfur-free and apparently compact and continuous chromium oxide scales grown in atmospheres containing small amounts of sulfur appear to be permeable to SO_2.

These experiments were conducted in atmospheres containing CO_2, CO and SO_2 or H_2S. The importance of the development of the `critical microstructure' is made even clearer by the results of Lin (21,22,), who also addressed the problem of corrosion in environments similar to those believed to exist in fluidized-bed combustors. In these experiments, a gas of a specific oxygen activity was first passed over a mixture of calcium sulfate and calcium oxide, or calcium sulfide and calcium oxide, with the aim of developing a sulfur activity in the gas phase similar to that in the combustor. As is usual, the controlled oxygen partial pressure was first developed by using a mixture of CO and

CO_2; this is the general choice, since these are the gas species that are found in actual combustion systems. Using this approach, microstructures were developed typical of those seen in corroded alloys within fluidized-bed combustors: chromium-rich sulfides develop in the metal, and rapid non-protective oxidation can result.

However, careful examination of specimens from actual combustors (The Grimethorpe PFBC and the Nova Scotia Power Company' Point Tupper AFBC) had revealed that for some alloys (notably Incoloy 800H) under certain circumstances, showed that internal carburization was present in the alloy ahead of the sulfidation front. Lin demonstrated that this internal carburization did not result from the carbon present in the alloy, but had to have derived from the environment. To determine whether this had any effect on the reaction, Lin compared atmospheres having the same oxygen activity (and, presumably, sulfur activity, although this could not be measured directly) using CO/CO_2 and H_2/H_2O initial gas mixtures. The results showed that in the absence of carbon there was no internal penetration of sulfide. Chromium sulfide did form, but at the oxide/metal interface only. Clearly, the critical microstructure is developed first by the internal carburization; the carbides are then sulfidized to form the internal sulfides, eventually leading to the microstructures reported by Choi and others.

Another important point concerning the effect of second oxidants is that reported by Perkins (23). His investigation was concerned with metals and alloys in gasification environments, and he reported that rapid non-protective corrosion occurred in atmospheres having oxygen and sulfur activities lying in the oxide-stable region, but close to the oxide/sulfide boundary. He determined the location of a boundary between regions where protective and non-protective corrosion took place, and this was a line approximately parallel to the oxide/sulfide boundary, but approximately two orders of magnitude higher in oxygen activity. He referred to this as a "kinetic boundary". These results were confirmed by Natesan (24). In an endeavor to determine the nature of the processes involved in the breakdown in protection, Perkins formed an initial protective chromium oxide under highly oxidizing conditions, and then switched to an atmosphere which approached the kinetic boundary conditions. He found that if the initial oxide was formed in the complete absence of sulfur, the oxide would remain protective even on the oxygen-poor side of the kinetic boundary for long periods. However, if the oxide was formed in the presence of even a very low activity of sulfur, the oxide would lose its protective capabilities immediately the conditions reached the kinetic boundary. There was no apparent difference in the oxide. However, at the time this work was done, techniques such as scanning Auger spectrometry were not available. It seems probable that the oxide grain boundaries were decorated with sulfur as a result of slow desorption of sulfur released as the oxidation process removed oxygen from chemisorbed SO_2, and that this is responsible for the rapid breakdown, presumably by allowing rapid transport of some species along the decorated grain boundaries. However, it must be remembered that the overall growth rate of the oxide was essentially the same in both cases, so the sulfur-decorated grain boundaries hypothesized here do not apparently enhance the transport of oxygen or chromium. This is an important result which should be looked at again in more detail.

Neither Perkins nor Natesan appear to have taken any account of the possible involvement of carbon in the reactions, although both generated their low oxygen activities using CO/CO_2 mixtures, at relatively high carbon activities. This aspect also needs further study.

To summarize: the initiation of breakaway corrosion induced by a second oxidant, sulfur being the example discussed here, appears to involve the generation of a critical microstructure, which corresponds to a relatively coarse distribution of sulfides within the metal. These coarse internal sulfides oxidize in situ, to tie up the element which would otherwise form the protective oxide. Other oxidants, such as carbon in the cases discussed here, may serve to generate some aspect of the critical microstructure. In addition, it appears that very small amounts of sulfur can produce an important, but subtle, change in the protective oxide which renders it more susceptible to degradation.

Practical Manifestations of Mixed Oxidant Corrosion

In coal combustion and conversion systems there are several examples of components which are subjected to severe mixed oxidant corrosion. The systems of interest include:

(a) coal gasification systems;

(b) pulverized coal boilers; and

(c) fluidized bed combustors.

(a) Coal Gasification Systems.

The object of a coal gasification system is to convert coal into a gaseous product, which in the cases of interest here is to be used as a fuel in some later process, normally integrated with the gasifier. The object of introducing this gasification step, instead of burning the coal directly, is that it is relatively easy to remove the incombustible mineral matter, the sulfur, and the fuel-bound nitrogen from the gas, so the fuel gas produced is essentially free from environmentally harmful constituents. It is also easy to use the gaseous fuel in a combined (gas turbine + steam turbine) cycle, which results in an efficiency improvement. The simplest way to do this would be to partially oxidize the coal to CO, but in practice it is usual to use steam in addition to oxygen or air to generate hydrogen and methane as well. The gasification reaction is endothermic, so some carbon must be burnt to CO_2 to provide the heat to drive the reaction; this represents an efficiency penalty, but this is more than compensated for by the use of the combined cycle.

The product gas before cleaning thus contains CO, CO_2, H_2, H_2O, CH_4, NH_3, and H_2S as major species, together with N_2 if the gasifier has been blown with air rather than oxygen, and a number of minor constituents. If the gas is at equilibrium, oxygen, sulfur, and carbon activities can easily be calculated simply from the relative proportions of the atomic species: it is not necessary to know the gas composition, provided the amount of hydrogen tied up as ammonia is small.

The gas as produced is hot: the exact temperature depends on the gasification process used, but for an entrained gasifier such as the Texaco or Shell gasifiers it is approximately 2000°C; for a fixed bed gasifier such as a Lurgi gasifier it is somewhere in the range 400 - 1000°C, depending on the fuel. For the high-temperature gases it is important from the point of view of system efficiency to recover as much of that sensible heat as possible. The metallic parts of the system at risk from mixed oxidant corrosion are those which experience the gas before the clean-up systems: in the majority of systems now in use the clean-up systems operate cold, and so the final stage before them is a quench cooler. Before that, there may be heat exchangers to cool the gas to some intermediate temperature. The materials for these heat exchangers are a problem, and most current designs limit the maximum surface metal temperature to 400°C or so, because of the problem of mixed oxidant attack (25). In an integrated combined cycle system, the overall efficiency would be improved if the steam could be superheated in this heat exchanger. The steam temperature would then be in the range 538 - 565°C (assuming present-day steam conditions: this temperature might go as high as 650°C for advanced steam conditions presently being considered). The temperature of the external metal surface in contact with the hot gas depends on the thermal flux, the wall thickness, and the different heat transfer coefficients, but will be between 50 and 100°C higher than the steam temperature. The only other potential high-temperature metallic components that would be in contact with the hot dirty gas are instrumentation wells, such as thermocouples; and these can probably be avoided in practical systems since the temperatures can be inferred from measurements taken elsewhere. Much early research was done using metal temperatures much higher than this, in some cases up to 1000°C: this work is now probably of only academic interest.

Estimation of the activities of the oxidants experienced by these cooled metal surfaces is very difficult. First, there is no clear evidence that the hot gas itself is at equilibrium. Gas samples collected in different ways give analyses which are inconsistent with an equilibrium composition at any temperature, although this may be an artefact of the sampling system. Even if it is, it is not clear that the reactions in the gas boundary layer at the metal surface will be rapid enough to re-establish an equilibrium characteristic of the temperature at the metal surface. However, this is the assumption customarily made in isothermal laboratory studies. On this basis, it appears that at temperatures of 650°C, the oxygen activity is of the order of 10^{-20}Pa, the sulfur activity 10^{-5}Pa, and the carbon activity 0.8. A further problem is the effect of total system pressure. Gasifiers used in integrated combined cycle systems typically operate at elevated pressures, of the order of 2.5 MPa (gasifiers used to make producer gas as a chemical feed stock may operate at significantly higher pressures, but this is outside the scope of this paper). The sulfur and oxygen activities are, at least to first order, independent of total system pressure, but the carbon activity is not; and clearly kinetic processes involving the gaseous boundary layer at the metal surface or gaseous transport through pores in the corrosion product will be expected to show a dependence on overall pressure. Furthermore, effects related to single gaseous species, such as HCl, will also show an effect of total system pressure. Relatively little testing has been done at elevated pressures, but it is now generally acknowledged that increasing the pressure does appear to increase the severity of attack (25,26). In laboratory testing, Choi (20) has shown a dependence of sulfidation-oxidation attack on the SO_2 partial pressure independent of the oxygen or sulfur partial pressures which he attributes to a gaseous transport process through pores in the scale.

For the oxygen and sulfur species alone, these activities lie in the Cr_2O_3 - stable region, but essentially at the kinetic boundary as defined above; they are probably within the Fe - stable region, but close to the Fe/FeS boundary. Severe sulfidation-oxidation corrosion has been experienced for virtually all candidate materials and coatings under these conditions. Austenitic stainless steels appear to have the best resistance for alloys, and coextruded tubes with an austenitic outer layer and a ferritic inner layer offer good prospects for moderate temperature capability improvements in evaporators.

Bakker and Perkins (25) have shown that an important part of the mechanism of the attack in gasifiers is the initiation of pitting by condensates formed on the cool surfaces when the unit is down. These hydrolyze the sulfides on the metal surface, producing highly acidic solutions, which pit the metal surface. On reheating, these pits appear to initiate high-temperature attack. This process is referred to as "down-time corrosion". The addition of small amounts of vanadium to the austenitic alloys greatly reduces this form of attack. The same authors believe that chromium diffusion coatings offer the best prospects for resistance, and that these too should contain a small percentage of vanadium. Aluminum diffusion coatings have been shown to be ineffective in this application, and in fact systems depending on the formation of Al_2O_3 scales are probably inappropriate at these low temperatures.

(b) Pulverized Coal Boilers.

In contrast to gasifiers, coal combustors are designed to be oxidizing overall: it is important to achieve as complete combustion of the coal as possible, and typically the overall combustion process is designed to have an excess oxygen content of at least 5% over that required to convert the carbon to CO_2 and the hydrogen to H_2O. However, locally the conditions may differ from this. The pulverized coal is injected pneumatically through burner nozzles,together with the preheated combustion air. The effect is that within the boiler vessel a fireball is established, and within this region the conditions are inhomogeneous, with oxygen, coal volatiles, and devolatilized coal char particles coexisting. Since equilibrium is not established, it is not possible to define activities of the different species in this region. If the conditions in the boiler become different to those assumed in the design, this region may contact the walls: this is sometimes called flame impingement. This can arise as a result of degradation of the burners, or as a result of changes in the combustion characteristics of the fuel. What happens is that a deposit forms on the relatively cool walls which contains mineral matter from the coal together with uncombusted char particles. The mineral matter is also incompletely converted to ash, and specifically the deposit may contain pyrite FeS particles. Large pulverized coal boilers are typically of water-wall construction, with tubes connected by membranes. In U.S. practice, the tubes are vertical, typically of approximately 60 mm outside diameter, with 10 - 15 mm membranes separating them. The preheated water enters the bottom of the wall, and boils towards the top of the combustion zone. The tubes on which the deposit forms are thus essentially filled with water close to the boiling point. The pressure in the system is typically 17 MPa, at which the boiling point is 350°C. The tube outer surface temperature is thus approximately 400°C maximum. Figure 8 illustrates the deposit, and the temperature gradient through the deposit and the metal (27).

Under these circumstances, the tubes may suffer accelerated wastage. This is called water-wall corrosion. Early studies attributed this to a molten salt accelerated attack, involving a complex iron alkali sulfate (28), but in fact such a species is not stable in these reducing conditions. Later, the severity of the attack was attributed to chlorine-containing species (NaCl, HCl, or $FeCl_x$), largely on the basis of U.K. experience (29). However, it is now generally agreed that the wastage is oxidation-sulfidation due to the local low oxygen and associated high sulfur activities, which should be essentially the same as those found in gasifiers (30). Nevertheless, the detailed mechanism of the wastage is still a matter of controversy, particularly in regard to the effect of chlorine (31).

Pulverized coal boilers also suffer accelerated wastage of the superheaters. This is generally agreed to be caused by the presence of a molten alkali-containing sulfate at the metal surface. Molten salt accelerated corrosion ("hot corrosion") is generally outside the scope of this paper, although it has been shown that some aspects of molten sulfate-induced corrosion may in fact be sulfidation-oxidation attack (32).

(c) Fluidized-Bed Combustors.

An attractive alternative to the pulverized coal boiler for the direct combustion of coal is the fluidized-bed combustor. In this, a bed which typically consists of coal ash, a sulfur acceptor such as limestone, and spent acceptor (calcium sulfate) with a mean particle size of approximately 0.8 mm is fluidized by air blown up through the bed from a distributor plate which also supports the bed. The bed is initially heated using an auxiliary burner, and when its temperature exceeds 400°C coal is injected, either below the bed or above it. The coal burns efficiently in the bed at a relatively low temperature, because of the excellent solids-gas contact in a fluidized bed. The bed temperature rises to a steady-state value of 850°C or so, since this optimizes the coal combustion efficiency and the sulfur capture efficiency. The temperature is maintained by balancing the heat input from the coal and the heat removal. Some of the heat is removed from the bed by the combustion gases and the excess air, and this heat is then removed by gas-pass heat exchangers, as in a conventional boiler. However, because of the relatively low combustion gas temperature, more heat exchanger surface would be required than in a conventional boiler. Additionally, heat is removed by a heat exchanger immersed in the bed: this takes advantage of the excellent heat transfer characteristics of a fluidized bed. It is this in-bed heat exchanger that may suffer oxidation-sulfidation attack.

The bed may be operated at atmospheric pressure (AFBC) or at elevated pressure (PFBC): in the latter case the exhaust gases are expanded through a gas turbine which drives the unit compressor, and in combined cycle mode also drives a generator. Most commonly, the in-bed heat exchanger is part of the steam circuit, and generally has both evaporator and superheater circuits. In some cases, the unit is used to heat air, which may then be expanded through a gas turbine. In the former case the metal surface temperatures in the bed will be up to a maximum in the range 400 - 450°C for the evaporator and up to 650°C for the superheater. In the latter case the maximum metal surface temperature will be close to the bed temperature. In both cases, there may be some uncooled support structures.

Stringer and Whittle pointed out several years ago (33) that the coexistence of calcium oxide CaO from the fresh acceptor and calcium sulfate $CaSO_4$ from the spent acceptor implied a restricted range of possible sulfur activities in the bed, and that in the presence of combusting coal particles local low oxygen activities could be accepted, which would generate high sulfur activities corresponding to the CaO-$CaSO_4$-CaS triple point. The thermodynamic relationships are illustrated in Figure 9. A further point is that oxygen probe measurements had indicated a rapid fluctuation in oxygen activities at a fixed location in the bed (34) from close to 2×10^4 Pa to 10^{-8} Pa with a frequency in the range 0.5 - 2 Hz. This result has been confirmed many times. The details of the environment and the corrosion mechanism have been reviewed in detail elsewhere (35).

Some examples of severe sulfidation-oxidation breakaway corrosion have been reported, and some alloys are very sensitive to the attack, particularly Incoloy 800 H, which as it happens is a leading candidate on mechanical strength grounds for the tubes in air-heater applications. However, if the air feed and fuel feed are carefully controlled, locations in the bed can be identified which have high oxygen partial pressures virtually the whole time; and Rocazella and Wright (36) have shown that under these circumstances Incoloy 800 H appears to behave perfectly well. As has been mentioned above, it has been shown that in a number of circumstances the severely corroded materials are also carburized ahead of the internal sulfidation front, and Lin has demonstrated that this carburization is a necessary part of the corrosion process, contributing to the development of the critical microstructure. It is possible that the mechanical wastage that is sometimes observed within fluidized beds (37) may also contribute to the corrosion by removing or damaging the external protective oxide layer. The rapid fluctuation in the oxygen activity may be reflected in a rapid fluctuation in sulfur activity in the opposite sense, but this has not been shown. It has been demonstrated that fluctuating oxygen activities do not appear to affect the corrosion (38).

Summary

Heat resisting alloys depend for their resistance to oxidation on the formation of a stable external protective oxide layer. In the presence of second oxidants, this layer may become unstable, leading to rapid non-protective oxidation. Usually, this breakdown can be attributed to either a destruction in the protective properties of the oxide as a result of the formation of short-circuit transport paths through the oxide, or to a destruction of the oxide-metal interface integrity; or an internal tying-up of the element required to form the protective oxide. In this latter case, it is not-sufficient to tie up the oxide-forming element as a dispersed second phase, since if this phase is finely distributed the alloy will still be capable of forming an external protective scale provided the internal dispersed phase is less stable than the oxide. The important step is the internal oxidation of the second phase in situ, and this appears to require a coarse distribution of the second phase: there may be other requirements, and the overall process is here termed the development of the critical microstructure. It is shown that in at least one set of circumstances of practical interest, a third oxidant is important in the development of the critical microstructure, although it does not itself play any other major role in the breakaway corrosion.

In coal combustion and conversion processes, there are several examples of sulfidation-oxidation corrosion which limit temperatures of application of materials, or restrict materials selection. In some cases these appear to be breakaway processes, which limits the usefulness of short-term testing for life prediction.

References

1. J. Stringer, G. Metcalf and M. L. Nicholson, "The High-Temperature Oxidation of a Commercial Titanium-Molybdenum Alloy" J. Less-Common Metals, 4 (1962), 69-77.

2. A. Atkinson, "Grain Boundary Diffusion in Oxides and its Contribution to Oxidation Processes" AERE R 12404 (Materials Development Division, Harwell Laboratory, Oxfordshire, U.K. 1987).

3. M. E. El-Dahshan, J. Stringer and D. P. Whittle, "The Effect of Carbon on the Hot-Corrosion of Cobalt-Base Alloys" Cobalt, No. 57 (1972), 182-191.

4. J. Stringer, D. P. Whittle, I. G. Wright, V. Nagarajan and M. E. El Dahshan, "A Novel Method for Designing SiO2 - Forming Alloys" Proc. 8th Intl. Cong. on Metallic Corrosion (DECHEMA, Frankfurt, FDR, 1981), 655-661.

5. V. Nagarajan, I. G. Wright, and J. Stringer, "A Dispersed Reservoir Approach to the Development of Protective Scales on High-Temperature Alloys" Proc. 12th International Plansee Seminar: ODS 6 (1989) 333-352.

6. J. Stringer, P. S. Corkish and D. P. Whittle "Oxidation of Two-Phase Alloys" Stress Effects and the Oxidation of Metals, ed. J. V. Cathcart (Met. Soc. of AIME, New York, New York, 1975), 75-93.

7. J. Stringer, "High Temperature Corrosion of Superalloys" Materials Science and Technology 3 (1987) 482 - 493.

8. A. W. Funkenbusch, J. G. Smeggil, and N. S. Bornstein, Met. Trans., Vol. 16A (1985) 1164 - 1166.

9. J. G. Smeggil, A. W. Funkenbusch, and N. S. Bornstein, Met. Trans., Vol. 17A (1986) 923 - 926.

10. J. G. Smeggil, "Some Comments on the Role of Yttrium in Protective Oxide Scale Adherence", Mat. Sci. and Eng., Vol. 87 (1987) 261 - 265.

11. J. L. Smialek, Met. Trans., Vol. 18A (1987) 164 - 167.

12. P.Y. Hou and J. Stringer, "A Study of the Apparent Detrimental Effect of a Surface-Applied HfO_2 Coating on the Oxidation Behavior of Alloys" submitted to J. Electrochem. Soc.

13. P Hancock, Proc. 1974 Gas Turbine Materials in the Marine Environment Conf., ed. J. W. Fairbanks and I. Machlin, Metals and Ceramics Information Center Report MCIC-75-27 (Battelle, Columbus, Ohio, 1974) 225 - 236.

14. P. L. Hemming and R. A. Perkins, "Thermodynamic Phase Stability Diagrams for the Analysis of Corrosion Reactions in Coal Gasification/Combustion Atmospheres" EPRI Publication FP-539 (Electric Power Research Institute, Palo Alto, California, 1977)

15. G. H. Meier, N. Birks, F. S. Pettit and C. S. Giggins, "Thermodynamic Analyses of the High Temperature Corrosion of Alloys in Gases Containing More Than One Reactant" High Temperature Corrosion ed. R. A. Rapp (National Association of Corrosion Engineers, Houston, Texas, U.S., 1983) 327 - 335.

16. J. Stringer and D. P. Whittle, "High Temperature Oxidation and Corrosion of Metals and Alloys" Proc. 1st Petten Colloquium on Advanced High Temperature Materials: Technological and Industrial Aspects. Also in Rev. Int. Htes. Temp. et Refract. 14 (1976), 6-20.

17. J. Stringer, "High Temperature Corrosion in Mixed Gaseous Oxidants" Metals and Energy - Proc. 33rd Ann. Conf. of the Australasian Inst. Met. (University of Auckland, New Zealand, 1980), 143-146; see also J. Stringer, "Corrosion and Erosion in Complex High Temperature Environments" Behaviour of High Temperature Alloys in Aggressive Environments , I. Kirman, J. B. Marriott, M. Merz, P. R. Sahm, and D. P. Whittle (eds.) (The Metals Society, London, U.K., 1980) 739 - 756.

18. D. R. Holmes, R. B. Hill, and L. M. Wyatt (eds.) "Corrosion of Steels in CO_2" (British Nuclear Energy Society, London, England, 1974).

19. S.- H. Choi and J. Stringer, "The Breakaway Corrosion of Fe-Cr Alloys in Atmospheres Containing Sulfur and Oxygen" High Temperature Corrosion of Materials and Engines for Energy Systems and Turboengines, ed. R. Streiff, J. Stringer, R. C. Krutenat and M. Caillet (Elsevier Sequoia, Lausanne, Switzerland, 1987) also Mat. Sci. and Eng., 87 (1987) 237 - 242.

20. S. - H. Choi, "High Temperature Corrosion of Iron-Chromium Alloys in Sulfur-Containing Atmospheres" Ph. D. Thesis, Lawrence Berkeley Laboratory, University of California, July 1985; Rep. LBL 20011, 1985 (Lawrence Berkeley Laboratory, University of California, Berkeley, California, 1985).

21. K. Mark, J. Stringer, J. S. Lin and D. A. Stevenson, "Oxidation-Sulfidation Corrosion in Fluidized Bed Combustor Environments" High Temperature Materials Chemistry II, eds. Z. A. Munir and D. D. Cubicciotti (Electrochemical Society, Pennington, New Jersey, 1983) 360-370.

22. J.-S. Lin, D. A. Stevenson, and J. Stringer, "The Role of Carburization in the In-Bed Corrosion of Alloys in FBC" Proc. 1987 Int. Conf. on Fluidized Bed Combustion, ed. J. P. Mustonen (American Society of Mechanical Engineers, New York, N. Y., 1987) 656 - 662.

23. R. A. Perkins and S. J. Vonk, "Materials Problems in Fluidized-Bed Combustion Systems. IV - Corrosion Chemistry in Low Oxygen Activity Atmospheres" EPRI Report FP 1280 (Electric Power Research Institute, Palo Alto, California, 1981).

24. K. Natesan, "Alloy Performance in Coal Gasification Environments" Materials for Coal Gasification, ed. W. T. Bakker, S. Dapkunas, and V. Hill (ASM International, Metals Park, Ohio, 1988) 51 - 60.

25. W. T. Bakker, R. A Perkins, and J. Van Lierre, Mat. Performance, 24 (1985) 9 - 17.

26. M. A. H. Howes, "Corrosion of Alloys in Simulated Coal Gasification Environments at Elevated Temperatures" Materials for Coal Gasification, ed. W. T. Bakker, S. Dapkunas, and V. Hill (ASM International, Metals Park, Ohio, 1988) 9 - 18.

27. A. J. B. Cutler and E. Raask, Corr. Sci., 21 (1981) 789

28. W. T. Reid, External Corrosion and Deposits: Boilers and Gas Turbines (Elsevier, New York, New York, U.S.A., 1971).

29. A. J. B. Cutler, C. J. Grant, J. W. Laxton, D. D. Price and C. G. Stevens, in Corrosion Resistant Materials for Coal Conversion Systems ed. D. B. Meadowcroft and M. I. Manning (Applied Science Publishers, London, U.K., 1983) 159.

30. E. Latham, D. B. Meadowcroft, and L. Pinder, "The Effects of Coal Chlorine on Fireside Corrosion" in Ref. 32.

31. J. Stringer and D. Banerjee (eds.) Proc. Intl Conf. on Chlorine in Coal (Elsevier Sequoia, Lausanne, Switzerland) in press.

32. J. Stringer and M. E. El Dahshan, "The Role of Sulphur in Hot Corrosion" Proc. 1974 Conf. on Gas Turbine Materials in the Marine Environment, eds. J. W. Fairbanks and I. Machlin (Metals and Ceramics Information Center Report MCIC-75-27, 1975), 161-182.

33. J. Stringer and D. P. Whittle, "High Temperature Corrosion of Metals and Alloys in Fluidized Bed Combustor Environments" Proceedings International VGB Conference on Corrosion and Deposition in Power Plants, Essen (1977), Paper 22, 20 pp.

34. M. J. Cooke, A. J. B. Cutler, and E. Raask, J. Inst. Fuel, 57 (1972) 153.

35. J. Stringer, I. G. Wright, and W. T. Reid, "Materials of Construction" Chapter 8 in Atmospheric Fluidized Bed Combustion: A Technical Source Book ed. S.-E. Tung and G. C. Williams (U. S. Department of Energy DOE/MC/14536-2544, Morgantown Energy Technology Center, Morgantown, W. Va., Jan. 1987) 8-1 - 8-169.

36. M. A. Rocazella and I. G. Wright, "The Corrosion Behavior of Some Candidate Heat Exchanger Alloys in the AFBC Environment" in High Temperature Materials Problems in Coal Combustion and Utilization ed. I. G. Wright (The Electrochemical Society, Pennington, New Jersey, 1983).

37. J. Stringer, "Erosion/Corrosion in FBC Boilers" in Advances in Material Technology for Fossil Power Plants ed. R. Viswanathan and R. I. Jaffee (ASM International, Metals Park, Ohio, 1987) 319 - 326.

38. K. Natesan, High Temp. Tech., 4 (1986) 193 - 206.

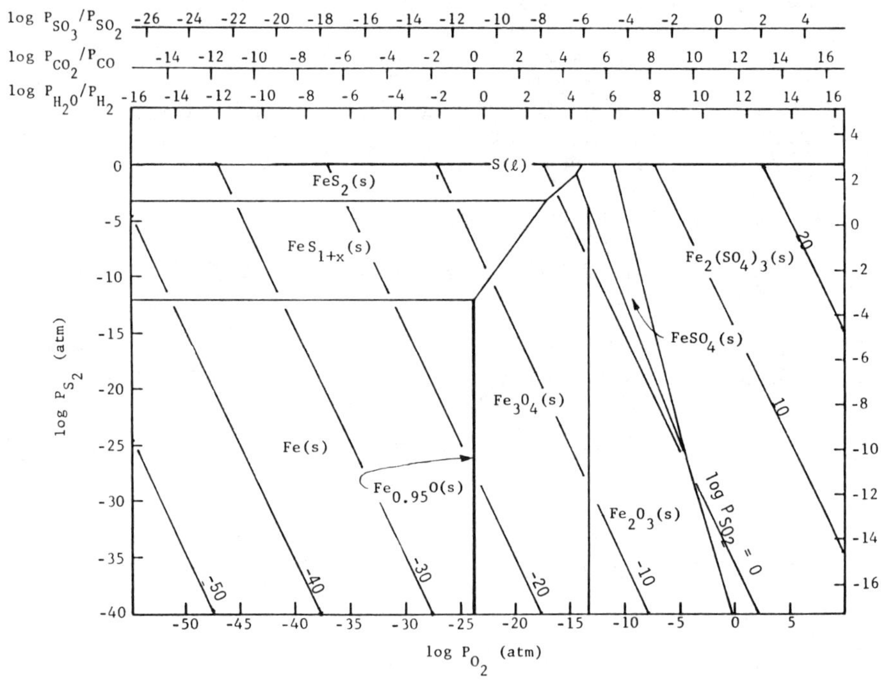

Figure 1. The thermodynamic phase stability diagram for the Fe-O-S system at 894 K (621°C). From Perkins and Vonk (14).

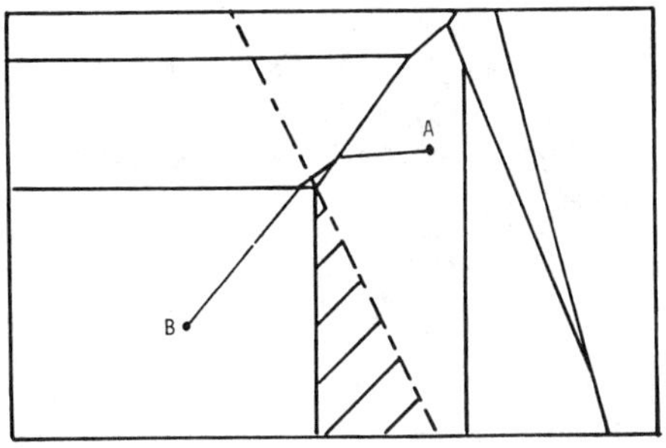

Figure 2. A simplified thermodynamic phase stability diagram, illustrating the concept of the reaction path.

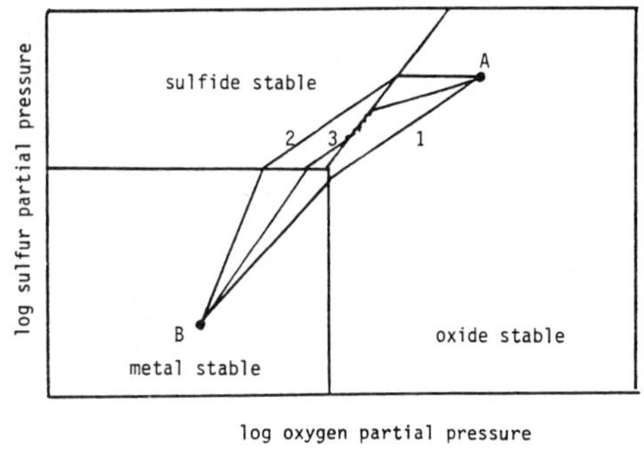

Figure 3. Examples of reaction paths corresponding to different reaction product distributions. Path 1 corresponds to the situation where only an oxide forms; in the case of Path 2 a single-phase sulfide layer forms between the outer oxide and the metal; for Path 3 a two-phase oxide + sulfide layer liers between an outer single phase oxide and an inner single phase sulfide.

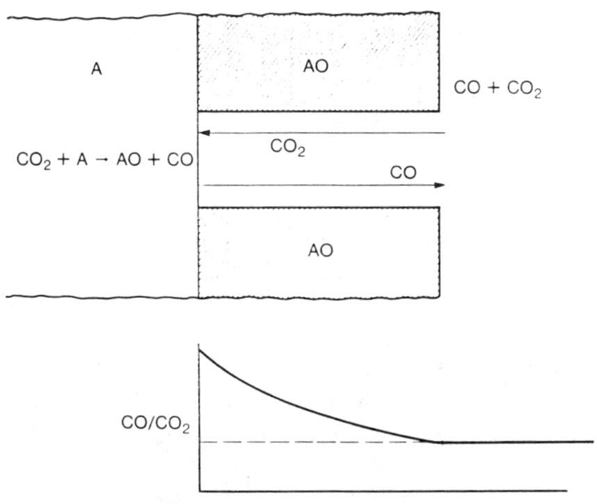

Figure 4. A schematic showing the increase in carbon activity possible if oxidation occurs by the gaseous diffusion of CO_2 through pores in the oxide towards the metal surface, and the migration of the product CO outwards.

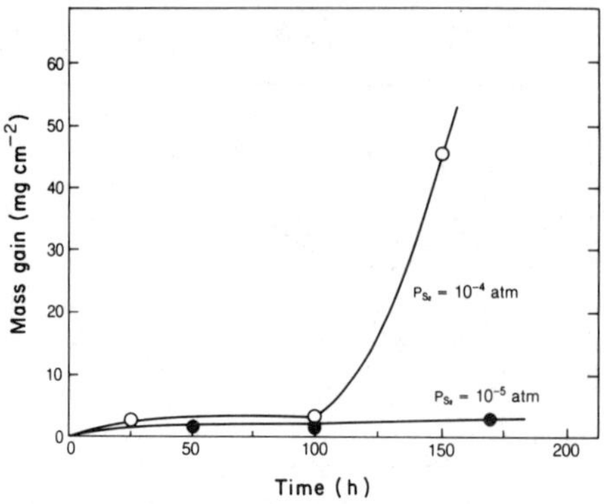

Figure 5. Breakaway corrosion of Fe - 20%Cr alloys at 1223 K in two CO_2 - CO - SO_2 - Ar atmospheres. The oxygen partial pressure was $10^{-6.2}$ Pa in each case; the sulfur partial pressures were 1 and 10 Pa. From Choi and Stringer (19).

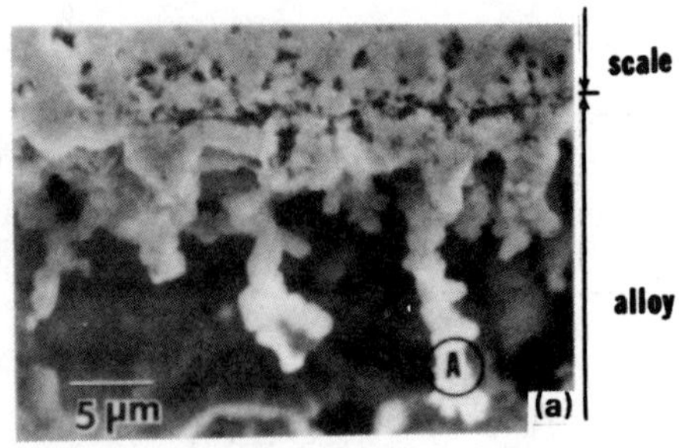

Figure 7. A scanning electron micrograph of an oxide intrusion from a Fe - 20%Cr alloy that is in the condition represented by Figure 6(d). The metal has been dissolved away in bromine-methanol. Dispersive X-ray analysis reveals sulfur at point A; otherwise, only chromium is detected. From Choi and Stringer (19).

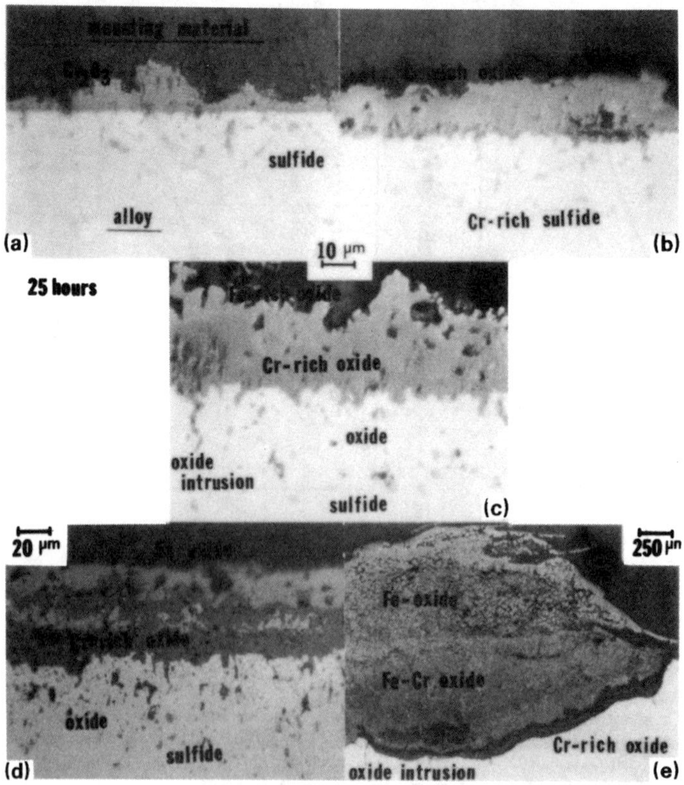

Figure 6. The evolution of the microstructure in a Fe - 20%Cr alloy undergoing breakaway. The conditions are those shown for the upper curve in Figure 5. The exposure times are: (a) 25 h; (b) 50 h; (c) 100 h; (d) and (e) 150 h. Note the differences in magnification. From Choi and Stringer (19).

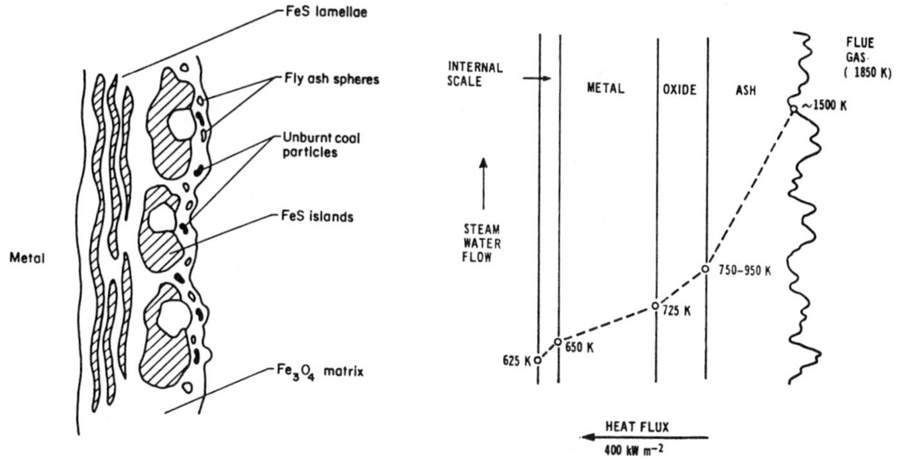

Figure 8. A schematic of the deposit formed on a water-wall tube that is undergoing rapid wastage. From Cutler and Raask (27).

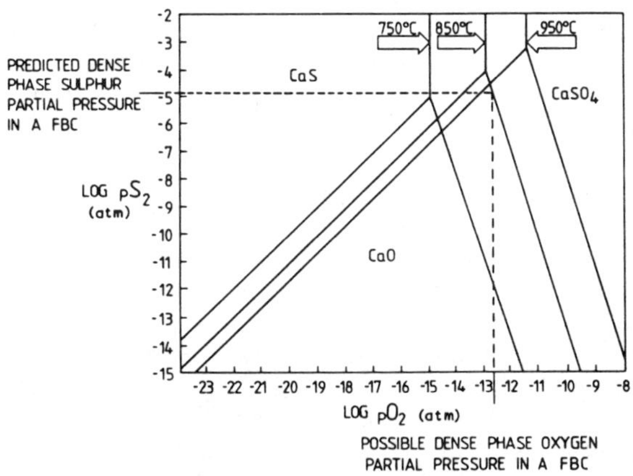

Figure 9. The thermodynamic phase stability diagram for the Ca - O - S system at three temperatures.

Oxidation of boiler tube steels in steam

P. Mayer and A.V. Manolescu
Ontario Hydro, Research Division, Toronto, Ontario, Canada, M8Z 5S4

The rate of formation and composition of oxide scales on fossil fired boiler tube steels exposed to steam at elevated temperatures are influenced by composition and metallurgical state of the metal, aggressivity of environment, exposure time and temperature, strains due to oxide growth, temperature changes and applied stresses. The relationships between these factors are complex but sufficiently understood to allow design and operation changes to eliminate a problem arising from excessive formation and exfoliation of these scales in steam generating and handling equipment. In this paper, accumulated knowledge on behaviour of steels in steam as well as recent advances in the understanding of the phenomenon are discussed in detail. Practical measures to eliminate consequences of steamside scale growth and exfoliation are reviewed.

Introduction

The extent and morphology of oxide scales formed on steel surfaces are governed by a number of factors. The most important of these include properties of the metal, environment and time of exposure, temperature and strains due to oxide growth, temperature changes and applied stress. Though much of this information is available, it is not usually considered in boiler design. The traditional approach is to follow classical guidelines for oxidation resistance at the temperature expected at any given location in a steam circuit. It is assumed that oxide scales will form on the surfaces of boiler tubes and steam pipes, reach an equilibrium thickness and stay intact throughout the service life of the unit. It has been found, however, that the cracking of oxide scales under certain conditions can lead to accelerated growth and then exfoliation (spalling-off or breaking away from the surface). These processes seldom cause tube failure, since the scales are usually very thin. It is only when the exfoliated scales are entrained in the steam that they cause problems, such as erosion of turbines and blockage of boiler tubes (1).

Extensive solid particle erosion (SPE) of turbines has occurred at many fossil fired stations (2). Figure 1 illustrates typical extent of the problem. Although this turbine damage has not caused any forced outages or extensions of maintenance outages, minor and major weld repairs of the tubing have been necessary. Exfoliation of scales from the internal surface of ferritic steel of superheater and reheater tubing and from the steam pipes upstream from the turbine has been identified as the main source of erosive particles (3). Therefore, there is considerable interest in determining the causes of steamside oxide exfoliation and in developing methods for lessening its effects on turbines.

Therefore, the aims of the work described in this paper were as follows: (a) to summarize the information available on the growth and morphology of oxide scales formed in steam on ferritic steel under boiler operating conditions, (b) to discover what insights into the scale formation mechanism can be gained from such investigation, and (c) to review various preventative and remedial measures to minimize SPE.

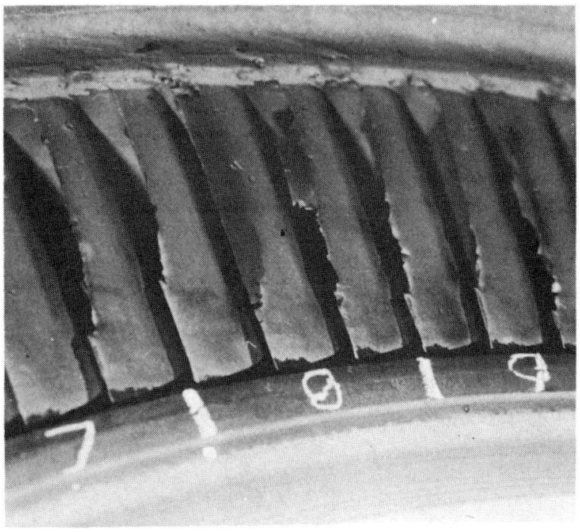

Figure 1. Turbine Damage Due To Solid Particle Erosion

Oxide Scale Growth and Morphology

The composition of oxide scale on iron varies with the steam temperature. At high temperatures, the scale consists of wustite (FeO) at the metal-oxide interface, magnetite (Fe_3O_4), and haematite (Fe_2O_3) at the oxide-steam interface. At lower temperatures only magnetite and haematite form. In the superheater and the reheater circuits of fossil fired boilers, ferritic alloys (See Table I) are used at locations where the design temperature is below that at which wustite is stable. Therefore, only the low-temperature oxide phases, Fe_3O_4 and Fe_2O_3, are commonly observed in steam-grown scales from operating boilers. Magnetite is the predominant phase; haematite occurs at the oxide-steam interface, but only as a thin and discontinuous oxidation layer (4). The haematite, therefore, has very little effect on oxidation rate (4). The magnetite portion of the scale exhibits simple duplex (double layer) morphology commonly observed in the laboratory, or laminated structure seen in plant scales. This morphology is characteristic for pure iron, and for a wide range of dilute iron alloys as well as ferritic chromium-molybdenum steels used in boilers that are used in steam circuits below 560°C. The inner layer of these scales contains virtually all alloying elements and impurities in metallic or oxidized form.

In some instances, the scales that form on steam-exposed surfaces exhibit multiple laminations. This scale consists of a thick magnetite layer at the steam-scale interface paired with a thicker layer consisting of several pairs of iron-chrome spinel and magnetite layers (laminations) and through-scale cracks which are partly or completely filled with magnetite.

TABLE I

Superheater and Reheater Materials
Exposed to Steam in Fossil Fired Boilers

MATERIAL	DESCRIPTION
SA-210-A1	Medium Carbon Steel (0.27 wt % max Carbon)
SA-209-T1A	0.25 wt % max Carbon - 1 wt % Molybdenum
SA-213-T22	2-1/4 wt % Chromium - 1 wt % Molybdenum
SA-213-T9	9 wt % Chromium - 1 wt % Molybdenum

Simple Duplex Scale

The results of many investigations indicate that the rate of scale formation (metal loss) is inversely proportional to time of steam exposure for a wide variety of ferritic alloys. The data obtained in laboratory or in operating boilers can be approximated to the parabolic law of the form:

$$h^2 = k_p t \tag{1}$$

where h = metal loss (mm)
t = time of oxidation (h)
k_p = parabolic rate constant ($mm^2.h^{-1}$).

The parabolic rate constant - time relationship is shown in Figure 2 summarizing the data by Manning et al (5), Grobner (6) and Mayer et al (7). Although the metal loss rates are similar for all materials, the increase of corrosion resistance with the chromium content is easily noticeable.

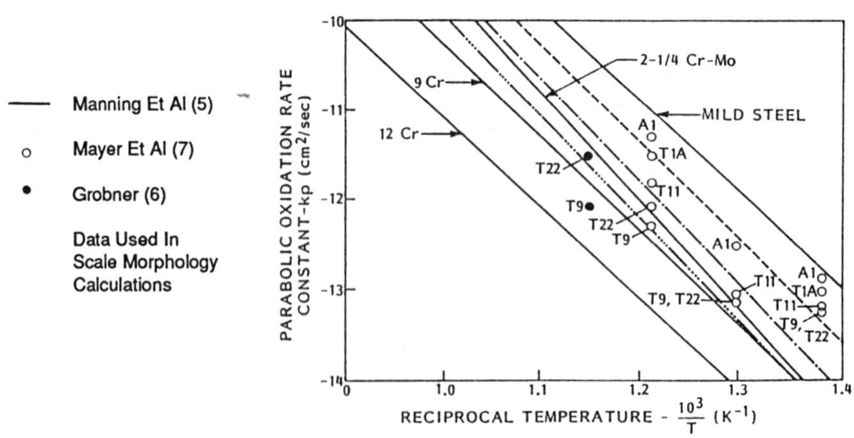

Figure 2. Oxidation Rate Of Ferrous Alloys In Steam

The simple duplex morphology of the scales is illustrated in Figure 3. The layered structure was revealed by selective etching of the cross-selectioned scale in the Clarke's solution.

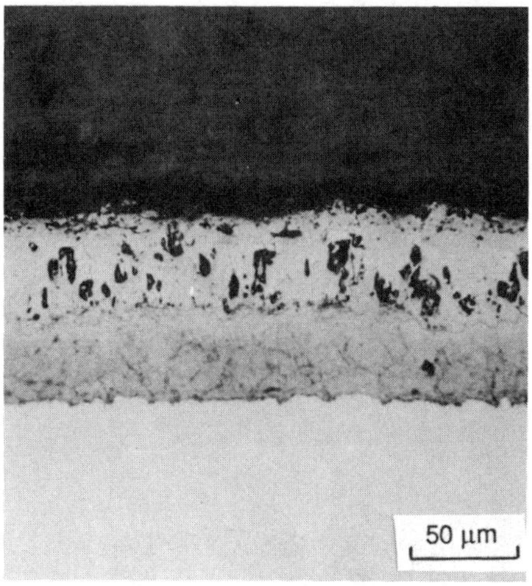

Figure 3. Typical Steamside Scale Morphology Observed On Boiler Tubes

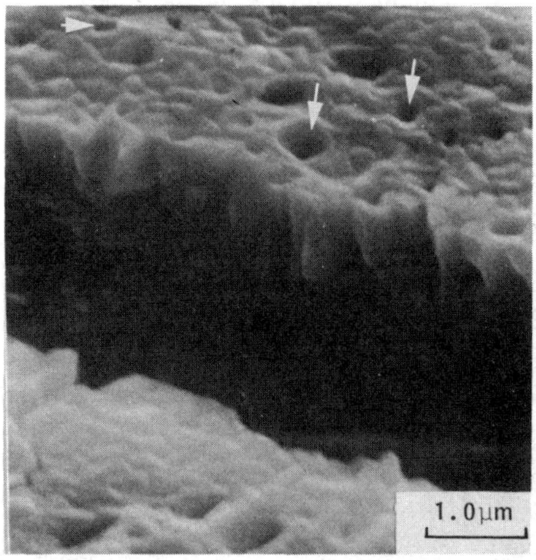

Figure 4. Fractured Scale Formed On T22 Material After 30 Minutes Oxidation In 550°C Superheated Steam (The Arrows Show Pores Emerging From The Outer Surface Of The Scale)

The structural characteristics of the scales on mechanically fractured specimens were examined in a scanning electron microscope. Figures 4, 5, 6, and 7 illustrate the typical morphological features of the scales. The inner layer consists of fine grains without defined crystal facets and irregularly shaped pores. The pores appear to be arranged in rows parallel to the oxide-metal interface (see Figure 12). The outer layer crystals are large and columnar. The pores in this layer are elongated channels, mostly at grain boundary triple points. These features are observed in scales on all alloys regardless of composition, temperature or time of exposure to steam.

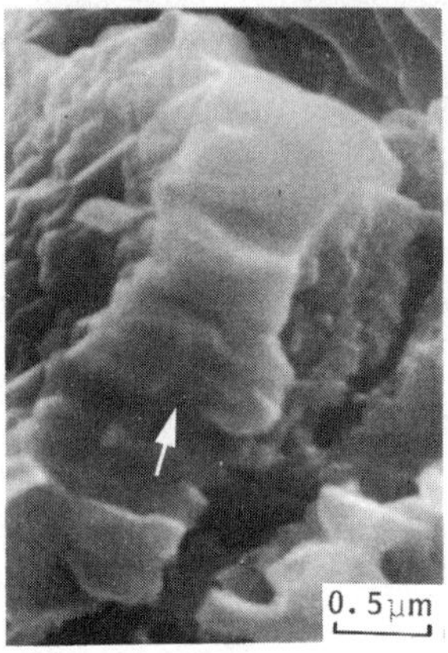

Figure 5. Fractured Oxide Scale Formed On T22 Material After 30 Minutes Of Oxidation In 550°C Steam (The Arrows Show Void In The Columnar Outer Magnetite Layer)

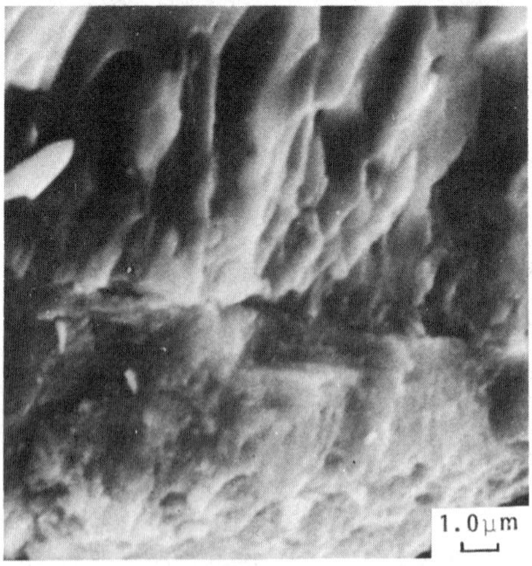

Figure 6. Fractured Scale Formed On T22 Material After 400 Hrs Oxidation In 550°C Superheated Steam

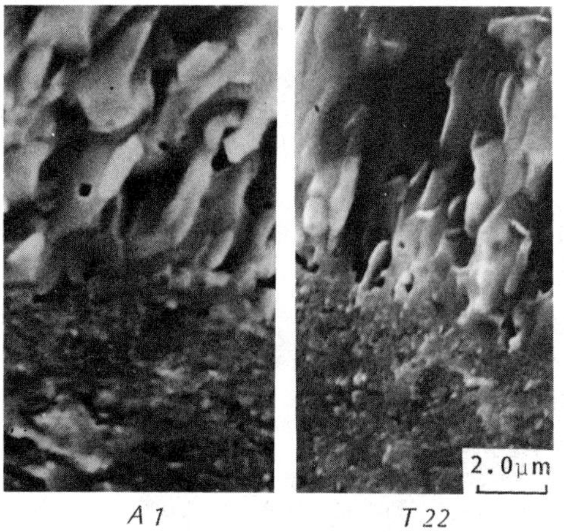

Figure 7. Fractured Oxide Scale Formed On A1 And T22 Materials After 1000 Hrs Oxidation In 550°C Superheated Steam

The compositional analysis using energy dispersive X-ray analysis (Figure 8) was limited by the resolution power of the instrument. It was impossible to establish any variations of the chromium concentration in the inner layer. Outer layer was found to be devoid of any alloying elements.

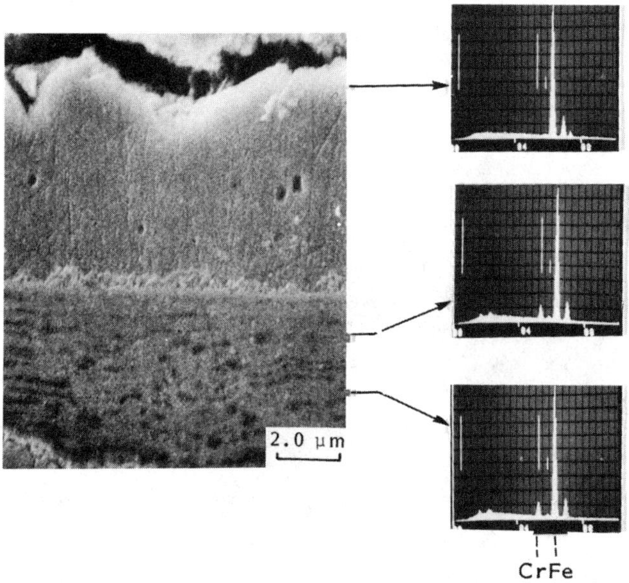

Figure 8. Energy Dispersive X-ray Analysis Of Chromium Content In The Scales Formed On T22 Alloy After 400 Hours Oxidation In 550°C Superheated Steam

The relative proportions of the two layers have been found to be constant with steam exposure time and temperature. However, the ratio of inner and outer layer thicknesses is dependent on the chromium content of the alloy. Figure 9 shows this relationship.

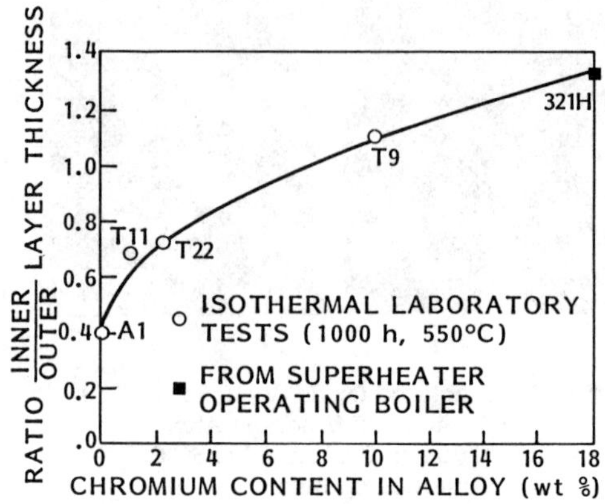

Figure 9. Ratio Inner/Outer Layer Thickness As A Function Of Chromium Content In Boiler Tube Alloys

At present, there are several accepted theories to explain of duplex magnetite formation on iron and dilute ferritic steels. These are shown schematically in Figures 10 to 14.

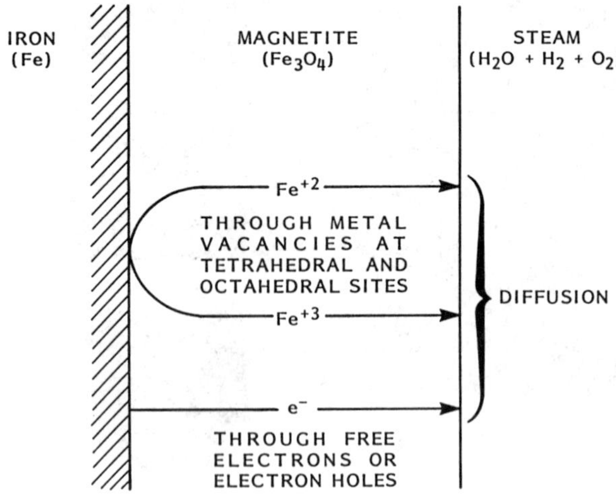

Figure 10. Schematic Representation Of Diffusion Process By Solid State Transport Through Scale During Oxidation Of Iron And Dilute Alloys According To Hauffe (5)

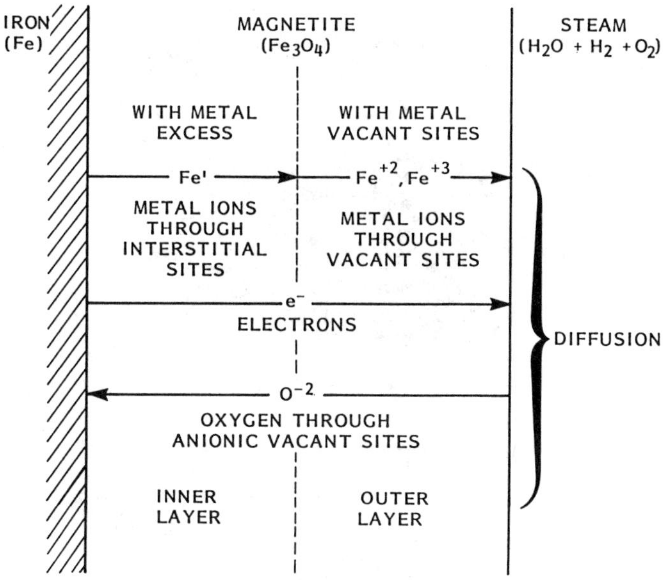

Figure 11. Schematic Representation Of Diffusion Process By Solid State Transport Through Scale During Oxidation Of Iron And Dilute Alloys According To Effertz Et Al (6)

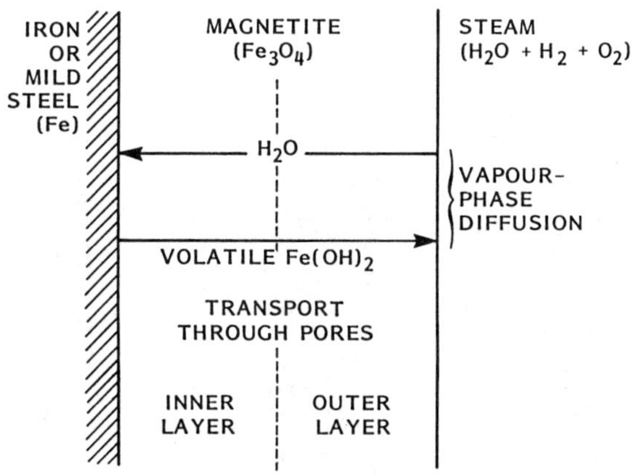

Figure 12. Schematic Representation Of The Diffusion Processes By Vapour-Phase Transport During Oxidation Of Iron And Dilute Iron Alloys According To Surman Et Al (7)

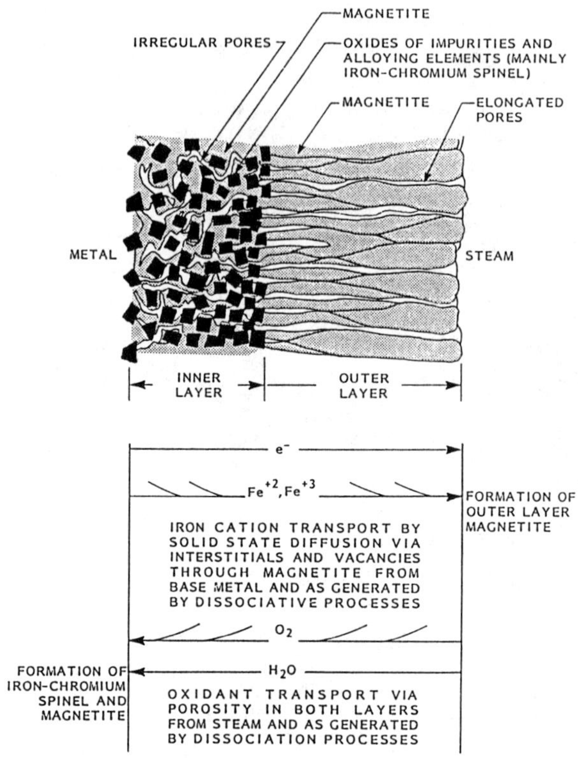

Figure 13. Schematic Representation of Processes Occurring During Oxidation Of Cr-Steel In Steam At High Temperature

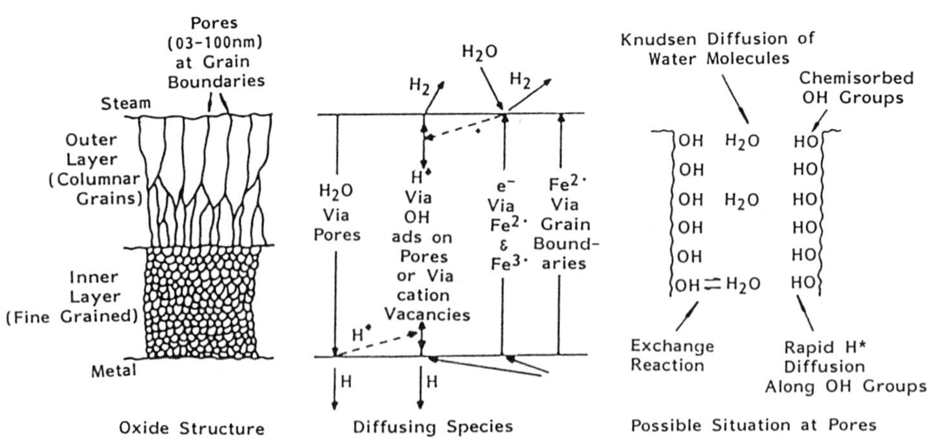

Figure 14. Latest Model Of The Mechanism Of Oxidation Of Ferritic Steels By Steam (17)

The first class of theories is based on a solid-state diffusion transport mechanism through the oxide scale; this was originally developed by Hauffe (8), and modified recently by Effertz et al (9). Oxygen molecules produced by dissociation of steam are adsorbed on the surface of the oxide, which is always present on steel at those conditions. The next step involves dissociation of the oxygen molecules, ionization of oxygen atoms and chemisorption of the oxygen ions. The necessary electrons are provided by the metal and are transported by diffusion through the oxide layer to the oxide-steam interface. The chemisorbed oxygen ions are incorporated in the spinel lattice, forming a new layer of magnetite with a high concentration of metal defects. At the metal-oxide interface, metal ions and electrons are transferred from the metal lattice into the scale and occupy interstitial positions in the spinel lattice of magnetite. Oxygen ions from the scale enter the uppermost layers of the iron lattice and start to form a spinel lattice. In this way, two layers of scale can grow from the original metal surface: on is growing outward and the other inward. Both consist of magnetite, but the outer layer is rich in iron vacant sites, while the inner one is characterized by iron excess and oxygen vacant sites. The inner layer also contains most of the alloying elements from the steel. The boundary between the inner and outer layer, representing the original position of the steel surface, is usually porous, because diffusion along vacant sites (outer scale) is faster than diffusion along interstitial sites (inner scale).

The second theory, advanced by Surman et al (10, 11), is based on an analogous process of iron and steel corrosion in a high temperature aqueous environment (10, 11, 12). This theory postulates that iron transport occurs by vapour-phase transport of volatile $Fe(OH)_2$ from the metal-oxide interface to the outer layer crystals through pores and cavities in the scale. The outer layer crystals act as "sinks" for deposition of Fe_3O_4. The volume created by volatization of iron is filled by growth of oxide nuclei to form a porous inner layer. The morphology of the scale is determined by competition between volatilization and growth of oxide nuclei. This mechanism, illustrated schematically in Figure 12, could produce a high flux of iron ions outwards, resulting in scale-metal separation.

Both of these transport processes and mechanisms of duplex scale formation are consistent with the observed duplex scale morphology. The growth of the scale can occur at metal-oxide and oxide-steam interface, so that the formation of two layers is possible. The reaction at metal-oxide interface facilitates the formation of alloying-element-bearing oxide phases. In most cases, these accumulate in the inner layer because of their high thermodynamic stability and low diffusion rates of the magnetite. The porosity of the inner layer can be attributed to volatilization of iron, vacancy condensation or interruptions in formation of chromium rich spinels because of chromium depletion in the alloy.

In our findings, the presence of interconnected network of pores in the scales suggest that the growth mechanism involves transport of iron to the steam-oxide interface, and access of oxidant to the metal-oxide interface. As a result, the scale grows simultaneously at both interfaces, a fact confirmed by marker studies (13) and oxidation tests (14) in ^{18}O-containing steam followed by autoradiography facilitated by means of the reaction: $^{18}O(p,n)\ ^{18}F$. The autoradiography (14) also indicates gaseous diffusion of oxygen in pores.

There are only two possible mechanisms that can account for these observations. These are the vapour phase transport (10, 11), and the mixed solid-gas-phase transport (15) based on oxide dissociative mechanism. The morphological development of the scale and the kinetics of formation can be explained using both of these models. But the observation that the outer surfaces of the oxides grown in the laboratory (7) differ in appearance from those usually associated with growth from vapour (10) favours the later mechanism. The surface (see Figure

4) is compact and dense with pores emerging from the oxide. The oxide grain boundaries are clearly visible since they form valleys around each grain. This type of surface is usually associated with solid-state diffusion transport (9). Because this observation does not agree with the vapour-phase transport mechanism, the mixed solid-gas phase transport model was applied by Mayer and Manolescu (7) to account for the scale formation on boiler tube alloys in steam.

As previously mentioned, the duplex oxide morphology on commercial purity metals and dilute alloys develops as a result of restricted plastic flow of the scale during formation. The plastic flow is hindered by the presence of immiscible oxides of the impurities and alloying elements in the inner part of the scale. The nuclei of these oxides also prevent full contact between the magnetite and the metal. Stresses and fast diffusion of iron cations away from the metal-oxide interface, cause voids and fissures to form, and lead to scale separation from the metallic core. This results in the dissociation of the outer part of the scale. Because of the anisotropic nature of the oxide, the dissociation leads to an instantaneous destruction of the scale compactness. This in effect enables an inward diffusion of oxidant molecules. The growth of scale occurs as a result of a simultaneous and counter-current diffusion of iron and oxidant. Alloying elements, mainly chromium, do not take part in the process of matter transport through the scale.

As a result of outward lattice diffusion of iron cations and electrons, the outer layer develops a characteristics morphology. Usually, the scale is composed of long intergrown columnar crystals perpendicular to the metal-oxide. The outer layers of scales investigated in this work show this texture. This texture is indicative of the anisotropic character of the oxide growth, since there are considerable differences between the formation rate of different crystallographic planes. Similar anisotropy may be expected in the dissociation process of such an oxide. The planes of slowest growth, ie, along the grain boundaries perpendicular to the surface of the steam-oxide interface, should dissociate at the highest rate. On the other hand, the dissociation process should be slowest on planes parallel to the surface of the metal. Therefore, the most favoured locations for the dissociation to occur are the grain boundary triple points. The result is formation of pores. The detailed morphological investigations of laboratory and plant scales (7) indicate that the majority of the pores in the outer-type layer are at grain boundary triple points. Some pores were also observed in the middle of the grains. The existence of these cannot be easily explained. The growth of the oxide on the outer scale surface favours "healing" of the channels. however, a permanent closure is continually retarded by the dissociation process, which regenerates the paths of inward oxidant transport (16).

The inner layer is formed by inward oxidant diffusion through discontinuities in both the outer and inner layers. The inner layer on chromium-containing steels consists of a more or less dense agglomeration of iron-chromium spinel particles with iron oxide in between. Since the outer layer is formed by cation diffusion outward, the metal ions are removed from the space between these particles. These tend to move together and accumulate at the interface between the outer and inner layers due to plastic flow of the magnetite part of the scale. Therefore, this plastic settlement or "collapse" of the inner layer scale is a function of chromium content of the alloy.

Results illustrated in Figure 9, show that the ratio of inner and outer layer thicknesses increases with increasing chromium content in the alloy. The increase is less pronounced at high chromium concentrations and tends to level off to a limiting value. It was suggested/14/ that this value is reached when the density of the iron-chromium spinel particles in the inner layer is so high that hey touch each other and prevent plastic flow of the magnetite entirely.

The salient points of this model developed by Mayer and Manolescu (7) are summarized schematically in Figure 13. The mass transport by solid-state diffusion of iron outwards and oxidant access to the metal-scale interface through an interconnected network of pores results in the development of duplex scale morphology.

Most recently Tomlinson and Cory (17) have further refined the model proposed by Mayer and Manolescu (7) and confirmed that both solid state and gas phase diffusion processes are involved. Based on the results of their experiments and previous workers (7, 8, 9, 10, 11) it was postulated that:

(1) Reaction occurs only at the gas/oxide and metal/oxide interfaces.

(2) Fe ions and electron diffuse outwards from the metal/oxide interface to the gas/oxide interface.

(3) All hydrogen formed at the metal/oxide interface diffuses through the metal.

(4) All hydrogen formed at the gas/oxide interface is emitted into the steam.

(5) All oxygen is transported across the oxide layer to the metal/oxide interface as H_2O molecules diffusing through micropores or small fissures in the oxide.

(6) Protons can diffuse across the oxide layer in both directions. Two possible routes are envisaged for proton transport: via OH groups adsorbed on micropore surfaces or via cation vacancies.

(7) The ratio of electron to proton transport across the oxide layer rises with increase in temperature.

A mechanism based on all seven assumptions is shown diagrammatically in Figure 14. This mechanism can account for the experimental results at higher steam partial pressures and at various temperatures.

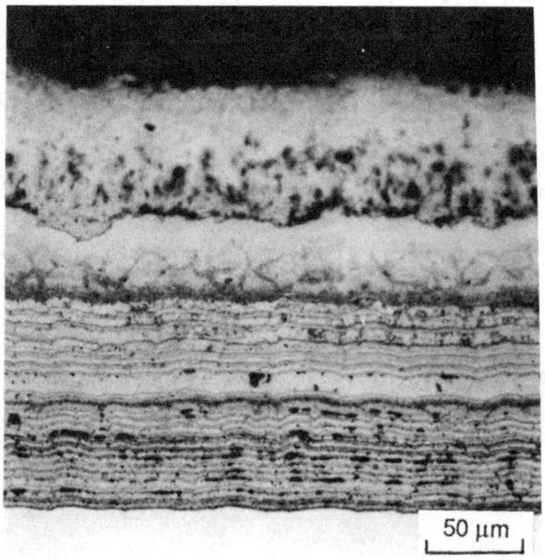

Figure 15. Typical Laminated Steamside Scale Morphology Observed On Boiler Tubes

Laminated Scales

The data reported in technical literature (1, 8) and examination of boiler tube samples removed from operating units (2, 7) indicate that steamside scales frequently have characteristic laminated morphology. These scales consist of several pairs of magnetite and iron-chrome spinel layers. Typical morphology of oxide scale observed on ferritic materials is shown in Figure 15. This laminated scale formed on a section of T-22-alloy tube located close to the outlet from a reheater. The estimated steam temperature at that location was about 500°C; the metal temperature is not known.

The SEM examination of those scales (Figures 16 and 17) illustrate the general features of the laminated scale. The outer magnetite layer consists of columnar grains with elongated pores mostly at the grain boundary triple points. However, some pore channels were observed in the middle of the grains. The inner laminated layer is composed of alternating layers of equiaxed and columnar crystals. The structure of the columnar crystals is the same as that of the outer magnetite layer. Energy dispersive X-ray analysis confirmed that this is magnetite with no alloying elements in it (Figure 18). Large amounts of chromium were detected in the layers of equiaxed crystals. Although it was not possible to analyze each individual grain, it is believed that these crystals are aggregates of grains with widely differing compositions and not a uniform layer of iron-chrom spinel as previously reported.

To date, it has not been possible to reproduce this type of laminated scale under controlled laboratory conditions. Consequently, all formation models have been inferred from examinations of boiler-tube scales. Forrest et al (18) and Grobner et al (14) reported observations suggesting that the formation of laminated scales is associated with through-scale cracking. The schematic description of this process is given in Figure 19. Forrest et al (18) also reported evidence for scale separation due to vacancy condensation at the metal-oxide interface.

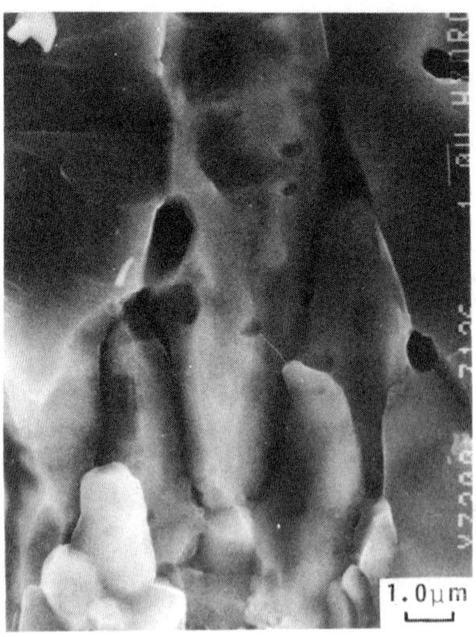

Figure 16. Fractured Oxide Scale Formed On T22 Boiler Tube During Operation (Outer Layer)

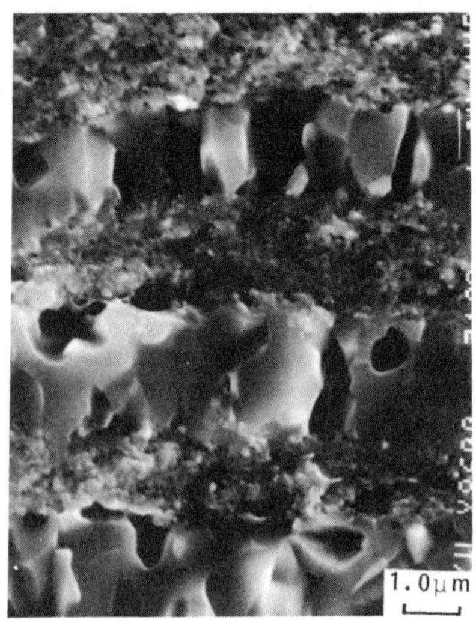

Figure 17. Fractured Oxide Scale Formed On T22 Boiler Tube During Operation (Laminations)

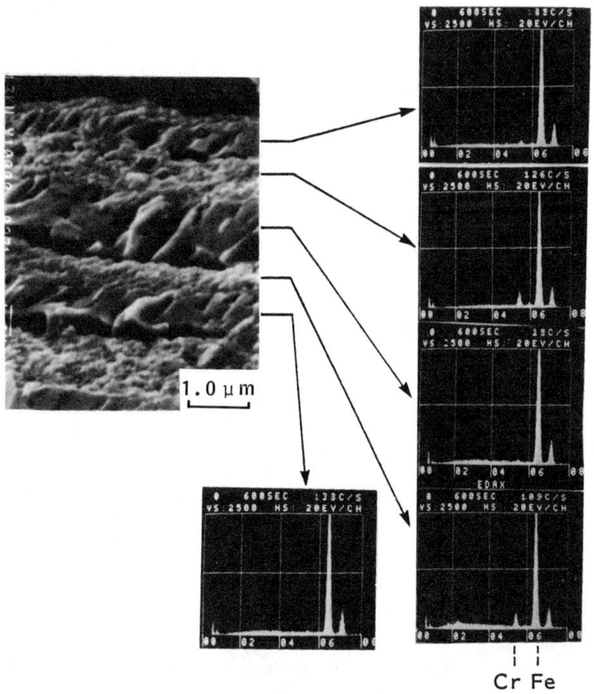

Figure 18. Typical Energy Dispersive X-ray Analysis Of Chromium Content In Laminations Formed On Boiler Tube Scales

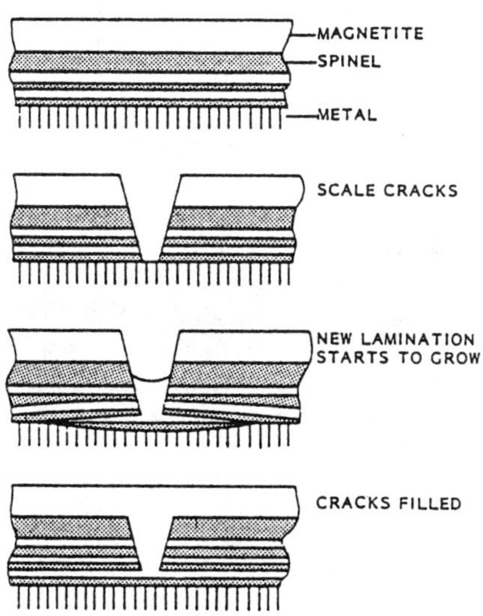

Figure 19. Schematic Illustration Of The Stages In The Growth Of Multilayer Scales By The Process Of Through-Scale Cracking. Reference (18)

Manning et al (19) proposed that laminated scales may be the result of periodic scale disruption when the oxide is unable to withstand the strains imposed upon it by the distension of the tube substrate. When a critical strain energy is exceeded, through-scale cracks occur, initiating the formation of a further lamination between the original oxide and the metal. This process relieves the elastic strain energy in the overlaying oxide layers. Two layer growth continues with inner layer formation at the metal-oxide interface and further thickening of the outer layer at the steam-oxide interface. When the strain energy limit is again exceeded, further laminations form in the same way. A numerical evaluation of this stored-energy model was successful in explaining the morphological details of scales observed on tubes removed from operating boilers (7).

Exfoliation of Scales

The release of oxide scales from internal surfaces of superheaters and reheaters and associated metal loss are the result of many factors. The difficulties in identifying those that have important influences are demonstrated in the paragraphs that follow.

Armitt et al (1) have identified the inability of oxide scales to withstand the strains imposed upon them as the cause of the exfoliation. Their treatment, the most comprehensive available today, recognizes the major source and types of strains and makes quantitative estimates of the levels of strain from each source. Some strains are due to the physical and chemical processes during oxide formation. Examples are epitaxial mismatch between oxide and metal, oxide density changes with stoichiometry and substrate elongation due to oxygen or carbon dissolution. Volume changes and mass transport associated with oxide formation or transformation may lead to strain where oxidation is non-uniform or where it occurs on curved surfaces. Time-dependent deformation processes act to reduce the elastic strains within the scales. External factors can cause additional strains. Examples are cooling strains from expansion mismatch, strains arising from heat flux removal, thermal shock and mechanical straining of the substrate. Fracture of the scales is considered to be the primary mode of stress relief and a function of the total elastic strain level in the oxide. In this treatment, it was possible to sum up algebraically all estimated strains from all sources and make allowance for stress relief by both time-dependent and time-independent deformation processes. Since the fracture of iron oxides usually occurs at $\approx \sigma_y/10$ for standard test geometries, the analysis largely concentrates on fracture in response to rapidly applied strains.

The results indicate that the consequences of exceeding the strain-tolerance of protective scale can be manifested in several ways. In compression, the whole or part of the scale may spall; in tension, through-scale cracking can cause enhanced localized corrosion and result in the subsequent formation of laminated scales. When the oxide is subjected to sustained deformation, linear oxidation rates can result from repeated scale fractures. Many aspects of the scaling behaviour of Fe-C steels have been rationalized and the failure mode map shown in Figure 20 has been constructed.

Grobner et al(4) have critically assessed the influence of various factors contributing to the propensity to exfoliate by comparing theoretical considerations with field observations. The significance of scale thickness, laminated structure, composition geometry, and heat-flux and thermal-cooling gradients was discussed in detail. Examination of the relationships between the exfoliation propensity and scale thickness revealed that the relationship proposed by Armitt et al(1) does not appear to have general validity(4, 20). While the strains associated with increasing scale thickness may explain exfoliation from austenitic materials, it does not appear to be the cause of exfoliation in ferritic materials. There is no clear evidence that laminated

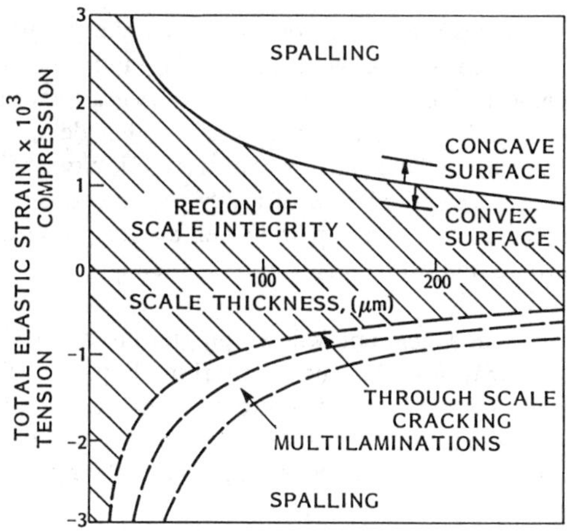

Figure 20. Oxide Failure Modes As A Function Of Thickness And Impressed Strain Reference (1)

scale formation is the cause of exfoliation, but conditions which promote this process may also cause exfoliation. Accumulation of voids at the metal-oxide interface and between the laminations can create planes of weaknesses. Heat flux has been identified as a potentially significant factor in the exfoliation processes, but it is noted that exfoliation is common in non-heat absorbing sections of the steam circuits. Based on examination of tube smaples removed from different sections of the boiler and from operating units with different lengths of service (7), it was possible to establish a definite relationship between exfoliation propensity and severity of thermal transients during operation. Reheater tubes that are exposed to more severe thermal gradients on start-up than are superheater tubes show far more exfoliation. Steam piping and headers, which retain their heat for some appreciable time after shutdown, may have similar temperature differentials on hot restarts, and also show severe exfoliation.

Preventive and Remedial Measures

There are several approaches to alleviating the problem of steamside scale exfoliation and turbine erosion. These include design and operating procedure changes, use of corrosion and exfoliation resistant materials, coatings, surface treatments of low alloy materials, and periodical scale removal by chemical cleaning.

The ultimate choice of the method to control turbine damage from oxide exfoliation may include one or a number of the solutions described. The method must be selected on the basis of detailed evaluation of each unit design, age, present and future operating modes. This evaluation must include cost benefit analysis of the proposed solution relative to turbine repair or replacement option(2).

Design and Operating Procedure Change

The damage of the turbine can be significantly reduced by mechanical removal of the exfoliated scale present in the steam circuit before introducing steam to the turbine(21). It has been known for years that steam blowing will remove particulates and preoperational cleaning procedures involving blowing to atmosphere have been used on essentially every utility unit. The purpose of preoperational steam blows is to remove weld beads, slag and other foreign material and construction debris which would be carried to the turbine. The important feature of any steam blowing program is to develop a force on particles equal to that developed during full load operation. Since the force on particles depends on the mass-velocity of the steam, the piping system must be sized to handle the required flow of lower pressure steam used in the blowing. Steam blowing would be expected to remove particulates broken loose during shutdown. If these particles bypass the turbine, major damage can be avoided.

It would be impractical to carry out a steam blowdown program to atmosphere each time a unit was brought on line. A blowdown procedure using a turbine bypass to the condenser can be employed. The connections, piping, valving and controls for such a system are costly to backfit in a system originally designed without such a feature in mind. A steam blowdown arrangement could readily be included in a new unit. Systems of this type are being used on units in Europe with reported good results.

Corrosion and Exfoliation Resistant Materials

It is known that increasing the chromium content of steel allows improves an alloy's oxidation resistance. Increasing the chromium content of an alloy well above 18% used in 300 series materials which have been used in superheaters and reheaters would be expected to significantly reduce the probability of turbine or boiler damage associated with oxide exfoliation(21). Constructing the steam circuits of a large utility unit with high chrome alloys could achieve the desired results but at a significant cost, provided they would be acceptable under the boiler code.

Coatings

One of the most promising options to alleviate the problem of exfoliation is treatment of the internal surfaces of reheaters and superheaters by coatings to reduce the rate of the scale formation and the propensity to exfoliation. The type of coating depends on the surface condition and location of application. There are basically three types:

- coatings for new or replacement installations which could be applied to clean surfaces away from the boiler;

- coatings for existing installations for treatment of clean surfaces, ie, acid cleaned in situ;

- coatings for existing installations for treatment of oxidized surfaces without any preparation to improve their resistance to corrosion and exfoliation.

The first two types of coatings have already been developed and are available commercially. Babcock and Wilcox (Alliance, USA)(22) have developed and tested chromizing treatment for new and replacement installations. Foster Wheeler Development Corporation (Livingston, USA)(23) and Ontario Hydro (Research Division) have developed and are currently field testing chromating and borating treatments for in situ application. The borating treatment, which is potentially the most cost effective way to overcome the problem, is currently under development at Ontario Hydro (Research Division). This program (No 112-G-277) is funded by the Canadian Electrical Association and final report will be available in May 1990.

Chemical Cleaning

One approach to controlling the effect of scale exfoliation is to periodically completely remove the oxide layer before it grows to a thickness where extensive exfoliation occurs(22). The critical period depends on the mode of the unit operation. Complete scale removal can be accomplished using chemical cleaning methods. Although the technical problems associated with chemical cleaning have been known for some time, with adequate care this operation can be performed safely.

Conclusions

The conclusions drawn from this work are summarized below:

- The major morphological features of duplex and multilaminated steamside scales are similar. The duplex scales, observed in laboratory isothermal tests and on thick tubes removed from an operating boiler, are composed of outer magnetite and inner chromium enriched layers. The magnetite layer has a pronounced texture. The oxide grains are elongated in a direction perpendicular to the metal surface. The pores observed in this layer are channels, located mostly at grain boundary triple points. The topographical features of the scale-steam interface indicate directional growth by solid-state diffusion. The inner layer consists of equiaxed crystals of iron-chrom spinel and magnetite. The porosity is irregular and appears to be aligned in parallel rows with the metal interface. The ratio of inner/outer layer thickness is independent of time but increases with increasing content of chromium in the alloy.

- The multilaminated scales are composed of alternating layers of magnetite with columnar structure and chromium-rich layers of equiaxed crystals. Usually a thick magnetite layer at the steam-scale interface is paired with a thinner iron-chromium spinel layer; several pairs of thinner layers (laminations) exit under this outer pair. The shape and extent of porosity is similar to that observed in duplex scales.

- A number of preventive and remedial measures to minimize or effectively eliminate minimize the exfoliation problem resulting in erosion of turbines is available.

References

1. J. Armitt, R.D. Holmes, M.I. Manning, D.B. Meadowcroft and E. Metcalfe. "The Spalling of Steam-Grown Oxide from Superheater and Reheater Tube Steels." EPRI Report TPS-76-655, February 1978.

2. R.B. Dooley, F.M. Somerset, A.C. Davidson, V.R.D. Miller and A.V. Manolescu. "Ontario Hydro's Experience with Solid Particle Erosion of Steam Turbines," presented at the Workshop on Solid Particle Erosion sponsored by ASME/ASTM/MPC Task Group, May 1980, Atlanta, GA.

3. T.H.Gladney and H.S. Fox. "Oxides Erode Turbine Blades." Electrical World, December 30, 1963.

4. P.J. Grobner, C.C. Clarke, P.V. Andreae and W.R. Sylvester. "Steamside Oxidation and Exfoliation of Cr-Mo Superheater and Reheater Steels." Paper No 172, presented at Corrosion 80, a NACE Conference, Chicago, March 1980.

5. M.I. Manning and E. Metcalfe, "Oxidation of Ferritic Steels in Steam," Proceedings of International Conference held by BNES and ICE, London, May (1977).

6. P.J. Grobner. "Scaling of Chromium-Molybdenum Steels in Steam and Air." Climax Molybdenum Company Report No S-4311, February 1978, Ann Arbor, MI.

7. P. Mayer and A.V. Manolescu. "The Role of Structural and Compositional Factors in Corrosion of Ferritic Steels in Steam," NACE Publication N. 6, p 368-379, 1981.

8. K. Hauffe. "Oxidation of Metals," p 285, Plenum Press, 1965, New York.

9. P.H. Effertz and H. Meisel. "Verzunderung Warmfesten Stahle in Hochdruckdampf nach Langen Betriebzeiten." Der Machinenshaden, 55 (1971), No 1, p 14.

10. P.L. Surman and J.E. Castle. "Gas Phase Transport in the Oxidation of Fe and Steel." Corrosion Science, 9 (1969), p 771.

11. P.L. Surman. "The Oxidation of Iron at Controlled Oxygen Partial Pressures - I. Hydrogen/Water Vapour." Corrosion Science, 12 (1973), p 113-124.

12. E.M. Field, R.C. Stanley, A.M. Adams and D.R. Holmes. "The Growth, Structure and Breakdown of Magnetite Films on Mild Steel." Presented at the Second International Congress on Metallic Corrosion, New York (March 1963).

13. E.C. Potter and G.M.W. Mann. "Mechanism of Magnetite Growth on Low Carbon Steel in Steam and Aqueous Solutions up to 550°C." Presented at the Second International Congress on Metallic Corrosion, New York, (March 1963).

14. T. Ericsson. "Stratified Oxide Scale Growth on Two Cr-Ni Steels Oxidized in High-Pressure Steam at 800°C." Oxidation of Metals, 2, (1970), p 173.

15. S. Mrowec and T. Weber. "Gas Corrosion of Metals." Published for NBS and NSF, Washington, DC, Warsaw, 1978.

16. G.B. Gibbs and R. Hales. "The Influence of Metal Lattice Vacancies on the Oxidation of High Temperature Materials." Corrosion Science, 17 (1977), p 487.

17. L. Tomlinson and N.J. Cory. "Hydrogen Emission during the Steam Oxidation of Ferritic Steels: Kinetics and Mechanism." Corrosion Science, 29 (1989), p 939-965.

18. J.E. Forrest and P.S. Bell. "The Relationship Between the Deformation of Substrate and Scale in the Mechanisms of Multilayer Oxide Formation on Low Alloy Ferritic Steels." Paper presented at the European Symposium on the Interaction between Corrosion and Mechanical Stress at High Temperature, Petten, The Netherlands, May 1980.

19. M.I. Manning and D.B. Meadowcroft. "Effects of Tube Creep Strains on Laminated Scale Formation in Ferritic Pressure Tubing." Central Electricity Research Laboratories Report No. RD/L/R2012, April 1980, Leatherhead, England.

20. T.A. McNary. "Exfoliation of Steamside Scale in Reheater Tubing of Croloys 2-1/4, 5 and 9." Babcock and Wilcox Alliance Research Centre, Report No 7724 (1966).

21. R.L. Attwood. "Controlling Solid Particle Erosion of Turbines." Paper No 83, presented at Corrosion 80, a NACE Conference, Chicago, March 1980.

22. P.L. Daniel, R.H. Emanuelson, J.M. Jevec and A.J. Blazewicz. "Steamside Oxidation Resistance of Chromized Superheater Tubes." Ibid, Paper No 173.

23. I.M. Rehn. "Corrosion Problems in Coal-Fired Boiler Superheater and Reheater Tubes: Steamside Oxidation and Exfoliation - Development of Chromate-Conversion Treatment." EPRI Program CS-1812, Research Project No 644-1, April 1981.

High-temperature degradation of advanced gas turbine engine materials and coatings*

P.C. Patnaik
R&D, Orenda Division, Hawker Siddeley Canada Inc., Mississauga, Ontario, Canada, L4T 1A9

ABSTRACT

High temperature oxidation is one of the predominant mechanism of material and coating degradation in gas turbine engines. Components in the hot end of the compressors undergo some oxidation under mechanical fretting wear or fatigue conditions. Turbine components such as nozzle guide vanes and blades with complex cooling configurations undergo severe oxidation and generally protected by high temperature coatings. Performance of these component materials such as titanium alloy, mechanically alloyed and directionally solidified superalloy materials with or without a protective coating is discussed from an oxidation point of view. Influence of the thermo-mechanical fatigue cracking on the turbine blade oxidation behaviour is presented.

INTRODUCTION

Materials for structural applications in the advanced gas turbine engine now range from titanium alloys in the hot end of the compressor to nickel base superalloys in the combustor as well as in the high and low pressure turbines. Majority of the engine components are now used in the coated condition with few exceptions. For example in the hot end of the compressor, rotor blades (Temp. 450°C- 600°C) are normally not protected except in the dovetail region. On the other hand, turbine components which operate at much higher temperatures (700°C-1100°C) are often protected by means of a high temperature coating. In the compressor, the basic function of the coating is to provide protection against fretting wear and fretting fatigue at the operating temperature, whereas in the turbine, coatings are used to provide adequate high temperature oxidation and corrosion resistance to the components. In this paper, the influence of oxidation and fatigue on the life expectancy of these components are discussed.

DEGRADATION OF TITANIUM ALLOY COMPONENTS

The predominant degradation modes of bare titanium alloys in the hot end of the compressor are fretting wear and/or fretting fatigue. Fretting wear is the removal of material from contacting surfaces through fretting as a result of low-amplitude oscillating sliding. A reduction in fatigue life is achieved due to fretting surface damage. Both fretting wear and fatigue are readily seen in dovetail region of the compressor blade and the processes are augmented in the presence of oxidation.

The influence of oxidation on the fretting wear and fatigue properties have been studied by Bill (1). The fretting wear for pure titanium as a function of temperature is shown in Fig. 1(a) indicating that the wear volume increased with temperature and they decreased drastically beyond 550°C. The fall in wear rate at high temperature was attributed to the change in growth kinetics of the oxide film from logarithmic to parabolic. This has also been observed in titanium alloys such as Ti-6Al-4V used frequently in the gas turbine, where wear volume is considerably reduced at higher temperatures (Fig.1(b)). However, under fretting fatigue conditions no improvement in fatigue properties are obtained in Ti-6Al-4V alloys as a result of the thick oxide produced by parabolic oxidation. The requirement for having better fretting fatigue properties is to have what's called as a "Glaze Oxide" with a spinel structure formed at high temperature and under sliding conditions (2). Unfortunately titanium alloys do not develop this type of oxide.

In practice titanium alloy compressor blades are protected at their dovetails by an anti-fret coating such as Cu-Ni-In accompanied by a solid film lubricant of MoS_2. Since the operating temperature of the blades is between 450°C-500°C and the Cu-Ni-In coating is not suitable for this temperature, severe degradation is observed. The blades in the dovetail showed signs of wear, fretting, coating delamination, oxidation and spalling(3). The deformation associated with unusually high loading conditions imposed on the blade dovetail caused the coating extruded and cracked. Better coatings or alternate surface modification techniques are therefore required to protect these components at the dovetail.

From a fretting wear or fatigue point of view, protection in the oxidizing atmosphere at moderate temperatures can be obtained either through the use of coatings or by surface modifications to produce the desired glaze oxide layer as discussed previously. The aim here is to have low friction and wear values at the operating temperature. Specifically for titanium alloys, implantation of Ba ions led to the formation of $BaTiO_3$ which has a pervoskite structure. This resulted some reduction in the coefficient of friction but the oxide layer was very thin (2). Also, implantation of Bi ions led to very low coefficients of friction on titanium alloys and reported very low wear rates (Fig.2). However effect of these ions on fretting fatigue properties has not been studied. In terms of coatings, it is suggested that chromium containing alloys or coatings under fretting conditions are capable of producing at high temperature a form of oxide which can exhibit low friction and wear. The oxide film even if it breaks is capable of reforming itself (2).

DEGRADATION OF NICKEL BASE SUPERALLOY COMPONENTS

Turbine Nozzle Guide Vanes

The first stage nozzle guide vane (NGV) is the component that directs the hot gas from the combustor towards the first stage of rotor blades. These are static components and designed on the basis the material stress-rupture lives and thermal fatigue properties. Typical materials used in the NGV are nickel based superalloys such as IN 738, René 80, MA 754, and MA 6000. Of particular interest is the MA 754 which is a mechanically alloyed material and is used currently as nozzle airfoil material. To date, this alloy is being used in the uncoated condition.

Fig. 3 shows a typical service exposed nozzle guide vane. The two airfoils are made from MA 754 and the bands from MM-509. The component is cooled by the circulation of cooling air internally. The MA 754 airfoil have been seen to undergo severe oxidation as they are used in the uncoated condition. Metallography carried out on different regions of the airfoil showed signs of oxidation to different degrees. The leading edge region showed signs of heavy oxidation with the formation of external and internal oxides (Fig.4). The alloy MA754 is said to be a lean Ni alloy containing about 20 wt.%Cr and 2.5 wt.% of Fe. Trace amount of Al and Ti total up to 1 wt. % are also present. The Y_2O_3 dispersoid is about 0.6 wt.% and is usually present in the form of $YAlO_3$. The oxidation products consist of an outer oxide layer rich in Ni and Fe and an inner layer containing Cr rich internal oxides. This is evident in the microprobe traces for elements Ni, Cr, Fe and Ti as shown in Fig. 5. The depth of internal oxidation zone ranged from 40-60 micron in the leading edge of the vane. On the other hand, a single external oxide layer consisting of spinel $NiCr_2O_4$, NiO and Fe oxide phases was observed in the region closer to the air cooling holes (Fig. 6). The morphology of the surface oxide was of nodular appearance within the cooling holes which is not exposed to the hot gas directly (Fig.7). The morphological appearance changed from mixed nodular to massive oxides within the air cooling holes which are partially exposed to the hot gas (Fig.8).

The vane airfoil has also been seen to be subjected to thermal fatigue loading conditions resulting from rapid temperature changes. Thermal fatigue cracks were particularly observed in the trailing edge region as well as near the cooling holes where thermal strains are quite high. This is expected as the trailing edge region is much thinner than the leading edge and therefore can heat or cool up faster than the rest of the component. Figure 9 shows typical thermal fatigue cracking as observed in the MA 754 airfoil trailing edge. Multiple crack branching was also observed.

In summary, it can be pointed out that a single nozzle guide vane can experience oxidation and thermal fatigue damage to a varying degree depending on the localized temperature distribution in the component.

Turbine Blades

The high pressure turbine (HPT) blades examined in this investigation were service exposed directionally solidified René 80. This component is coated with a Codep B-1 type of diffusion alumnide coating. The predominant mode of failure of the blade was catastrophic oxidation and radial cracking at the blade tip, specially between the mid-chord and trailing edge region (Figure 10). The tip was heavily oxidized and the cracks had a broad mouth (1-3mm wide) towards the concave or pressure surface. Some of these cracks extended up to the convex surface ie., across the entire trailing edge thickness. Extensive loss of substrate alloy in the tip cracks by oxidation followed by scale evaporation and spalling contributed towards widening of these cracks. These tip cracks ranged from 0.1 to 6mm in length and propagated right into the substrate in the downward or radial direction. Figure 11 shows a typical cross-section through one of these cracks. Some of these cracks originated at the DS grain boundary and some far away from it ie., within the grain, suggesting that the grain boundary channels of the DS alloy did not have any major impact on the material degradation process.

The general morphology of the oxidized layer at the blade tip and in a growing thermal fatigue crack was quite similar. The outer layer scale contained Ti and Cr rich oxides and the inner layer contained small elongated internal oxide particles rich in Al and Ti (Fig. 12). Elemental analyses below the internal oxidation zone revealed the presence of a gamma prime depleted layer as illustrated in Fig.12. Also observed in these blades were shrinkage porostities which when open to the surface led to oxidation. This is shown clearly in Fig.13. A mechanism has been recently proposed by Oldfield (4) proposing the failure of a turbine blade due to the presence of solidification shrinkage cavities open to the atmosphere. The mechanism takes into account of the initiation of crack beneath the cavity under thermo-mechanical fatigue conditions when the cavity is filled with an oxide. Oxide formation within the pore have been said to generate pressures exceeding the material yield strength leading to the formation of lips on the edge of the crack (Fig. 14).

The aluminide coating on the HPT blade consisted of a coarse grain β-NiAl structure in the outer layer and a diffusion zone containing TiC, $M_{23}C_6$ carbides and sigma phases. Two modes of coating degradation have been predominantly observed in the service exposed blades. They are respectively oxidation and thermo-mechanical fatigue cracking. Some interdiffusion of alloying elements between coating and substrate took place giving rise to sigma phase formation (Fig 15).

Oxidation

In the HPT blades investigated, aluminide coating provided moderate oxidation resistance in the regions where the coating was undamaged. In some regions localized coating oxidation attack resulted some coating delamination (Fig. 16).

Thermo-mechanical Fatigue (TMF)

The HPT blade experiences complex thermal and mechanical loadings during a typical mission of the aircraft. Temperature gradients and mechanical constraints during such complex cycles give rise to cyclic thermal stresses and therefore induce thermo-mechanical fatigue damage in the HPT blade. Since the aluminide coating on the HPT blade has a fairly high brittle to ductile transition temperature, the coating can be somewhat brittle at near service temperatures. Any TMF crack formed in the coating can then be propagated into the substrate alloy.

Figures 17(a) to 17(e) illustrate the thermal fatigue cracks present in various regions of an HPT blade. Depending on the location of the crack, minor to extensive substrate alloy attack has been observed. Figure 17(a) shows the initiation of a TMF crack in the coating whereas Figures 17(b) and (c) show attacked regions near the blade leading edge. It should be noted in Figure 17(d) that near the leading edge of the blade, the diffusion zones of the external airfoil coating and the internal coating on the cooling hole surface have completely overlapped each other. In other words, no substrate alloy was found to be present in between these two coating interfaces. AS a result of this and poor TMF properties of the coating, a crack that had initiated in this region of the coating propagated through the entire blade wall thickness. Gamma prime depletion zones beneath the interior oxidized surfaces were also observed in these cracks.

It is evident from the service behaviour of two superalloys MA 754 and DSR80H that a coating is required to protect the bare MA 754 and to better tailor the coating properties toward the specific requirement of DSR80H HPT blade. Thermal fatigue or thermo-mechanical fatigue leading to oxidation of substrate alloys seem to be the key issue in both of these components. Apart from designing coating compositions, it is only recently that issues related to thermal and thermo-mechanical fatigue properties of coatings have been addressed in a systematic manner (5).

SUMMARY AND CONCLUSIONS

Investigations on the service exposed hardwares of the gas turbine both from the hot end of the compressor as well as the high pressure turbine showed that fatigue and oxidation are the two predominant modes that govern the degradation of these components. In the hot end of the compressor where titanium alloy blades are used, fretting wear or fatigue assisted by oxidation of the dovetail pressure surface can limit the blade life. Coatings now used do not provide adequate protection to the dovetail. Surface modification techniques such as ion implantation or advanced coatings are being considered to provide resistance to dovetail fretting fatigue at elevated temperatures.

In the turbine end, nozzle guide vanes and blades suffer from oxidation, thermal and thermo-mechanical fatigue. Specially in the case of turbine nozzle, which is now used in uncoated condition, suffer from severe thermal fatigue damage and internal oxidation. Oxidation of the braze joint between the vane airfoil and the band can also limit the life since the braze contains significant amount of borides which are very prone to oxidation. This indeed has been observed. In the case of HPT blade, a number of factors govern its life. Material defects such as shrinkage cavities when close to the surface have been the sites for TMF crack initiation. Poor TMF property of the diffusion aluminide coating also led to crack initiation in the coating and subsequent propagation into the substrate. The coating properties must, therefore, be improved in this situation to resolve the cracking problem. Tip cracking to some extent can be resolved by better design of the blade in terms of providing better access to cooling air, although such an effort has not been able to solve the cracking problem. This leaves behind the only options to improve blade material and coating properties tailoring to service requirements.

ACKNOWLEDGEMENTS

The author wishes to acknowledge the financial support of the department of National Defence Canada for this work. Support from the structures and materials laboratory of the National Aeronautical Establishment, NRC is also greatfully acknowledged.

REFERENCES

1. R.C. Bill, ASTM STP-780, p.165-182, (1982) Philadelphia, PA, ASTM.

2. R.B. Waterhouse, AGARD-CP-461, p.8 (1989).

3. J.E. Eldes et. al, AGARD-CP-461, p.16 (1989).

4. W.Oldfield et al. Paper presented at the ASM Repair and Life Assessment of hot section components, Phoenix, Arizona April 17-19, 1990.

5. R. Mevrel, AGRARD-CP-461, p.9 (1989).

a.

b.

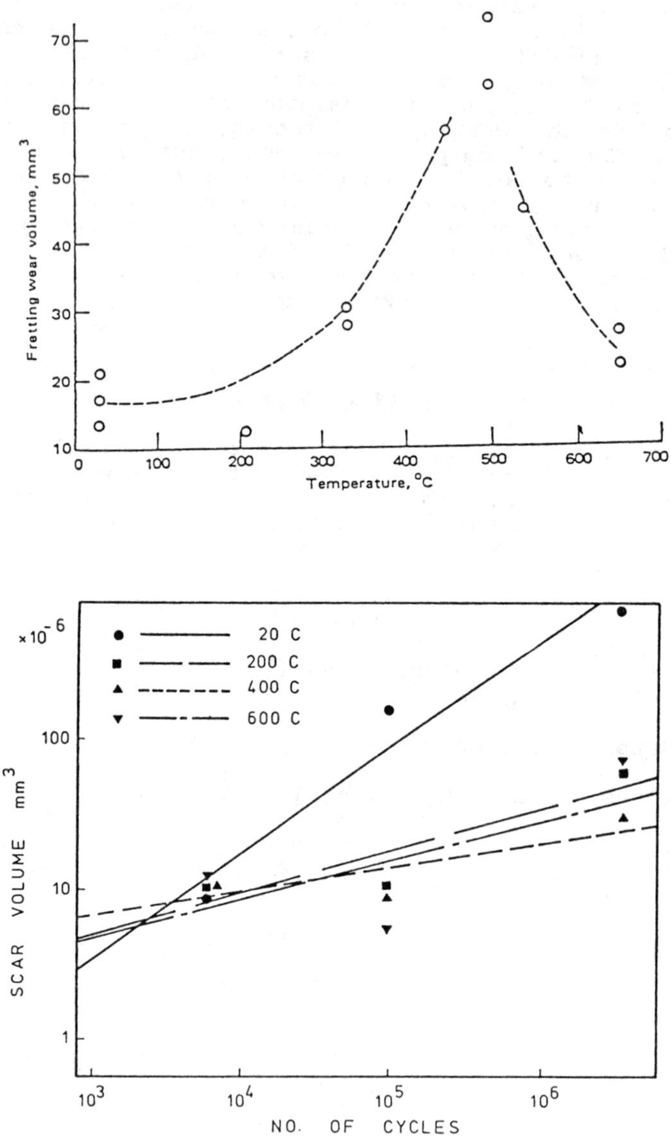

Fig.1 (a) Fretting wear of pure titanium vs. Temperature (1).

(b) Fretting wear of Ti-6Al-4V vs. temperature (2).

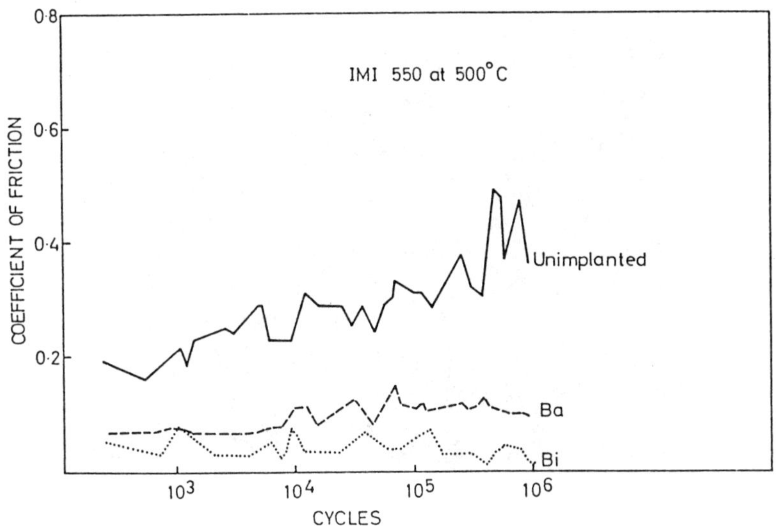

Fig.2 Effect of ion implantation on the coefficient of friction of IMI 550 at 500°C (2).

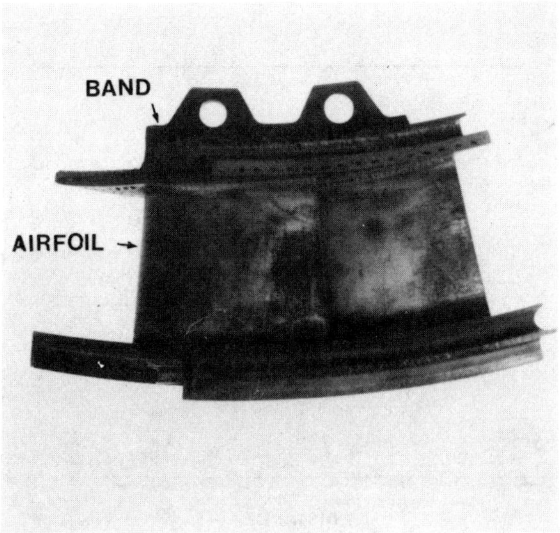

Fig.3 Photograph showing a typical high pressure turbine nozzle.

Fig.4 SEM micrograph of the cross section of a service exposed MA 754 turbine nozzle. The oxidation products consist of an outer layer of Ni and Fe rich oxide and an inner layer containing Cr rich internal oxides.

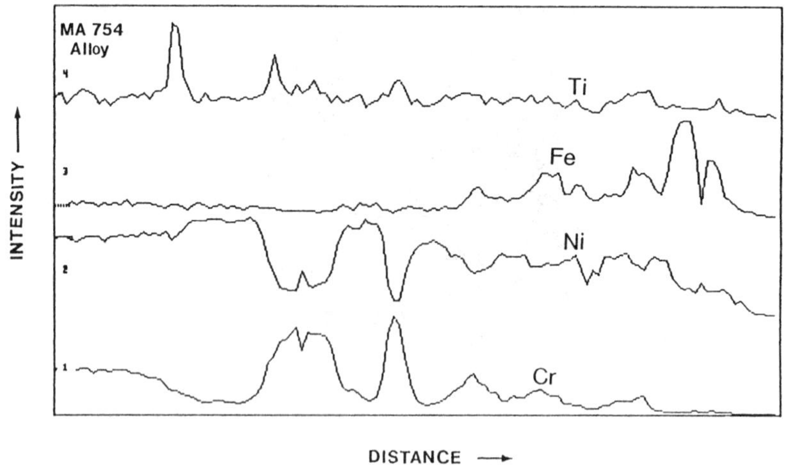

Fig.5 Elemental microprobe scan across the oxide layer shown in Fig.4.

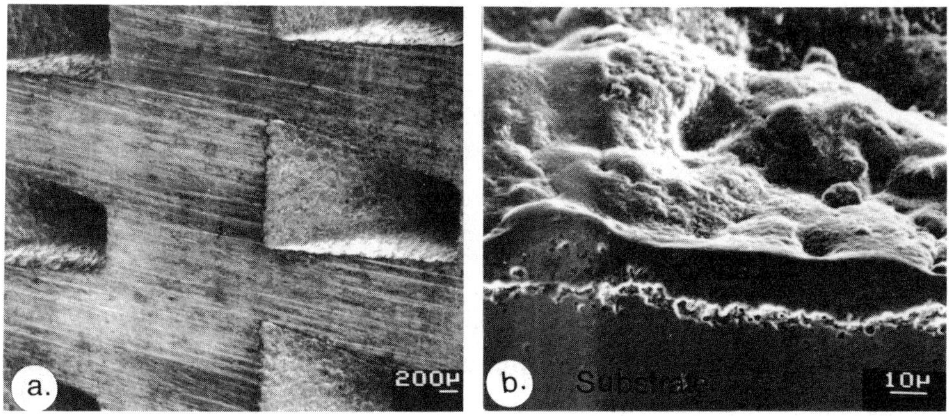

Fig.6 SEM micrograph showing (a) the air cooling hole region of the airfoil convex surface (b) a typical cross-section of the oxide layer near one of the cooling holes. Note very little internal oxidation in this region.

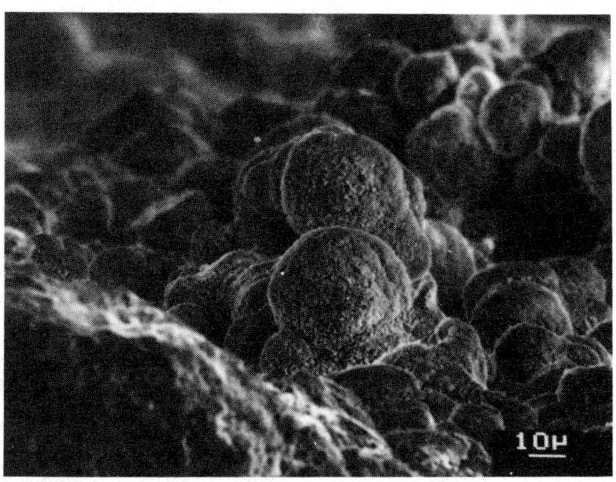

Fig.7 Nodular oxides rich in Ni as appeared in one of the cooling hole of the nozzle airfoil.

Fig.8 Morphology of oxides on the MA 754 airfoil (a) nodular appearance near the cooling hole (b) away from the cooling hole.

Fig.9 SEM micrograph showing (a) the cooling hole region of the vane airfoil (b) from which a thermal fatigue crack has initiated (c) further branching of this crack.

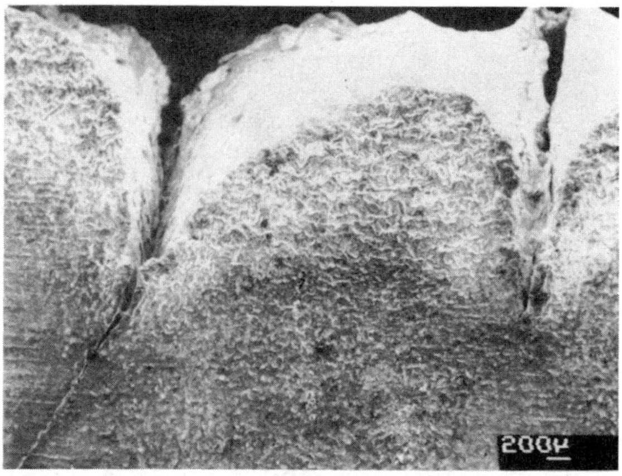

Fig.10 SEM micrograph showing severe trailing edge tip cracking, oxidation and loss of substrate alloy.

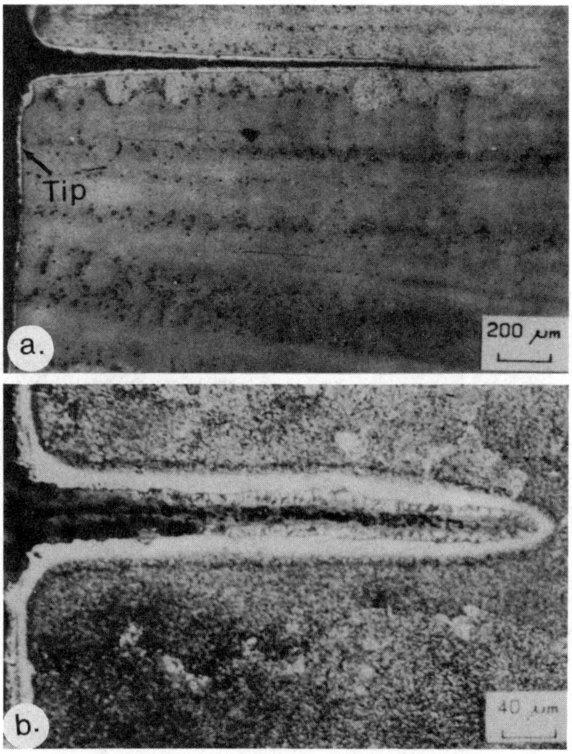

Fig.11 SEM micrographs showing a longitudinal cross-section of the tip crack.

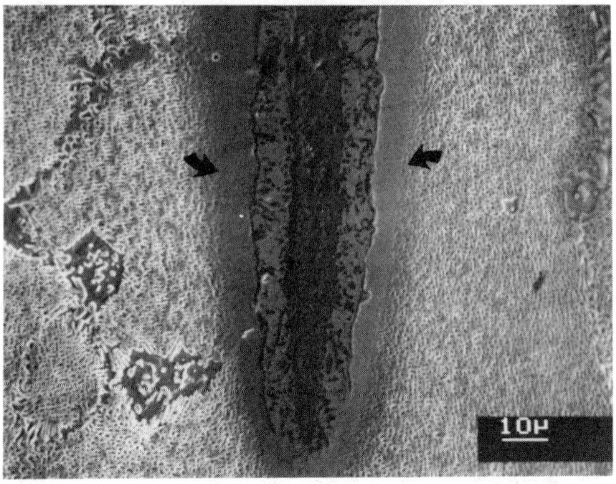

Fig.12 SEM micrograph showing the oxide morphology in the vicinity of a tip crack. A solute depleted zone is evident in this micrograph.

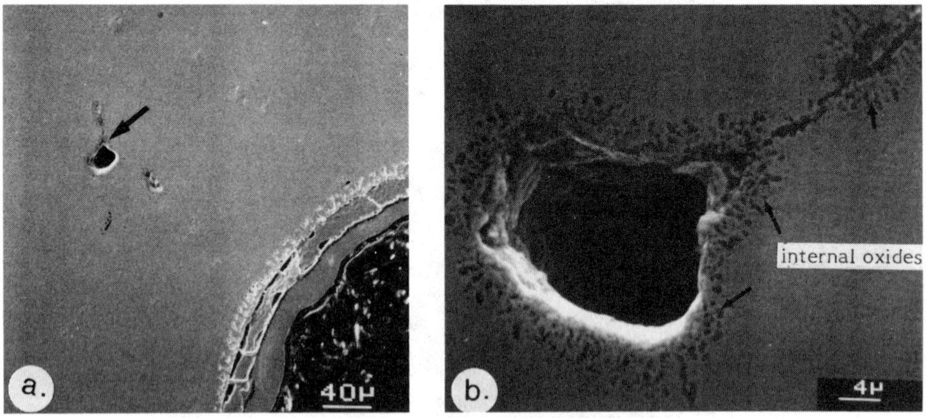

Fig.13 SEM micrograph (a) illustrating the presence of shrinkage cavities (b) oxidation within the cavity.

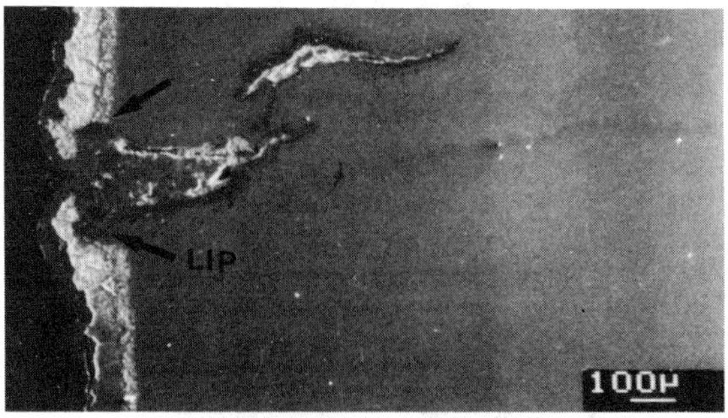

Fig.14 SEM micrograph illustrating the lip formation due to oxidation in a microcavity open to the surface.

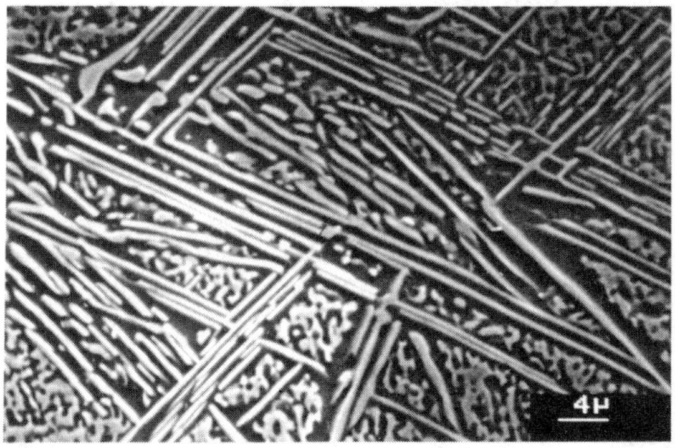

Fig.15 SEM micrograph showing the formation of needle like sigma phases just beneath the coating diffusion zone. This micrograph is taken from the sample in a region parallel to the coating surface.

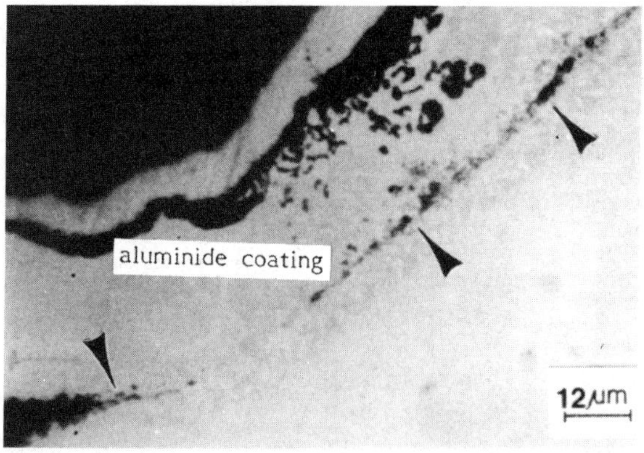

Fig.16 SEM micrograph illustrating oxidation attack of the coating resulting in a coating failure.

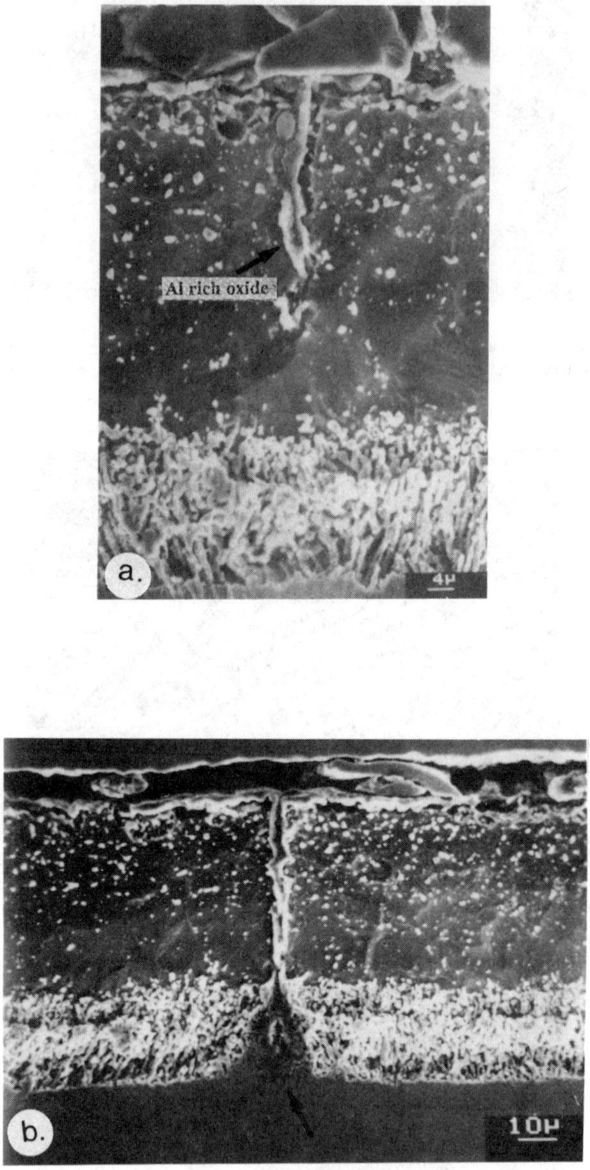

Fig.17 a) SEM micrograph illustrating the initiation of a TMF crack in the coating

b) SEM micrograph showing the propagation of a TMF crack through the entire coating thickness.

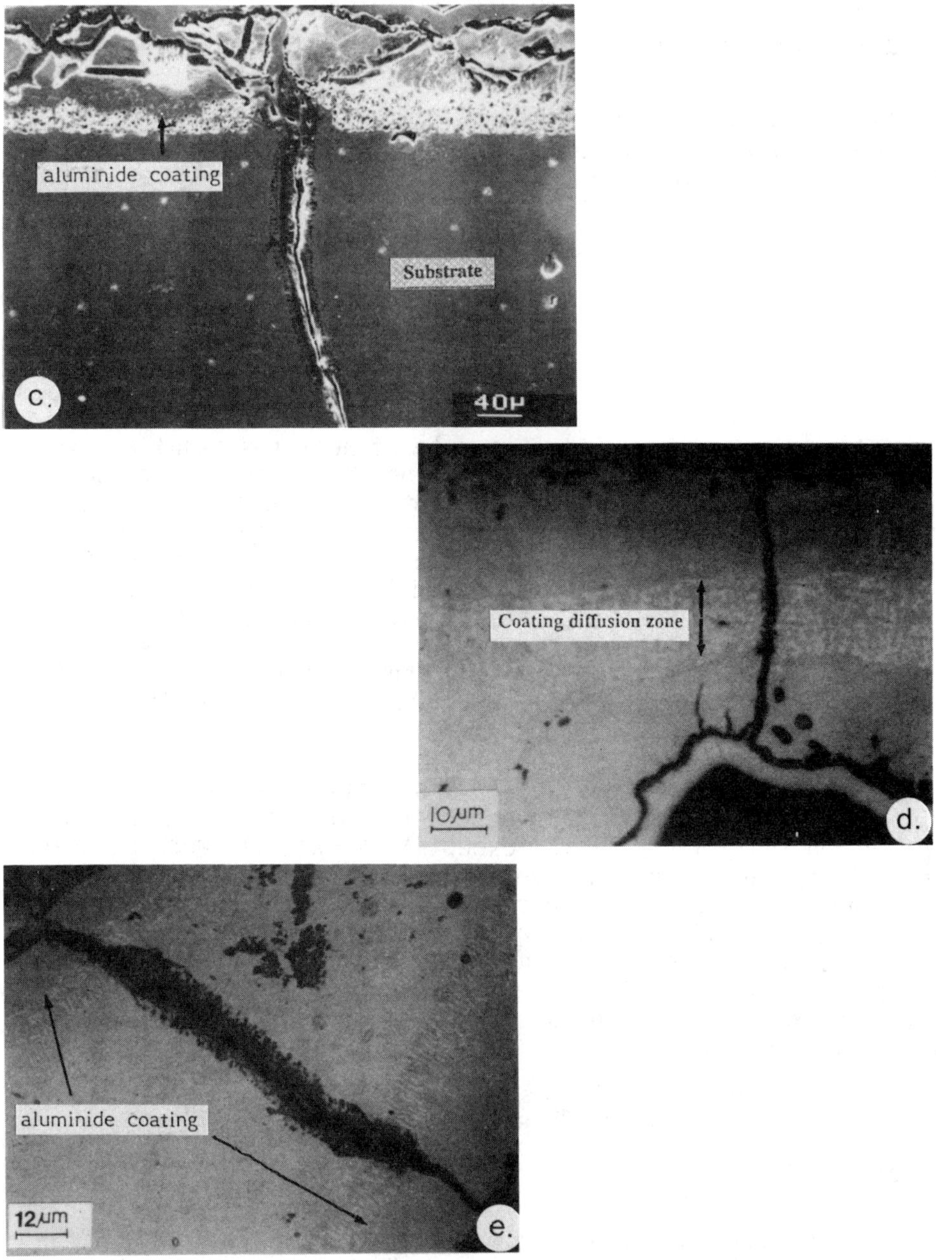

Fig.17 c) SEM micrograph showing further propagation of a TMF crack right into the alloy substrate.

d) Complete attack through a wall cross-section.

e) Overlapping of coating diffusion zone and cracking in this zone.

NaCl-induced high-temperature oxidation of Ni and Ni-Cr alloys

Yutaka Shinata and Yoshizumi Nishi
Department of Materials Science and Engineering, Mining College, Akita University, Akita 010, Japan

Introduction

High -temperature oxidation rate of heat-resistant alloys is frequently stimulated by the deposition of combustion products or impurities such as NaCl, SO_2, Na_2SO_4 and V_2O_5. The Na_2SO_4-attack of nickel-base superalloys used in gasturbine engine has been considered a very important problem and known as hot corrosion (1, 2). Deposition of Na_2SO_4 is caused by the reaction between sulfur in fuel and NaCl injected into the turbine as an aerosol of sea salt together with the intake air. Thus, hot corrosion induced by NaCl seems to be an important problem. The role of NaCl in hot corrosion by Na_2SO_4 and corrosion behavior in Na_2SO_4-NaCl mixture have been discussed(3-8).

The authours have examined the hot corrosion of Ni(9), Ni-Cr(10), and Ni-Cr-M alloys(11-14) systematically from the viewpoint of salt-composition dependence in Na_2SO_4-NaCl mixtures and showed that Ni-Cr alloys, except Ni-Cr-Al(11), were attacked severely by NaCl more than by Na_2SO_4 or their mixtures.

There has been a little report dealing with NaCl-induced hot corrosion(4, 15-17). In the present paper the oxidation behavior of Ni and Ni-Cr alloys with and without NaCl was examined under conditions below and above the melting point of NaCl from the view point to examine a role of Ni and Cr to the NaCl-induced high-temperature oxidation.

Experimental Methods

Specimen

Commercial Ni, and vacuum melted Ni-Cr binary alloys, i.e., Ni-5%, -10% and -20%Cr alloys were used in this experiment. TABLE 1 shows chemical compositions of the materials.

These Ni-Cr alloys were used after solution treatment to dissolve chromium carbide of grain boundary into matrix. After the heat treatment the alloys were cut and shaped into specimens 9 mm x 9 mm x 5 mm. These specimens were polished with SiC paper of grade 400, ultrasonically cleaned in alcohol, gauged

TABLE 1 Chemical Compositions of Specimens Used in This Work as Mass Percent.

	C	Si	Mn	P	S	Cu	Ni	Cr
Ni	0.01	0.06	0.26	——	0.002	0.01	99.6	——
Ni-5Cr	0.05	0.06	0.02	0.001	0.001	0.02	bal.	4.95
Ni-10Cr	0.06	0.06	0.02	0.001	0.001	0.01	bal.	10.24
Ni-20Cr	0.01	0.06	0.04	0.003	0.003	0.02	bal.	19.75

with a micrometer and determined the mass. Then the specimens were immersed cyclically in acetone, in which reagent-grade 99.9% NaCl was suspended, to coat their surfaces with the salt homogeneously. The amount of salt coating on the specimens was $0.2 \pm 0.01 kg \cdot m^{-2}$. When oxidation rate was measured, $0.1 kg \cdot m^{-2}$ NaCl was coated.

The NaCl-coated or -uncoated specimen was put in alumina crucible and oxidized in a electric furnace at a constant high temperature under static air for 57.6ks. Heating temperature was changed from 873 to 1373.

After the oxidation descaling was conducted by an electrolytical method in molten NaOH-Na_2CO_3 mixture at 873K and the corrosion loss was measured. Scales and cross section of the oxidized specimens were examined by X-ray and electron probe microanalysis, respectively.

Thermal analysis was conducted to examine an interaction between NaCl and Ni, NaCl and Cr. In the analysis DTA and TG curves were simultaneously obtained at a heating rate of $0.33 K \cdot s^{-1}$ with a Rigaku Thermoflex apparatus. As a reference for DTA, α-Al_2O_3 was used.

Results and Discussion

Corrosion by Na_2SO_4, NaCl and their mixtures

FIGURE 1 shows corrosion loss of Ni and Ni-Cr alloys after coating test in Na_2SO_4-NaCl mixtures at 1173K for 57.6ks. The corrosion loss of Ni is very large on corrosion by 100%Na_2SO_4 and decreases with increasing NaCl content, whereas that of Ni-Cr alloys increases gradually with NaCl content and shows a maximum value on corrosion by 100%NaCl. Both Ni and Ni-Cr alloys are oxidized acceleratedly by coating 100%NaCl and the effect of Cr-addition is not clear from these data. To examine which element is detrimental on corrosion by NaCl, an interaction between NaCl and Ni or Cr was measured by thermal analysis.

Thermal Analysis

FIGURE 2 shows DTA and TG curves. On the DTA curve of Ni

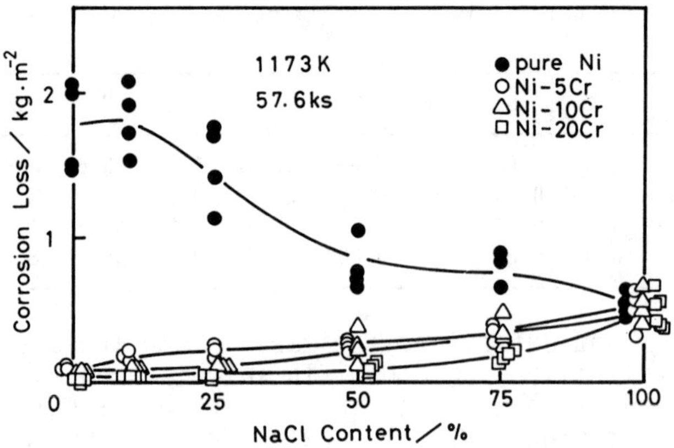

FIG. 1 Corrosion Loss of Ni and Ni-Cr Alloys After Coating Test in Na_2SO_4-NaCl Mixture at 1173K for 57.6 ks.

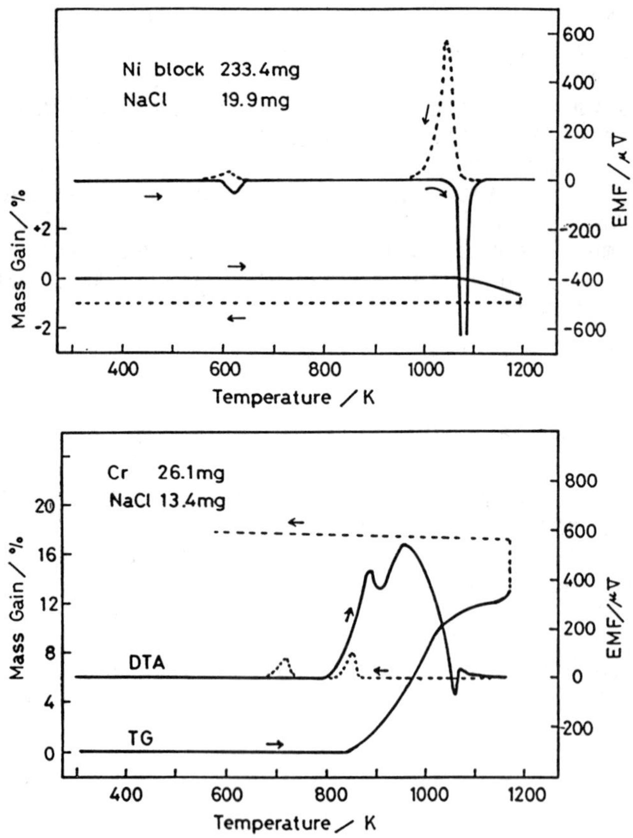

FIG. 2 DTA and TG Curves in Oxygen for Heating and Cooling of Ni and Cr.

two reversible peaks, i.e. endothermic on heating and exothermic on cooling, are observed at about 630K and 1070K. The magnetic transformation temperature of Ni is 626K and the melting point of NaCl is 1074K. Then the two peaks seems to correspond magnetic transformation and melting-solidification temperatures, respectively. On TG curve a little mass-loss is only observed after melting of NaCl. This is considered to be vaporization of NaCl because NaCl has high vapor pressure especially above the melting point.

In the case of Cr, a big change is observed on both DTA and TG curves even below the melting point of NaCl. On heating process big exothermic peaks with mass-gain starts from about 800 K. On cooling two small exothermic peaks are observed at about 850K and 700K. This irreversible reaction has been elucidated to be caused by oxidation of Cr and formation of Na_2CrO_4(18, 19).

From the results of thermal analysis it is clear that there is little interaction between NaCl and Ni below the melting point, whereas a big interaction between NaCl and Cr.

Oxidation Rate Below the Melting Point of NaCl

FIGURE 3 shows oxidation kinetics of Ni, Cr and Ni-20%Cr specimens oxidized with 0.1 kg・m^{-2} NaCl at 1023K which is about 50 K below the melting point. At this temperature, oxidation of

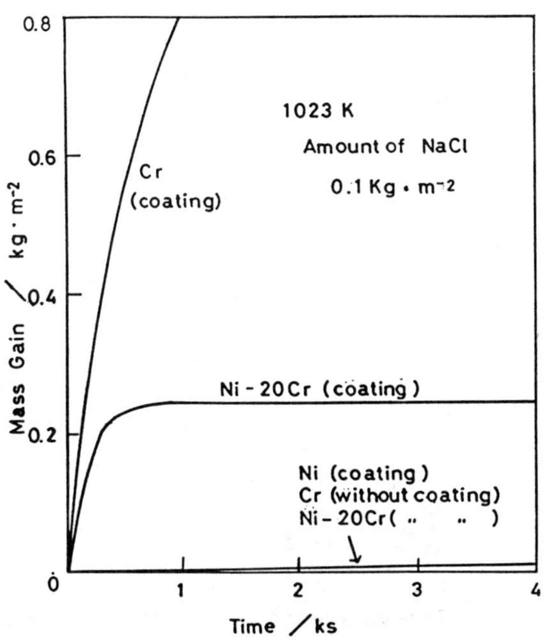

FIG. 3 Mass-Gain versus Time Curves for NaCl-Coated and Uncoated Ni, Cr and Ni-20%Cr Alloy at 1023 K in Flowing Air.

every specimen without coating NaCl occurs very slowly, and mass-gain is very small, too. The oxidation rate of NaCl-coated Ni is the same with that of uncoated specimen. On the contrary, the oxidation of Cr is markedly accelerated by coating NaCl. These behavior corresponds to the results obtained by thermal analysis. In the case of Ni-20%Cr alloy, the oxidation is also accelerated by NaCl, but the oxidation rate shows an intermediate value between Ni and Cr, as expected from the alloy composition.

From these result it is clear that Cr has a big influence to NaCl-induced oxidation of Ni-Cr alloy. To examine the effect of Cr more detail, corrosion-loss was measured by using Ni-Cr alloys which have a different Cr content, as a function of temperature.

Corrosion Loss Below the Melting Point of NaCl

FIGURE 4 shows corrosion losses of Ni and Ni-Cr alloys oxidized with coating NaCl for 57.6ks below the melting point of NaCl. As is obvious from the figure, corrosion loss of Ni at every temperature is smaller than that of Ni-Cr alloys and it increases with increasing Cr content.

From this result it is concluded that Cr, added to commercial Ni-base superalloys to improve corrosion resistance, is a detrimental element to NaCl-induced oxidation below the melting point.

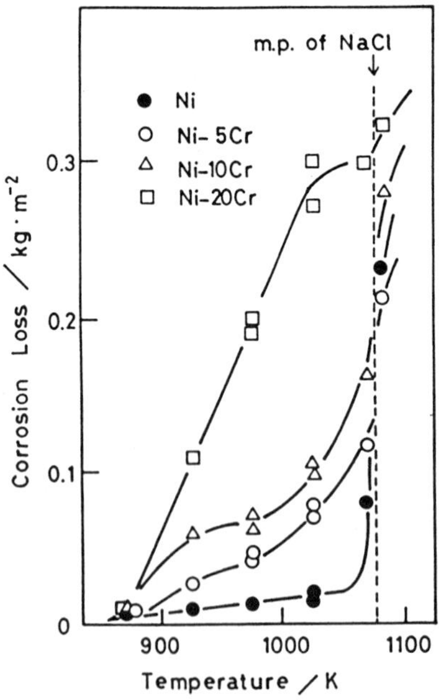

FIG. 4 Corrosion Losses of Ni and Ni-Cr Alloys Oxidized with Coating NaCl for 57.6 ks below the Melting Point of NaCl.

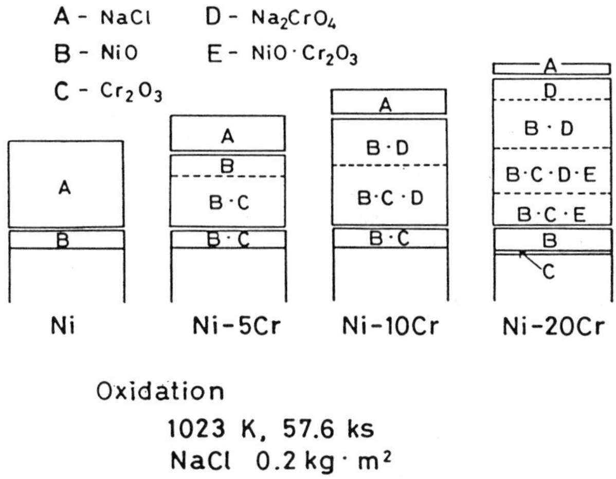

FIG. 5 Schematic Representation of Scales Oxidized at 1023 K for 57.6 ks with NaCl and Identified by X-Ray Diffraction.

Analysis of Scale

To decide the reason why corrosion of Ni-Cr alloys is accelerated by NaCl, scales formed on surface after oxidation with NaCl was examined by X-ray diffraction.

FIGURE 5 shows the result schematically for specimens oxidized at 1023K for 57.6ks. The depth profile was measured by polishing the specimens from surface to matrix metal with emery paper.

On Ni surface white NaCl remains without any reaction, and only NiO is identified. Whereas on surface of Ni-Cr alloys yellowish NaCl presents on outermost scale, and beneath the NaCl, NiO and Cr_2O_3 are identified. In the case of Ni-10%Cr and -20%Cr alloys, presence of Na_2CrO_4 is confirmed obviously. An amount of Na_2CrO_4 seems to be large in Ni-20%Cr alloy. This fact corresponds to the experimental result which Ni-20%Cr alloy showed a large quantity of corrosion loss.

This Na_2SO_4 as a reaction product, i.e., hexavalent compound of Cr , is considered to be formed by next reactions:

$$Cr + 2NaCl + 2O_2 \longrightarrow Na_2CrO_4 + Cl_2$$

$$Cr_2O_3 + 4NaCl + 1/2\ O_2 \longrightarrow 2Na_2CrO_4 + Cl_2$$

Protective scale may not be created, because Na_2CrO_4 forms eutectic melt with NaCl, and the eutectic temperature is about 850K as shown in FIG. 6 (19). When once Na_2CrO_4 is formed on surface, corrosion in molten Na_2CrO_4-NaCl mixture will occur eventhough an oxidation temperature is below the melting point of NaCl. The Na_2CrO_4 is formed more easy in high chromium alloy. This is the reason why Ni-20%Cr alloy showed the most severe corrosion.

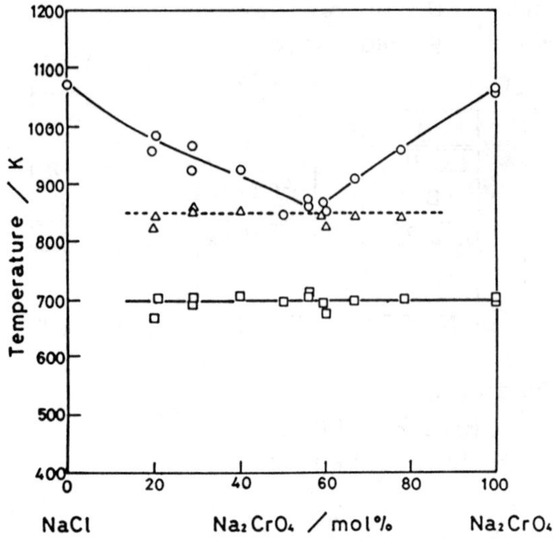

FIG. 6 Phase Diagram for the NaCl-Na_2CrO_4 System Obtained by the DTA Method(19).

Corrosion above the Melting Point of NaCl

As is obvious from FIG.4, just above the melting point of NaCl corrosion loss of Ni increases abruptly and approaches to that of Ni-Cr alloys. To examine this behavior more detail, cross section of a Ni specimen was observed.

FIGURE 7 shows the cross section of Ni oxidized slightly with coating NaCl at 1173K for 0.6ks. This temperature is about 100K higher than the melting point of NaCl. In NaCl coating on the specimen, layered and spherical NiO are observed, as identified by electron probe microanalysis. Some parts in NaCl layer contain dendritic NiO. It seems that these oxide were formed by dissolved Ni from matrix metal into NaCl solution. As an evidence of this assumption, presence of Ni^{2+} ion in NaCl was confirmed by the dimethylglyoxime method. Above the melting point of NaCl protective NiO scale is not formed and the oxidation rate becomes very fast.

FIGURE 8 shows corrosion losses of Ni and Ni-Cr alloys above the melting point of NaCl. From the melting point to 1173K, Ni and Ni-Cr alloys show nearly equal corrosion loss. This is because corrosion loss of Ni increases remarkably above the melting point as stated previously, whereas corrosion behavior of Ni-Cr alloys does not change markedly below and above the melting point. Until 1373K corrosion loss of Ni-20%Cr alloy does not

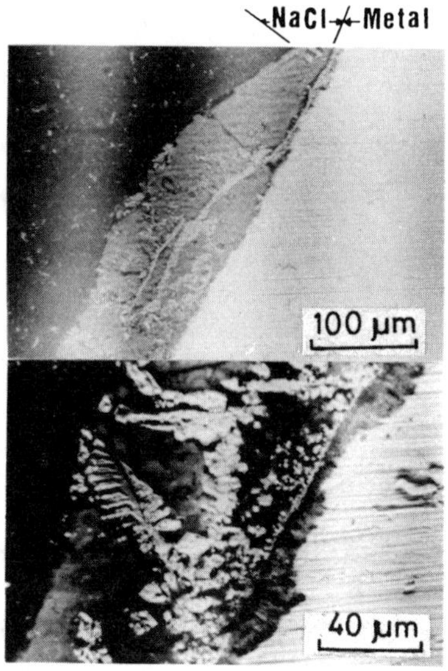

FIG. 7 Cross Section of Ni Oxidized Slightly with Coating NaCl at 1173 K for 0.6ks.

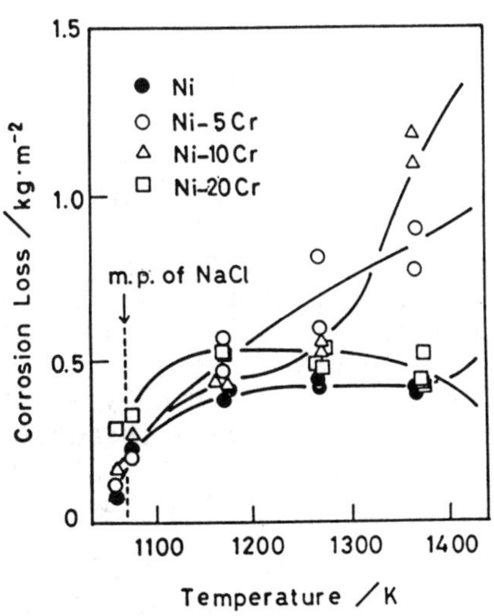

FIG. 8 Corrosion Loss of Ni and Ni-Cr Alloys Oxidized with Coating NaCl for 57.6 ks above the Melting Point of NaCl.

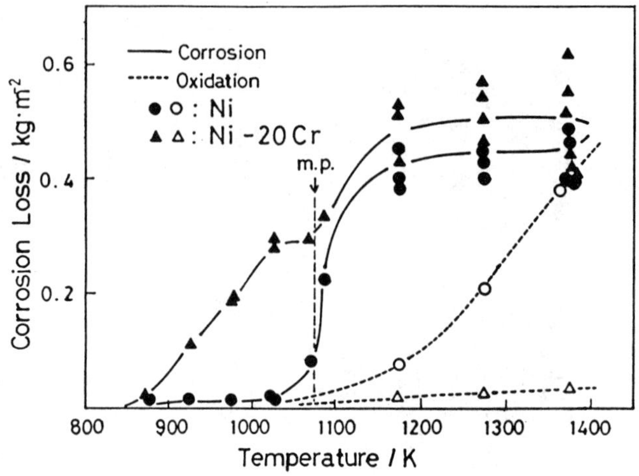

FIG. 9 Comparison of Corrosion Loss after Oxidation with (Corrosion) and without NaCl (Oxidation) as a Function of Temperature.

increase, but that of Ni-5%Cr and -10%Cr alloys increases with temperature. Then corrosion loss of low Cr alloys becomes exceed to that of Ni-20%Cr alloy at higher temperature. At high temperature vaporizing velocity of NaCl becomes very fast and corrosion behavior of alloys considered to be similar to that without NaCl.

Comparison of oxidation with and without NaCl.

FIGURE 8 shows a summarization of corrosion loss below and above the melting point of NaCl. In the figure, data of Ni-5%Cr and -10%Cr were omitted for simplicity and corrosion loss without NaCl was illustrated simultaneously as dotted line for comparison. Corrosion loss of Ni-20%Cr alloy without NaCl was calculated from mass-gain data after oxidation because oxide film covered surface tightly could not remove completely. Comparing these data it is clear that oxidation of Ni is accelerated only above the melting point of NaCl, whereas that of Ni-Cr alloy is accelerated at every temperature by the detrimental effect of Cr. At higher temperatures the influence of NaCl becomes small especially on corrosion of Ni.

TABLE 2 shows a summarization of accelerated oxidation of Ni, Cr and Ni-Cr alloy induced by Na_2SO_4 and NaCl. Here, accelerated oxidation means that oxide forming reaction of metals and alloys in air or O_2 is markedly accelerated by the presence of salt. Both Na_2SO_4 and NaCl can cause accelerated oxidation of Ni above their melting point. But concerning to Ni-Cr alloy there is a big difference between Na_2SO_4- and NaCl-induced oxidation. The

TABLE 2 Summerization of Accelerated Oxidation of Ni, Cr and Ni-Cr Alloy Induced by Na_2SO_4 and NaCl (9, 10, 19).

	Below Melting Point		Above Melting Point	
	Na_2SO_4	NaCl	Na_2SO_4	NaCl
Ni	×	×	**O**	O
Cr	×	**O**	×	**O**
Ni-Cr	×	O	×	O

Acclerated Oxidation
O : severe O : medium × : not observed

difference depends on the role of Cr as stated previously. For the Na_2SO_4-induced oxidation Cr is very effective element and not only Cr but also Ni-Cr alloy does not receive an acceleration. Whereas for the NaCl-induced oxidation Cr is detrimental even below the melting point of the salt, and acceleration occurs in oxidation of Cr and Ni-Cr alloy similarly at every temperature above the eutectic point of Na_2CrO_4 and NaCl.

Conclusion

High temperature oxidation of NaCl-coated Ni and Ni-5%Cr, -10%Cr and 20%Cr alloys was carried out and the oxidation behavior was examined by thermal analysis, measurement of oxidation rate and corrosion loss, X-ray diffraction analysis of scale and metallographic observation. Results obtained were summarized as follows:

(1) Below the melting point of NaCl, there was no interaction between Ni and NaCl. Then oxidation of Ni was not influenced by solid NaCl. But above the melting point, Ni dissolved into molten NaCl and protective dense oxide did not form. Then oxidation of Ni is accelerated by molten NaCl above the melting point.

(2) Accelerated oxidation of Ni-Cr alloys occurred even below the melting point of NaCl due to the detrimental effect of Cr in alloys. In the oxidation process, Cr reacted with NaCl and formed not only NiO and Cr_2O_3 but also hexavalent Cr compound, Na_2CrO_4. This Na_2CrO_4 and NaCl formed eutectic melt with low melting point(850K) and accelerated the oxidation reaction. Corrosion loss below the melting point of NaCl increased with increasing Cr content.

Above the melting point Ni-Cr alloys were oxidized acceleratedly, too. But the effect of Cr content was not so clear, because oxidation both Ni and Cr were influenced by NaCl.

References

1. R.A.Rapp, Mater. Sci. Eng., 87,319(1987).
2. A.Rahmel, Mater. Sci. Eng., 87,345(1987).
3. D.M.Johnson, D.P.Whittle and J.Stringer, Corros. Sci., 15, 721(1975).
4. Y.Bourhis and C.A.Barrett, J. Electrochem. Soc.,9,507(1975).
5. G.J.Santoro and C.A.Barrett, J. Electricchem. Soc., 125, 271(1975).
6. D.W.Mckee, D.A.Shores and K.L.Luthra, J. Electrochem. Soc., 125,411(1978).
7. D.N.H.Trafford and D.P.Whittle, Corros. Sci., 20, 497(1980).
8. S.Kameswari, Werkst. Korros., 40, 213(1989).
9. Y.Shinata, H.Kobayashi and Y.Nishi, Nippon Kinzoku Gakkaishi (J.Jpn.Inst.Met.), 49, 267(1985).
10. Y.Shinata, H.Kobayashi and Y.Nishi, Nippon Kinzoku Gakkaishi (J.Jpn.Inst.Met.), 49, 273(1985).
11. Y.Shinata and Y.Nishi, Nippon Kinzoku Gakkaishi (J.Jpn.Inst.Met.), 49, 995(1985).
12. Y.Shinata and Y.Nishi, Proc. JIMIS-3, Trans, Jpn. Inst.Met., 24, 385(1985).
13. Y.Shinata and Y.Nishi, Nippon Kinzoku Gakkaishi (J.Jpn.Inst.Met.), 50, 315(1986).
14. M.Hara and Y.Shinata, Nippon Kinzoku Gakkaishi (J.Jpn.Inst.Met.), 53, 771(1989).
15. A.U.Seybolt, Oxid. Met., 2, 119(1970).
16. H.E.Evans, D.A.Hilton, R.A.Holm and P.W.G.Simpson, Oxid. Met., 16, 327(1981).
17. Y.Shinata, F.Takahashi and K.Hashiura, Mater. Sci. Eng., 87, 399(1987).
18. Y.Shinata and Y.Nishi, Oxid. Met., 26, 201(1986).
19. Y.Shinata, Oxid. Met., 27, 315(1987).

Sulphidation of Vanadium in H_2S-H_2 atmospheres at temperatures between 1073 K and 1273 K

F.A. Elrefaie

Department of Metallurgical Engineering, Cairo University, Giza, Egypt

W.W. Smeltzer

Department of Materials Science and Engineering, and Institute for Materials Research, McMaster University, Hamilton, Ontario, Canada, L8S 4M1

Abstract

The sulfidation behavior of vanadium in H_2S-H_2 atmospheres, $10^{-4} \leq P_{S_2} \leq 10^{2}$ Pa, at a total pressure of 10^{5} Pa was investigated at temperatures in the range $1073 \leq T \leq 1273$ K. If the dependence of the parabolic sulfidation rate constant on sulfur pressure is represented by $k_p \propto P_{S_2}^{1/n}$, n increases with sulfur pressure. Monosulfide scales were formed at sulfur pressures less than 10^{-3}, 10^{-2} and 10^{-1}Pa at 1073, 1173 and 1273 K, respectively. At higher sulfur pressures, duplex scales were formed consisting of external fine-grain V_3S_4 and inner coarse-grain VS layers. Marker studies indicated that the growth of V_3S_4 and VS layers involve outward vanadium and inward sulfur diffusion, respectively.

Key words

V sulfidation, V_3S_4, VS.

Introduction

Limited information is available concerning the high temperature corrosion of vanadium in sulfur bearing atmospheres(1,2). Crousier and Laffitte(1) sulfidized vanadium in an H_2S-H_2 atmosphere at mole ratio 7 in the range 1123-1373 K; the reaction kinetics for growth of VS scales were characterized by two-stage parabolic rates, the rate in the final stages being most rapid. More recently, Taniguchi and coworkers(2) investigated the sulfidation kinetics in H_2S-H_2 atmospheres, $10^{-13} \leq P_{S_2} \leq 10^{-8}$Pa, at temperatures in the range 1123-1323 K. The kinetics were parabolic with growth of single-phase VS and V_3S_4 scales at low and high sulfur pressures, respectively. The isobaric two-phase thermodynamic equilibria were also determined and a tenative V-S phase diagram constructed(2-4) containing VS, V_3S_4, V_2S_3 and V_5S_8 phases. Accordingly, multi-phase sulfide scales can be expected when vanadium is sulfidized at high sulfur pressures. The objective of this investigation is, therefore, to gain additional knowledge on the sulfidation properties of vanadium by establishing the sulfidation kinetics and scale microstructures upon sulfidation in H_2S-H_2 atmospheres over a wide range of sulfur pressure, $10^{-4} \leq P_{S_2} \leq 10^{+2}$Pa.

Experimental

Vanadium, 99.7% pure, was supplied as turnings containing impurities of 0.06 Fe, 0.05 Si, 0.02 Al, 0.02 Mg, 0.005 Co and 0.063 Ni (wt.%). Cylindrical rods, 7.6 cm long and of 1.6 cm diameter, were produced by tungsten - arc melting of the turnings under argon gettered of oxygen by titanium. Disks 0.1 cm thick were sectioned from these rods, polished to 1μ diamond paste and cleaned in methanol.

These vanadium specimens were sulfidized at 1073, 1173 and 1273 K in flowing H_2S-H_2 atmospheres, $10^{-4} \leq P_{S_2} \leq 10^{+1}$Pa, maintained at 10^5 Pa in a balance assembly as previously described(5). Marker measurements were carried out by placing Pt wire, 25 μm diameter, around a vanadium specimen which was then sulfidized for 5 h at 1273 K and $P_{S_2} = 10^2$ Pa. Scale morphologies were examined by optical and electron microscopies. Sulfides were identified using X-ray reflection diffraction from the outer scale surface and at different depths upon abrading the scale using 600 SiC grit paper.

Results

Typical sulfidation curves obtained at 1173 K are illustrated in Fig.1. The parabolic type plots, $(\Delta W/A)^2$ vs time, demonstrated that vanadium sulfidized parabolically after an induction period of ~ 60 min as illustrated by the plots in Fig.2 of the kinetic data at 1073 K. The parabolic rate constants for growth of the sulfide scales given in Table I were obtained from the slopes of the linear sections of these types of plots.

Figure 3 illustrates microstructural features of the sulfide scales. As illustrated by the cross-sections, a to c, the scales were duplex at $P_{S_2} = 1$ and 10Pa. The external columnar fine-grain layer is thicker than the inner equiaxed coarse-grain layer. At lower sulfur pressures, a single-layer sulfide scale was formed as manifested by the SEM micrograph shown in Fig.3d for the scale formed at $P_{S_2} = 10^{-4}$Pa.

The reflection X-ray diffraction patterns from the surfaces of scales formed at the sulfur pressures $10^{+2} \leq P_{S_2} \leq 10^{-4}$Pa, and 1273 K are illustrated in Fig.4. The sulfide phases indentified at the three investigated temperatures of 1073, 1173 and 1273 K are summarized in Table II. At $P_{S_2} \geq 10^{-1}$Pa, monoclinic V_3S_4 with preferred (310) and (020) textures was only detected. At sulfur pressures close to that of the V_3S_4-$VS_{1\pm\delta}$ coexistence, diffraction lines were observed corresponding to that of monoclinic V_7S_8. With decreasing sulfur pressure, the diffraction lines characteristic of orthorhombic ($VS_{0.985}$) and tetragonal (V_5S_4) monosulfides appear in addition to those of monoclinic V_7S_8. That is, the V_7S_8 lines became weaker and those

TABLE I
Values of the Parabolic Sulfidation Rate constant at
$10^{-4} \leq P_{S_2} \leq 10^{+2}$Pa and at 1073, 1173 and 1273 K

P_{S_2}, Pa	k_p, $g^2\ cm^{-4} s^{-1}$		
	1073 K	1173 K	1273 K
10^{-4}	4.5×10^{-10} 4.8×10^{-10}	7.4×10^{10}	1.1×10^{-9}
10^{-3}	9.7×10^{-10}	1.3×10^{-9}	2.5×10^{-9}
10^{-2}	1.2×10^{-9}	2.4×10^{-9}	3.8×10^{-9}
10^{-1}	1.4×10^{-9}	2.7×10^{-9}	4.9×10^{-9}
1.0	1.5×10^{-9}	3.6×10^{-9}	7.0×10^{-9}
10^{+1}	1.5×10^{-9} 1.4×10^{-9}	4.5×10^{-9}	8.7×10^{-0}
10^{+2}	1.2×10^{-9}	5.0×10^{-9}	1.1×10^{-8} 1.2 x 10-8

TABLE II
Vanadium sulfide phases in the scales formed at
$10^{-4} \leq P_{S_2} \leq 10^{+2}$Pa and at 1072, 1173 and 1273 K

P_{S_2}, Pa	k_p, $g^2\ cm^{-4} s^{-1}$		
	1073 K	1173 K	1273 K
10^{+2}	V_3S_4	V_3S_4	V_3S_4
10^{+1}	V_3S_4	V_3S_4	V_3S_4
1.0	V_3S_4	V_3S_4	V_3S_4
10^{-1}	V_3S_4	V_3S_4	V_3S_4
10^{-2}	V_3S_4	$V_5S_4 + V_7S_8$	V_7S_8
10^{-3}	$V_5S_4 + V_7S_8$	V_7S_8	$V_7S_8 + VS_{0.9985}$
10^{-4}	$V_7S_8 + VS_{0.985}$	$V_7S_8 + VS_{0.985}$	$V_7S_8 + VS_{0.985}$

of orthorhombic $VS_{0.985}$ stronger with decreasing sulfur pressure. The formed orthorhombic sulfide phase exhibited a preferred (122) texture.

Reflection X-ray diffraction patterns obtained at different increasing depths through duplex sulfide scales, as illustrated in Fig.5, also indicated that the external sulfide layer was V_3S_4 and that the inner sulfide layer was composed of monoclinic V_7S_8, orthorhombic $VS_{0.985}$ and tetragonal V_5S_4. V_7S_8 was detected adjacent to the V_3S_4 layer; tetragonal V_5S_4 was formed at the scale/metal interface below the $VS_{0.985}$ sulfide phase. These results are summarized in Table III for the scale formed at $P_{S_2} = 10^{+1}$Pa and 1273 K. These results illustrate that the thickness of the V_3S_4 layer is relatively large and that the metal-deficient part of the monosulfide layer (V_7S_8) is of much smaller thickness than the sulfur-deficient part ($VS_{0.985}$ + V_5S_4).

The thickness dependences of the monosulfide type phases $VS_{1\pm\delta}$ obtained from reflection X-ray diffraction at various depths in the single-layer scale formed at 1073 K and $P_{S_2} = 10^{-4}$Pa are illustrated in Fig.6 and summarized in Table IV. These results also demonstrate that the thickness of metal deficient V_7S_8 is very small.

TABLE III

Vanadium sulfide phases present in scales at various depths x/x_L where x is the thickness of sulfide abraded to expose new sulfide surface to the X-ray beam and X_L is the total scale thickness.

x/x_L^*	Sulfide phases in duplex scales formed at $P_{S_2} = 10^{+1}$Pa and 1273 K
0.0	V_3S_4
0.1	V_3S_4
0.2	V_3S_4
0.3	V_3S_4
0.4	V_3S_4
0.5	V_3S_4
0.6	V_3S_4
0.7	V_3S_4 + V_7S_8
0.8	$VS_{0.985}$
0.9	$VS_{0.985}$ + V_5S_4
1.0	V

TABLE IV.
Vanadium sulfide phases present at various depths x/x_L in a single-layer scale.

x/x_L	Sulfide phases in single-layer scale formed at $P_{S_2} = 10^{-4}$Pa and 1073 K
0.0	$V_7S_8 + VS_{0.985}$
0.2	$VS_{0.985} + V_5S_4$
0.4	$VS_{0.985} + V_5S_4$
0.6	$VS_{0.985} + V_5S_4$
0.8	$V_5S_4 + V$
1.0	V

Figure 7 illustrates the position of two markers, one of which was spot welded to the sample surface prior to sulfidation, Fig.7b. These micrographs show that the markers, though highly deformed, are located near the V_3S_4-VS interface; apparently the outer columnar V_3S_4 layer grew by outward vanadium diffusion whereas growth of the inner layer containing monosulfide phases was supported by inward sulfur diffusion.

Discussion and Conclusions

The sulfidation kinetics were parabolic at the investigated sulfur pressures and temperatures. Figure 8 shows logarithmic plots of the parabolic rate constants versus sulfur pressure. When described by the relationship $k_p \alpha P_{S_2}^{1/n}$, the value of n decreases with increasing pressure at 1173 and 1273 K. At 1073 K, the parabolic rate constant exhibits a very small dependence on sulfur pressure over the high pressure range 10^{-2} to 10^{+2}Pa. Values of k_p extrapolated from this investigation to the higher sulfur pressure range utilized by Crousier and Laffitte(1) are ~ 2 orders of magnitude less than their reported values. On the other hand, the values of k_p determined by Taniguchi and coworkers(2) are in fair agreement with values interpolated from the results of this investigation as represented in Fig.8. The independence of k_p on sulfur pressure over the high sulfur pressure range as reported by these latter investigations was approximately valid in this investigation only at 1073 K.

Polycrystalline $VS_{1\pm\delta}$ type scales were formed at sulfur pressures less than 10^{-3}, 10^{-2} and 10^{-1}Pa at 1073, 1173 and 1273 K, respectively. At higher sulfur pressures, duplex scales consisting of an outer fine-grain V_3S_4 layer and an inner coarse-grain $VS_{1\pm\delta}$ layer were formed. Although Taniguchi and coworkers(2) reported growth of a single V_3S_4 layer when vanadium foils were

TABLE V
Values of sulfur pressure required for growth of the duplex VS/V_3S_4 scale and equilibration of $VS-V_3S_4$ phases.

Temperature, K	P_{S_2}, Pa (growth of duplex scale)	P_{S_2}, Pa (equilibrium VS - V_3S_4 sulfur pressure)
1073	$P_{S_2} \geq 10^{-3}$	5×10^{-6}
1173	$\geq 10^{-2}$	10^{-3}
1273	$\geq 10^{-1}$	10^{-1}

sulfidized within the stability range of this sulfide, growth of $V_3S_4/VS_{1\pm\delta}$ duplex scales observed in this investigation is more consistent with the equilibrium sulfur pressure-composition isotherms reported for the $VS-V_5S_8$ system(2-4). Table V summarizes these equilibrium sulfur pressures for the $VS-V_3S_4$ system and the sulfur pressures at which the duplex scales were found to form. Formation of this duplex scale at $p_{S_2} > 10^{-1}$Pa at 1273 K is consistent with the thermodynamic data; at lower temperatures, however, the transition from growth of a monosulfide type scale to a duplex scale occurred at sulfur pressures higher than the equilibrium values.

Arrhenius plots of the parabolic sulfidation rate constants are given in Fig. 9 and values of the activation energy calculated from the slopes of these plots are recorded in Table VI. The activation energy increases with sulfur pressure; the values are 11,000 and 22,000 cal/mole at $P_{S_2} = 10^{-4}$ and 10^2Pa, respectively. The activation energy value previously reported for growth of a single phase V_3S_4 scale at $P_{S_2} = 10$ Pa is 43,400 cal/mole whereas the value obtained in this investigation for growth of the $VS_{1\pm\delta}/V_3S_4$ duplex scale under the same condition is 24,000 cal/mole.

Although a monosulfide type layer was formed in the single and duplex scales, X-ray reflection diffraction established the presence of several $VS_{1\pm\delta}$ nonstoichiometric phases. Monoclinic V_7S_8 was formed in the outer region of the $VS_{1\pm\delta}$ layer with orthorhombic $VS_{0.985}$ underneath. Tetragonal V_5S_4 occurred in the layer above the scale/metal interface. According to the $VS-V_5S_8$ equilibrium system(2), hexagonal and orthorhombic monosulfides of different nonstoichiometry are stable forms at high temperature. V_7S_8 and V_3S_4 possibly exhibit epitaxial relationships since both phases are of monoclinic symmetry. The sequence of monosulfide phases also occurred in the

TABLE VI
Values of the activation energy for growth of sulfide scales at various sulfur pressures.

P_{S_2}, Pa	Activation energy, cal/mole
10^{-4}	11,500 ± 500
10^{-3}	11,500 ± 500
10^{-2}	15,400 ± 500
10^{-1}	16,800 ± 500
1.0	20,900 ± 500
10^{+1}	24,500 ± 700
10	32,500 ± 1,000

single layer scales formed on the 1.0 mm thick metal discs which is to be compared to the formation of hexagonal VS on 0.1 mm thick vanadium foils(2). Possibly these differences are associated with different stress in the two systems as the inert Pt markers placed on the discs were deformed by growth of the sulfide scale (Fig.7).

The marker experiments indicated that growth of the outer V_3S_4 layer involved outward V diffusion while growth of the inner monosulfide type layer involved inward S diffusion. These diffusion mechanisms are consistent with the point defect structures of V_3S_4 and $VS._{1\pm\delta}$. These sulfides are regarded as containing neutral point defects due to their metallic type electrical conductivities. Monclinic V_3S_4 is classified as a defective NiAS structure with V vacancies confined to every second metal layer and ordered within the layers(6,7). The inner sulfide layer is largely composed of S deficient $VS_{1-\delta}$ phases since the thickness of V deficient V_7S_8 phase is of much smaller thickness as indicated by the X-ray diffraction analyses. The V rich $VS_{1-\delta}$ exhibits a defective anionic lattice with S vacancies as the predominant point defects (8).

In light of these findings, the mechanism for growth of the duplex scales, as illustrated schematically in Fig.10, probably involves the following reactions. At the $VS_{1-\delta}/V$ interface, metal is incorporated into the monosulfide lattice by generation of sulfur vacancies, v_S^x:

$$V + S_S^x = VS + (v_S^x)_{VS} \qquad (1)$$

Thus, a sulfur flux is induced inward through the sulfur deficient part of the monosulfide layer. This inward sulfur flux is accompanied by an outward vanadium flux originating at the stoichiometric composition of the monosulfide phase. That is, sulfur and vanadium vacancies are annihilated at the stoichiometric VS composition. The appropriate vanadium fluxes via vacancies in the metal-deficient $VS_{1+\delta}$ phase and the V_3S_4 layer are partially established by the interfacial reaction:

$$4VS_{1+\delta} + (v_V^x) = V_3S_4 + (V_V^x)VS \quad (2)$$

The defect reaction leading to growth of the V_3S_4 layer at the scale/gas interface may be then represented by the reaction:

$$4H_2S = 4H_2 + 4S_s^x + 3(v_V^x)V_3S_4 \quad (3)$$

A schematic model representing growth of the monosulfide type scale is presented in Fig. 11. Vanadium is incorporated into $VS_{1-\delta}$ at the metal interface by generation of sulfur vacancies as given previously by reaction (1). At the gas/scale interface, sulfur is incorporated into $VS_{1+\delta}$ by

$$H_2S = H_2 + S_S^x + (v_V^x)_{VS} \quad (4)$$

Thus scale growth is supported at both its inner and outer interfaces by sulfur and vanadium transport, respectively, with the vanadium and sulfur vacancies being annulled at VS of stoichiometric composition according to the defect reaction:

$$v_V^x + v_S^x = V_V^x + S_S^x \quad (5)$$

Reactions (3) and (4) are to be regarded as simple approximations to the generation of vanadium metal vacancies in V_3S_4 and $VS_{1+\delta}$ by interaction with H_2S. One would expect a relationship of $k_p \alpha (P_{S_2})^{1/n}$ with n changing with sulfur pressure due to the growth of a monosulfide type scale at low pressure and a duplex $VS_{1\pm\delta}/V_3S_4$ scale at higher sulfur pressures. The limiting values of n at low and high pressures (Fig. 8), however, cannot be determined from these illustrative point defect equations for vanadium vacancies and their role in governing the diffusion of vanadium as one of the reactants.

Acknowledgements

The investigators were indebted to the Natural Sciences and Engineering Council of Canada for financial support of this research.

References

1. J. Crousier and M. Laffitte, C.R. Acad. Sci., 260, 6903 (1965).
2. M. Taniguchi, M. Wakihara, and Y. Kojima, JIMIS, 3, 78 (1982).
3. M. Wakihara, T. Uchida, and M. Taniguchi, Met. Trans., 913, 29 (1978).
4. T. Uchida, H. Hiroda, M. Wakihara, and M. Taniguchi, Met. Trans., 12B, 71 (1981).
5. G. Romeo, W.W. Smeltzer, and J.S. Kirkaldy, J. Electrochem. Soc., 118, 740 (1971).
6. M. Chevreton and A. Sapet, C.R. Acad. Sci., 261, 928 (1965).
7. S.H. Hold, R.J. Bouchard, and A. Wold, J. Phys. Chem. Solids., 27, 755 (1966).
8. H.F. Franzen, and T.J. Burger, J. Chem. Phys., 49, 2268 (1968).

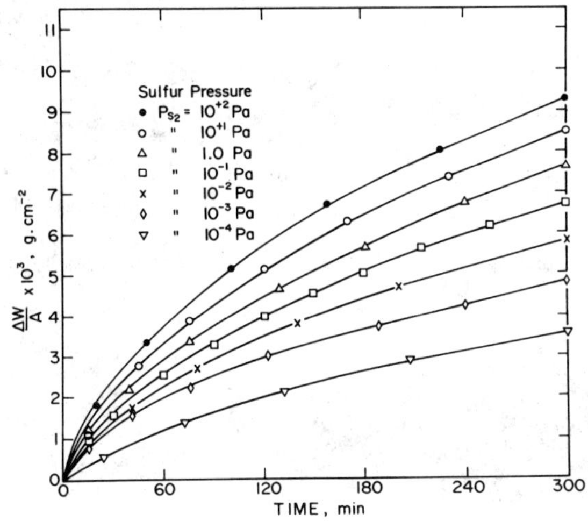

FIG.1. Sulfidation kinetics of vanadium at 1173K in H_2S-H_2 atmospheres at $10^{-4} \le P_{S_2} \le 10^{+2}$Pa.

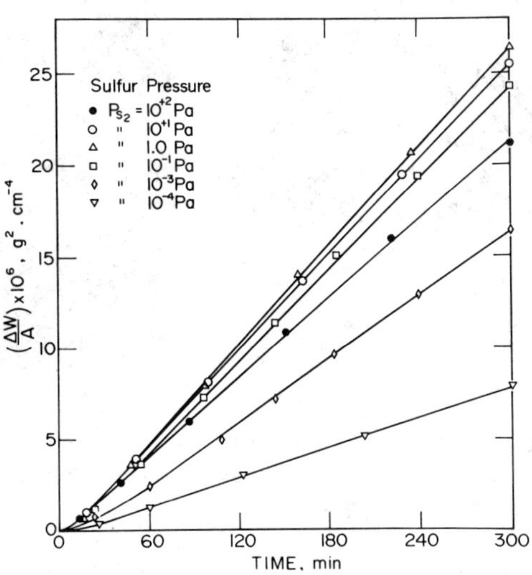

FIG. 2. Sulfidation kinetics of vanadium at 1073K in H_2S-H_2 atmospheres at $10^{-4} \le P_{S_2} \le 10^{2}$Pa plotted in parabolic form.

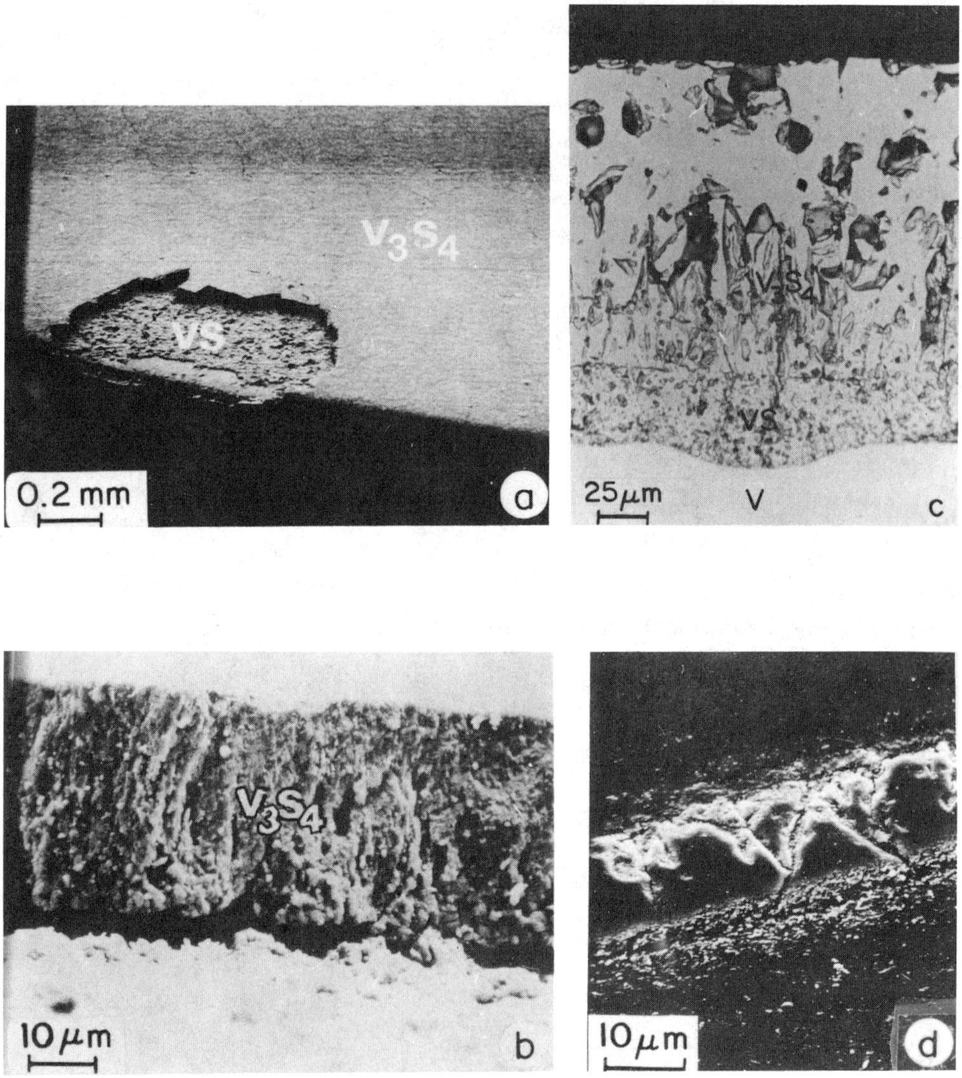

FIG. 3. Microstructural features of the sulfide scales: (a) a section broken out of outer V_3S_4 layer showing inner VS layer formed in duplex scale at 1173K and P_{S_2} = 1.0 Pa (x 55); (b) fracture cross-section of the scale in (a) illustrating columnar microstructure of outer V_3S_4 layer (x 1100); (c) cross-section of duplex VS/V_3S_4 scale formed at 1173K and P_{S_2} = 10Pa (x 400); (d) SEM image of single-layer $VS(V_7S_8 + VS_{0.985})$ scale formed at 1073K and $P_{S_2} = 10^{-4}$Pa (x 1350).

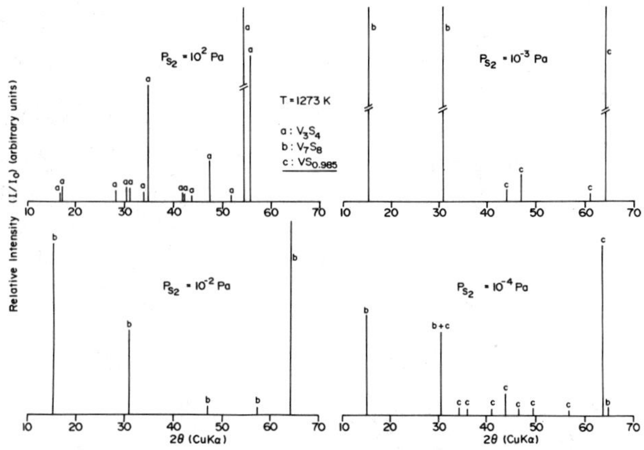

FIG.4. Reflection X-ray diffraction patterns from the surfaces of scales formed at 1273K and $10^{-4} \leq P_{S_2} \leq 10^{2}$Pa.

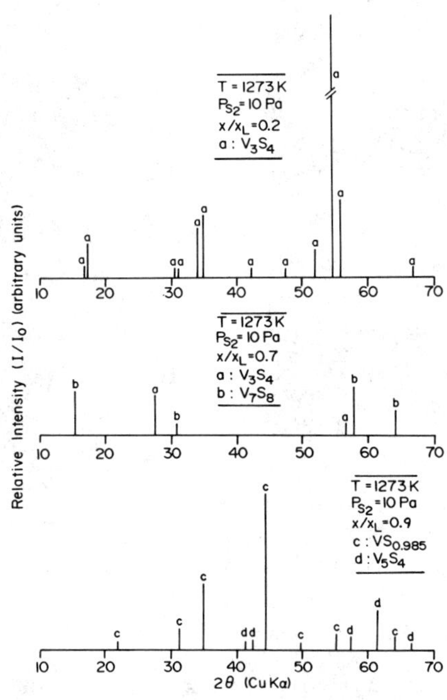

FIG.5. Relative X-ray intensities at scale depth fractions 0.2, 0.7 and 0.9 for scales formed at 1273K and P_{S_2} = 10 Pa.

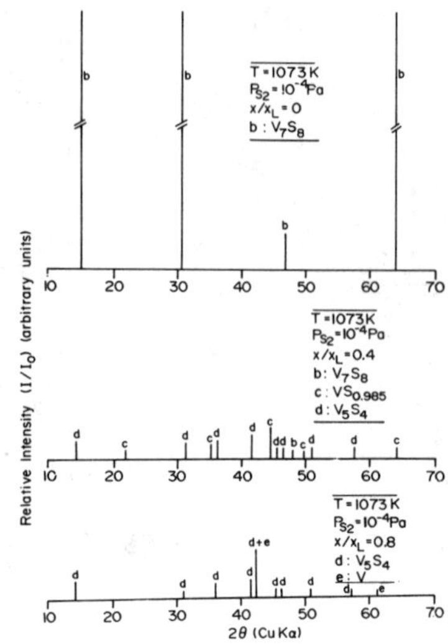

FIG.6. Relative X-ray intensities at scale depth fractions 0.0, 0.4 and 0.8 for scales formed at 1073K and $P_{S_2} = 10^{-4}$Pa.

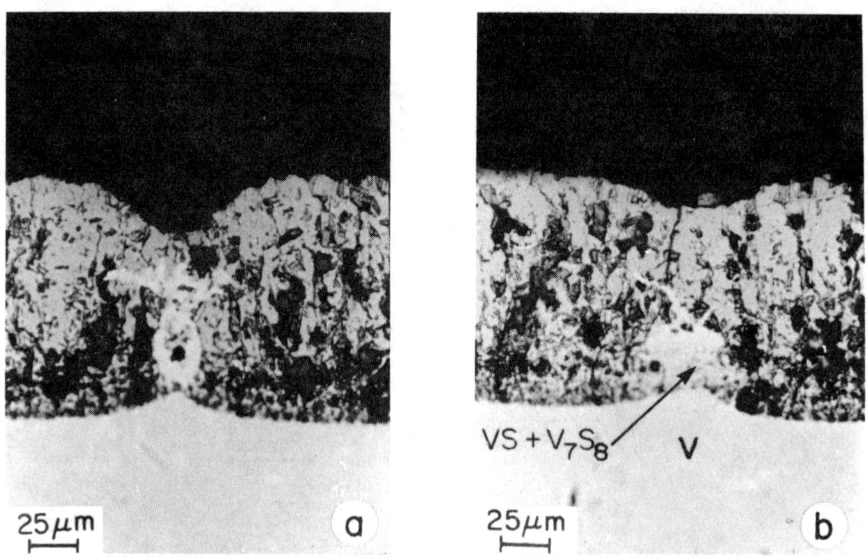

FIG.7. The positions of two Pt markers in the region of the VS/V_3S_4 interface of duplex scale after sulfidation of vanadium at 1273K and $P_{S_2} = 10^3$Pa: (a) Pt originally placed in contact with metal; (b) Pt originally spot-welded to metal surface.

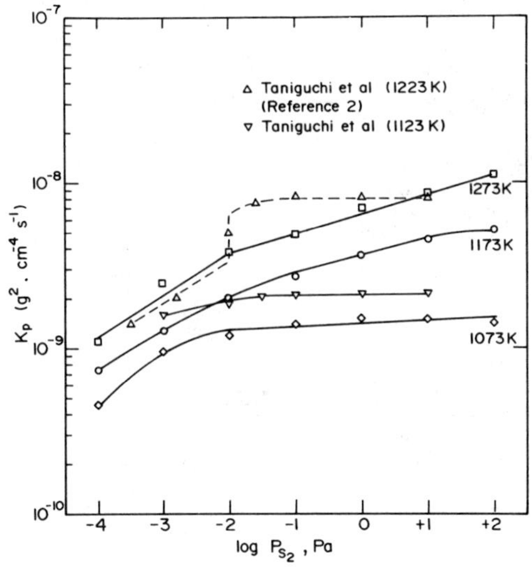

FIG.8. Plots of the parabolic sulfidation constant versus sulfur pressure in the form of k α $P_{S_2}^{1/n}$ at temperatures of 1073, 1173 and 1273K. Results from previous work (2) at 1123 and 1223K are also included.

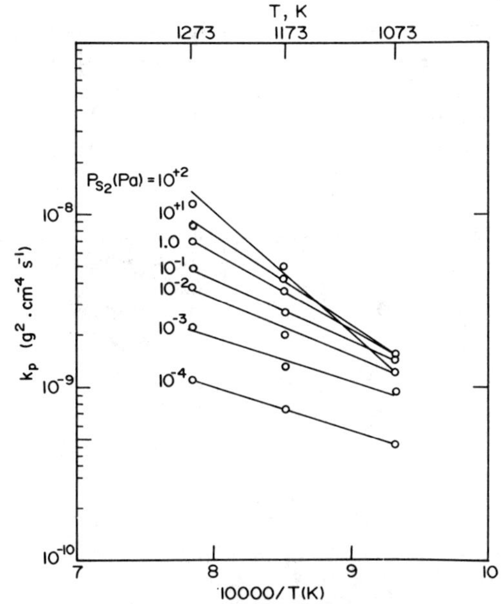

FIG.9. Arrhenius plots of the parabolic sulfidation constant at sulfur pressures ranging from 10^{-4} to 10^{2}Pa.

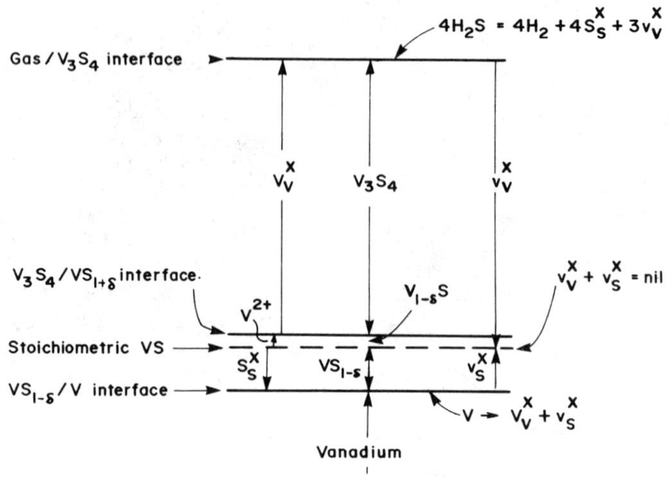

FIG.10. Schematic representation of growth mechanism for duplex VS/V_3S_4 scale on vanadium.

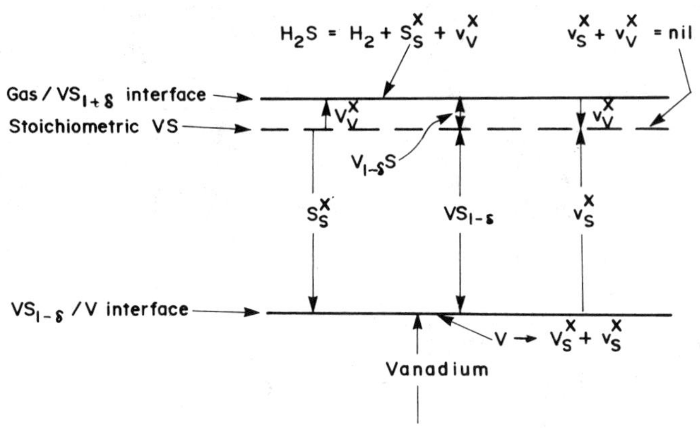

FIG.11. Schematic representation of growth mechanism for single-layer VS scale on vanadium.

Oxidation behaviour improvement on FeCrAl and FeCrNi alloys by surface applied ceria

J.P. Larpin, G. Aguilar, H. Buscail and J.C. Colson
Laboratoire de Recherches sur la Réactivité des Solides, URA 23, Faculté des sciences Mirande, B.P. 138, 21004 Dijon Cedex, France

Introduction

General interest on Fe-Cr and Fe-Cr-Al alloys in high temperature corrosion relies upon the formation of protective oxide scales (Cr_2O_3) (Al_2O_3) (1, 2, 3, 4, 5, 6).

Unfortunately, protective oxide films have a great tendancy to spallation under thermal cycling. Rare earth elements, as cerium, are known to present a beneficial effect on the oxide scale properties.

Still remain the difficulties to introduce active elements in the substrate as alloying components.

Other technics, as sol-gel methods have been proposed to realize local applications (1, 7, 8, 9, 10, 11, 12).

In order to confirm these results, a new type of cerium deposits has been carried out on two different commercial alloys, Fe-CrNi and Fe-Cr-Al.

Experimentals

The chemical compositions of the alloys are reported in Table I. Coatings have been applied by two different laboratories : Rhône Poulenc Corp. and the University of Franche-Comté.

Fe-Cr-Al

element	C	P	S	Mn	Si	Cr	Ni	Mo	Al	Mg
weight %	0.045	0.019	0.015	0.35	0.51	22.75	0.18	0.11	5.45	0.02

Fe-Cr-Ni

element	C	P	S	Mn	Si	Cr	Ni	Mo
weight %	0.021	0.022	0.013	1.44	0.48	17.25	11.8	2.1

TABLE I : Alloy composition.

The Fe-Cr-Al alloy was treated by dipping in a sol (specimens N° 1 to 6) while electrophoretic deposits have been achieved on Fe-Cr-Ni alloys (specimens A to E). Anyway, the deposit was sintered in air by firing at 650°C. Specimens, in the form of coupons, 10 mm x 10 mm x 2 mm were tested in air at 1150°C (150 Pa) and 980°C (400 Pa) for Fe-Cr-Al and Fe-Cr-Ni respectively.

Thermal cycling was performed by heating up the specimens at 980°C for one hour and then quenched to room temperature. This cycle was repeated 15 times successively. Oxide scale morphologies and partitioning of the different elements were determined by scanning electron microscopy (S.E.M.) and energy dispersive X-ray analysis (E.D.A.X.). Identification of the different phases was done by Xray diffraction (X.R.D.).

Weight gain measurements were followed by thermogravimetry using a Setaram MTB 10-8 microthermobalance.

Results

I - Kinetics

I-1 : Aluminum containing alloy (Fe-Cr-Al)

Weight gain curves are reported on FIG. 1. A transient stage is present during the first two hours, due to oxide nucleation and cerium oxidation achievement of the thickest cerium deposits.

Most of the deposits realized on this substrate did not give the expected good results. The N° 6 type specimens (dip coating in cerium polyacrylamid solution) show an important decrease of the growth rate compared to the blank specimens. But one should notice here that a great proportion of the mass gain was obtained during the transient stage.

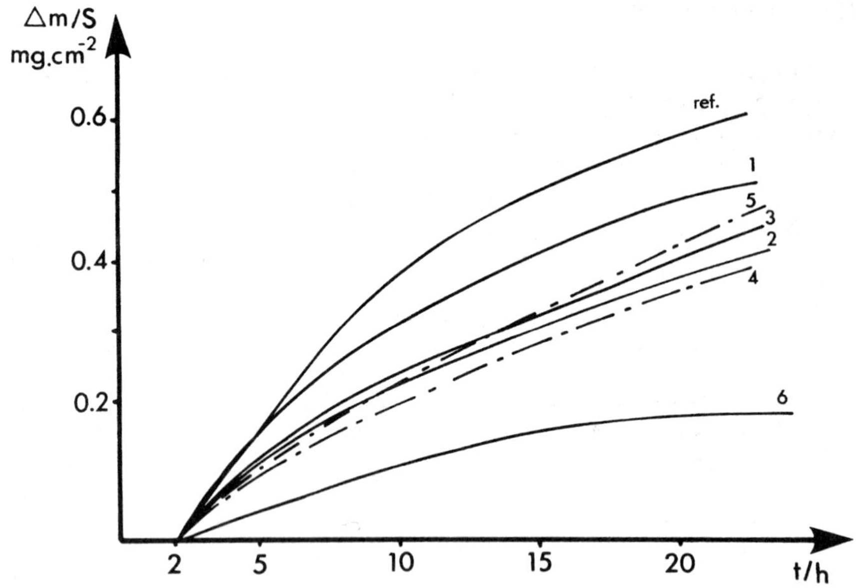

FIG. 1 : Isothermal oxidation of coated Fe-Cr-Al in air at 1150°C.

I-2 : Nickel containing alloy (Fe-Cr-Ni)

A summary of the kinetics results is given in Table II and FIG. 2. Several tests were performed on different metallographically abraded specimens. Surface statement does not seem to influence the registered mass gains. There was not any real transient stage observed with this alloy, mainly due to the relatively low thickness of the cerium deposit applied on the surface. This transient stage is however present on the kinetics curves of specimens, covered by the thickest coatings (specimens A and E). The linear plots $\Delta m/S = f(\sqrt{t})$ demonstrate that the oxidation kinetics follow the classic parabolic rate law which denote a diffusion controlled growth process.

specimen	kp1 ($mg^2.cm^{-4}.s^{-1}$)	kp2 ($mg^2.cm^{-4}.s^{-1}$)
blank	$2.6\ 10^{-6}$	$8.7\ 10^{-6}$
C	$1.6\ 10^{-6}$	$2.2\ 10^{-6}$
E	1.10^{-6}	
A	$8.6\ 10^{-7}$	
B	$4.2\ 10^{-7}$	
D	$1.6\ 10^{-7}$	

TABLE II : Parabolic rate constants.

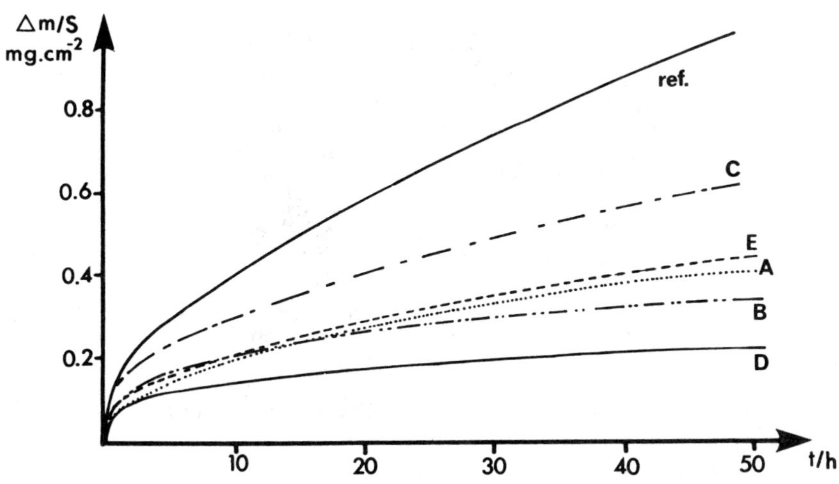

FIG. 2 : Isothermal oxidation of coated Fe-Cr-Ni in P(air) = 400Pa at 980°C.

II - Scale Morphology

II-1 : Aluminum containing alloy (Fe-Cr-Al)

II-1-1 : Morphology of the coatings :

The cerium oxide coating aspect is dependent upon the method used. Dipping in aqueous sol (specimen N° 1) did not permit a significant deposit on the surface. The Makishima method (specimen N° 2) gives thicker coatings but severe agglomerations of cerium oxide particles occurred which led to a flake - like surface morphology after firing (FIG. 3).

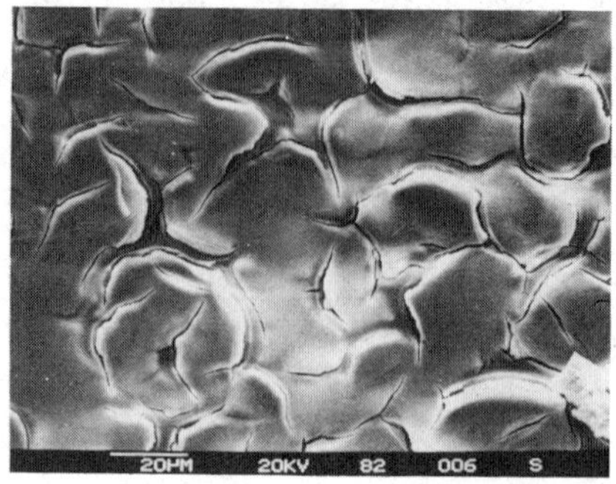

FIG. 3 : Fe-Cr-Al (specimen N° 2), flake-like surface morphology.

Some specimens were laser treated (specimen N° 3) in order to obtain a thicker homogeneous scale after a Makishima treatment (FIG. 4). A fine alumina scale containing a great proportion of cerium oxide was then observed on this specimen. Other specimens (N° 4 to 6) exhibit relatively thin coatings but they are quite inhomogeneous, the traces of a drop are observed on the borders of the specimen.

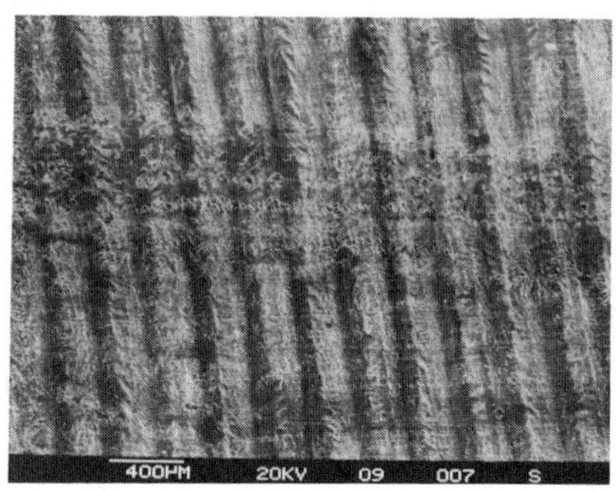

FIG. 4 : Fe-Cr-Al (specimen N° 3), laser treated.

II-1-2 : Morphology of the surface after oxidation tests :
A preliminary study was performed on the specimens in air under atmospheric pressure at 800°C during 15 hours. The oxide scale is thin composed with aluminum, chromium and iron : in every case, the coating is relatively thick (0,5 to 1,5 m), the cerium is located at the scale-gas interface. EDAX (FIG. 5) clearly shows that even a quite thick initial coating permits an alumina subscale to grow at the coating substrate interface. Concerning the laser treated specimens, a new alumina scale developped under the initial cerium containing one. In this case, some spallation occured during cooling, probably due to the different natures of the two alumina subscales. Spallation is greatly minimized when thin coatings were applied before testing. This confirms the assumption that the best protectiveness is obtained on alloys with the thinner uniform coatings as it was proposed by somé authors (13, 14).

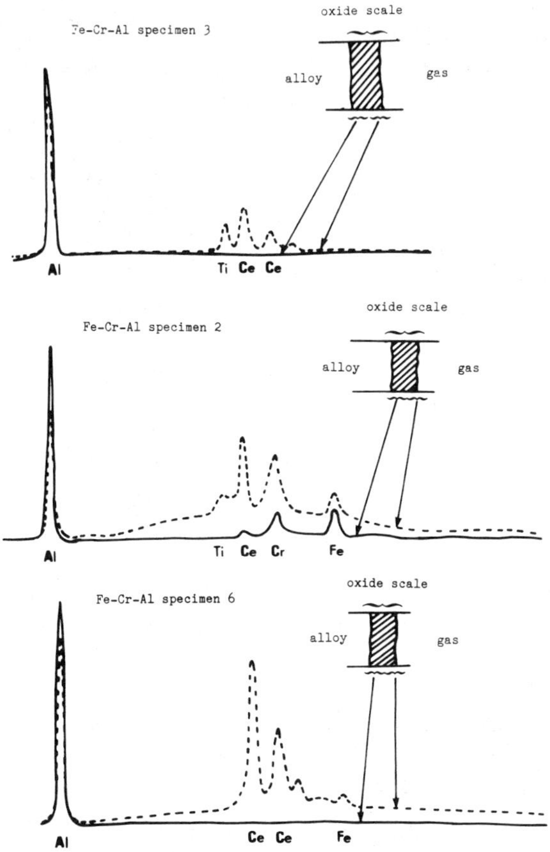

FIG. 5 : Energy dispersive X-ray analysis system (EDAX) of different coated Fe-Cr-Al.

II-2 : Nickel containing alloy (Fe-Cr-Ni)

II-2-1 : Morphology of the coatings

On this alloy, the coatings obtained were generally thin. Electrophoresis leads to much more uniform deposits than dipping only. Moreover, the consecutive thermal treatment did not need any drying period.

II-2-2 : Morphology of the scale after oxidation tests

Fe-Ni-Cr alloy oxidation has been already largely investigated (1, 2, 3, 15). The oxidation behaviour of this alloy seems to be dependent on minor

alloying elements as Mn, Si, or Mo present with a weight percentage higher than 0,5 %. Our observations on blank material after testing are in good agreement with the literature. A duplex compact oxide scale is formed ; its thickeness is relatively low after 50 hours of testing. The spinel phase $(Mn,Fe)Cr_2O_4$ is present in the outer subscale while the inner subscale contains mainly the chromia phase (Cr_2O_3) (FIG. 6). Spallation occured systematically during cooling, due to a rupture at the oxide substrate interface.
All the specimens, with and without coatings, always present inclusions located along grain boundaries interface (FIG. 6). These nodules contain a large amount of Si and Fe and could contribute to the so called "pegging effect". Finally, a depletion in chromium concentration is observed in the substrate close to the oxide alloy interface.

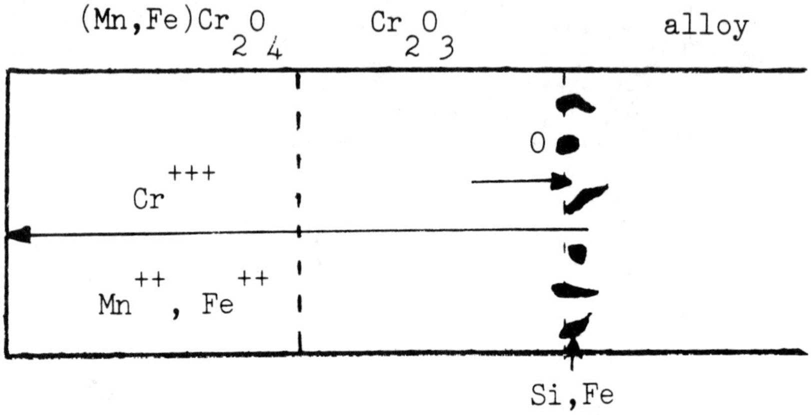

FIG. 6 : Schematic structure of the oxide scale on Fe-Cr-Ni alloy.

The scale morphology of coated specimens is quite similar to the one obtained on blank materials. The main differences observed are related to the absence of voids at the oxide-alloy interface which gives to the scale a better adherence. Thermal cycling was performed on these specimens to confirm such good results. As on the one hand we could observe that a 15 cycles test leads to a tremendous spallation of blank specimens, on the other hand no spallation was observed on coated specimens.

II-2-3 : Structural analysis (X.R.D.)

The identification of the phase, on blank materials, shows the presence of a spinel phase $(Fe,Mn)Cr_2O_4$ and chromia in the scale (FIG. 7). Coaled specimens present typical diffraction pics of chromia and ceria (which confirms the morphological observations) (FIG. 8). After testing in air, these specimens also exhibit a spinel phase. But the main differences between coated and uncoated specimens rely on the presence of another phase present after testing in air on blank materials. This phase is now identified as a Fe-Ni solid solution located below the oxide scale (FIG. 7).

Discussion

Although the beneficial effect of rare earth elements has been demonstrated as alloying component or coating, one should notice the different results obtained, depending on the substrate and the nature of the rare earth element itself.
During the last few years, numerous mechanisms were discussed, considering the different methods of reactive elements introduction or application (4, 7, 8, 15, 16, 17).

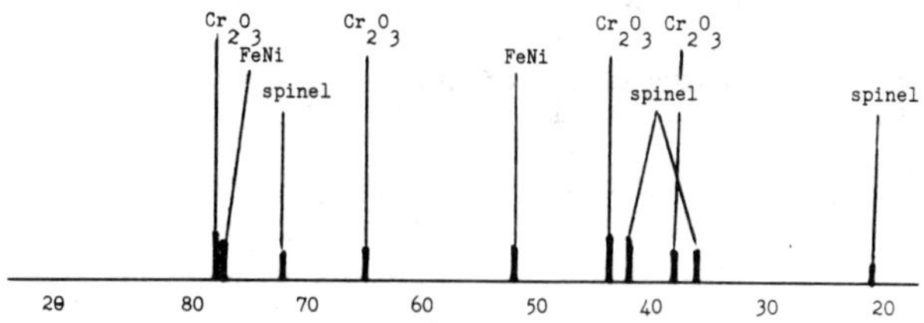

FIG. 7 : X-ray diffractogram after oxidation (blank Fe-Cr-Ni).

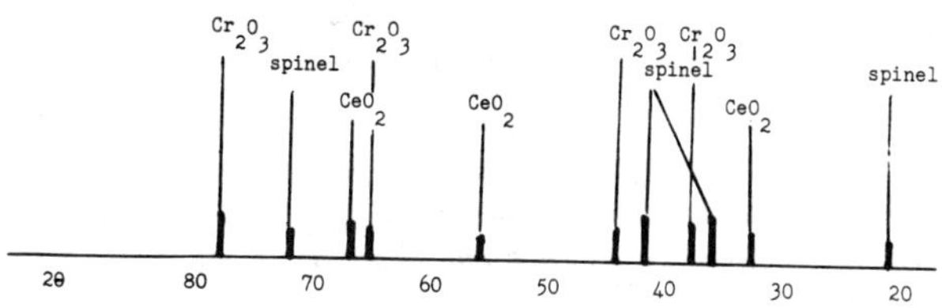

FIG. 8 : X-ray diffractogram after oxidation (Fe-Cr-Ni specimen B).

It has been attributed to largely different reasons, the following effects being proposed :

- Acceleration of the nucleation process of a continuous protective Al_2O_3 or Cr_2O_3 layer ;
- Modifications of the structure and mechanical properties of the scale;
- Keying effect due to physical and/or chemical bonds at the scale substrate interface ;
- Prevention of the sulfur segregation ;
- Change in the transport properties of the scale.

Moreover, reactive element location in the scale seems to be of a great interest. Some authors found the reactive elements localized at the oxide alloy interface (7, 16, 18) and some others at the oxide gas interface (8, 19) or finally disseminated regularly in the oxide scale (1, 17, 20).
In our study, cerium was localized in the oxide scale grown on FeCr-Ni alloy while the rare earth element was concentrated at the external interface in the case of the Fe-Cr-Al alloy.
Then, the alloy composition seems to play an important role. WEI (16) and NAGAI (7) have observed the good oxidation behaviour of alloys (Fe-28Cr-0.5Ce and Fe-20Cr-1Ce respectively) but spallation occured systematically as we have also observed on non-containing cerium alloys.
On the other hand, Ni-base alloy Ni-15Cr-1Ce (7) did not show any change on oxidation kinetics but spallation was considerably reduced. MASKAOUI (21) found similar results on the Fe-20Cr-6Al-0.025Ce alloy.
Finally, all the studies concerning alloys containing (or coated by) cerium demonstrate a better oxidation behaviour either under isothermal or cyclic conditions (1, 10, 14, 15, 17, 19).

In this paper, the efficiency of new methods of cerium oxide deposits have been tested in high temperature oxidation conditions. Testing performed on Fe-Cr-Al coated specimens generally show a bad oxidation behaviour. Coatings were thick, irregular and cracked.
This structure led to the developpement of an internal alumina scale under the initial coating. This subscale was 4 m thick after a 20 hours experiment at 1150°C.
The reactive element was then located at the coating oxygen interface or in the outer part of the alumina scale.
The laser treated material reacts similarly to other specimens. The supposed thick and homogeneous cerium containing coating permits in fact a new alumina scale to grow underneath. The different nature of the two subscales induces the spallation of the initial diffusion barrier.
A great improvement is observed for the coated Fe-Cr-Ni alloys. Electrophoretic methods seems to be a better technique compared to the classical sol-gel methods.
Non ionic surfactant additions provide a better adherence of the deposit. This leads to an increased behaviour of the substrate in oxidizing environment (D specimen).
In the mass, the scale morphology is not significantly modified. However, structural analysis tends to show that a metallic phase has disappeared. The lack of this phase could be responsible of the non-spallation of the scale.
The better corrosion behaviour of the coatings performed on the Fe-Cr-Ni alloy is related to a uniform and thin deposit (sample B and D). Moreover, the low thickness of the oxide scales indicates a slowing down of the growth processes.
Generally speaking, whatever the substrate tested here, the tendency to spallation is greatly reduced.
Scale adhesion is related to many parameters such as :

- the difference of Pilling Bedworth coefficients between the substrate and the scale,
- impurity segregation or vacancy coalescence at the scale-substrate interface,
- growth stresses and crack generation in the scale.

Still remain the difficulties to determine the mechanical properties of the oxide scale. However, reactive elements can largely modify impurity segregation and act as vacancy sinks.
Coated and blank materials both exhibit pegs at the scale substrate interface. It seems that pegs are not responsible for scale adhesion. Finally, spallation could be attributed to the depletion of chromium leading to the presence of the Fe-Ni phase observed at the scale-substrate interface after testing on blank specimens. Cerium also change the growth mechanism and strengthen the diffusion barrier properties of the oxide scale.
Our results demonstrate that uniformity and thickness of coatings are two determining factors of protection against high temperature corrosion. Electrophoresis is then the best technics to be used.

Conclusion

Our results confirm the beneficial effect of cerium concerning high temperature corrosion of the commercial alloys.
Usually, sol-gel technics as well as ion implantation (22, 23) have a beneficial effect on kinetics and scale adhesion.
We have clearly shown that the quality of the coatings is of great importance especially concerning their thickness and homogeneity.
Electrophoresis seems to be a very promissing technique. By contrast, thick cerium containing coatings even laser treated did not give any good result.

Morphology and composition of the corrosion scale were not drastically modified by the presence of cerium.
Nethertheless, it seems that the disparition of some phases present at the scale-substrate interface on blank materials could be responsible for the better adherence on coated materials.

References

1. M.J. Bennett, J. Vac. Technol. B 2, 800, (1984).
2. A.F. Smith, Werkst. Korros., 30, 100, (1979).
3. A.F. Smith, Werkst. Korros., 29, 441, (1978).
4. E. Lang, The role of active elements in the oxidation behaviour of high temperatures metals and alloys, Edited by LANG, PETTEN, the Netherlands, (1989).
5. G. Ben Abderrazik, F. Millot, G. Moulin and A.M. Huntz, J. Am. Ceram. Soc. 68, 307, (1985).
6. M. Skeldon, J.M. Calvert and D.G. Lees, Oxid. Met., 28, 109, (1987).
7. H. Nagai, Mater. Sci. Forum, 43, 75, (1989).
8. P.Y. Hou and J. Stringer, J. Electrochem. Soc., 137, 1836, (1987).
9. G.M. Ecer, R.B. Singh and G.H. Meier, Oxid. Met., 18, 55, (1982).
10. M.F. Stroosnijder, V. Guttmann, R.J.N. Gommans and J.H.W. de Wit, Mater. Sci. Eng., A 121, 581, (1989).
11. T. Fransen, P.J. Gellings, J.C. Fuggle, G. Van Der Laan, J.M. Esteva and R.C. Karnatak, Appl. Surf. Sci., 20, 257, (1985).
12. J.L. Whoodhead, Sci. Ceram., 169, (1980).
13. D.P. Moon, Oxid. Met., 32, 67, (1989).
14. V.A.C. Haanappel, T. Fransen, B. Geerdink, P.J. Gellings and M.F. Stroosnijder, submitted for publication in Oxid. Met.
15. R.L. Nelson, J.D.F. Ramsay, J.L. Woodhead, J.A. Cairns and J.A.A. Crossley, Thin Solid Films, 81, 329, (1981).
16. F.I. Wei and F.H. Stott, Corros. Sci., 29, 839, (1989).
17. A.T. Chadwick and R.I. Taylor, Proc. 5th I.C.M.C., 3, p. 381, (1984).
18. W.E. King, L.H. Park, J.L. Routbort and K.C. Goretta, Oxid. Met., 31, 217, (1989).
19. O.T. Goncel, Proc. 8th I.C.M.C., p. 681, (1981).
20. W.J. Quadakkers, H. Holzbrecher, K.G. Briefs and H. Beske, Oxid. Met., 32, 67, (1989).
21. M. Maskaoui, Thesis, Dijon, France, (1981).
22. S. Mrowec, J. Jedlinski and A. Gil, Mater. Sci. Eng., A 123, 169, (1989).
23. C.M. Cottel and G.J. Yurek, J. Electrochem. Soc., 134, 1871, (1987).
24. M.J. Bennett, B.A. Bellamy, C.F. Knights, N. Meadows and N.J. Eyre, Mater. Sci. Eng., 69, 359, (1985).

High-temperature oxidation properties of ODS MA754 alloy

K. Pryzybylski
Institute of Materials Science, Academy of Mining and Metallurgy, Krakow 30059, Poland

Commercial, oxide-dispersion-strengthened (ODS) alloy MA754 (Ni-20Cr-0.3Al-0.5Ti-0.6Y_2O_3), was oxidized at 1000°C in Ar-5%O_2 and in pure O_2, respectively. The oxidation behaviour of alloy that had similar composition (Ni-24Cr). Multiphase scales formed over alloy grain boundaries was mainly Cr_2O_3 with a thin layer of Ni-rich oxide at the scale/gas interface. The microstructures and compositions of the Cr_2O_3 and NiO layers formed on MA754 alloy were studied by STEM. Using EDX analysis, concentration profiles for Y and Ti across grain boundaries in the scales were measured, which demonstrated strong segregation of Y and Ti to the oxide grain boundaries. Particles of Y_2O_3 and mixed Y_2O_3-Al_2O_3 were found in the Cr_2O_3. The effects of the segregation of Y to oxide grain boundaries on the rates and mechanisms of oxidation, on the scale microstructures, and on scale adhesion are discussed.

(paper not available at time of printing)

Phase stability and reactions of chromite with carbon

M.V. Klein and R.G. Reddy
Department of Chemical and Metallurgical Engineering, University of Nevada-Reno, Reno, Nevada 89557, U.S.A.

Thermodynamic phase stability diagrams of Fe-C-O and Cr-C-O systems for several temperatures were constructed. The rate of reaction of $FeCrO_4$ with carbon was determined by weight loss measurement technique using TGA. The effect of temperature, Fe/C molar ratio and particle size on reaction rates were determined. The kinetics of the chromite-carbon reaction were modeled using a diffusion controlled model. Based on the deduced activation energy (1 Kcal) and inverse square dependence of chromite particle size, it was concluded that the rate of reaction of chromite with carbon was controlled by the diffusion process. The possible reaction mechanisms and scheme of reaction steps are discussed.

(paper not available at time of printing)

Authors' Index

Subject Index